Clean Technology and the Environment

Edited by

R.C. Kirkwood
Department of Bioscience and Biotechnology
University of Strathclyde
Glasgow

and

A.J. Longley
Engineering and Physical Sciences Research Council
Swindon

BLACKIE ACADEMIC & PROFESSIONAL
An Imprint of Chapman & Hall
London · Glasgow · Weinheim · New York · Tokyo · Melbourne · Madras

Published by
Blackie Academic and Professional, an imprint of Chapman & Hall,
Wester Cleddens Road,
Bishopbriggs, Glasgow G64 2NZ

Chapman & Hall, 2–6 Boundary Row, London SE1 8HN, UK

Blackie Academic & Professional, Wester Cleddens Road, Bishopbriggs, Glasgow G64 2NZ, UK

Chapman & Hall GmbH, Pappelallee 3, 69469 Weinheim, Germany

Chapman & Hall USA, One Penn Plaza, 41st Floor, New York, NY 10119, USA

Chapman & Hall Japan, ITP-Japan, Kyowa Building, 3F, 2–2–1 Hirakawacho, Chiyoda-ku, Tokyo 102, Japan

DA Book (Aust.) Pty Ltd, 648 Whitehorse Road, Mitcham 3132, Victoria, Australia

Chapman & Hall India, R. Seshadri, 32 Second Main Road, CIT East, Madras 600 035, India

First edition 1995

Typeset in 10/12 pt Times by Photoprint, Torquay, Devon
Printed in Great Britain by St Edmundsbury Press, Bury St Edmunds, Suffolk

ISBN 0 7514 0037 8

A catalogue record for this book is available from the British Library
Library of Congress Catalog Card Number: 94-67890

∞ Printed on acid-free text paper, manufactured in accordance with ANSI/NISO Z39.48–1992 (Permanence of Paper)

Clean Technology and the Environment

Preface

Many environmental problems resulting from atmospheric, land and water pollution are now widely understood. The combination of both improved technology and legislative pressure has led to a reduction in pollution from industrial practices in the West in recent years. However, sustainable development is dependent upon a new approach to environmental protection – clean technology.

This book is in two parts. The first explores the ecological principles governing the function of ecosystems, sustainability and biodiversity (Chapter 1) and the problems resulting from atmospheric pollution (Chapter 2), water pollution (Chapter 3) and land pollution (Chapter 4). For example, there is increasing international concern that the combustion of fossil fuels is leading to an increase in the levels of carbon, sulphur and nitrogen gases which pollute the atmosphere of our planet. The enhanced levels of carbon gases such as carbon dioxide may cause change in our global climate and, in turn, lead to flooding and loss of low-lying coastal regions. In addition, the deposition of sulphur and nitrogen oxides is believed to be the cause of 'acid rain' which has led to loss of fish stocks from upland lochs and damage to forestry plantations.

The problem of pollution of rivers, lakes and estuaries is widespread, occurring in association with human aggregation in towns and cities. It results from the release of sewage and other readily decomposable waste products (e.g. from farming or food industries) which lead to a biological oxygen demand (BOD), loss of dissolved oxygen and death of aerobes (oxygen-requiring organisms). A more insidious problem concerns the transmission of persistent compounds via food webs leading to the accumulation of chemical residues at the upper trophic (feeding) levels and possible death of carnivores such as birds of prey. Such 'biomagnification' can occur in the form of persistent inorganic and organic compounds found in the effluents from manufacturing processes or released from landfill sites; they include heavy metals, radioisotopes, certain persistent pesticides and polychlorinated biphenyls.

The difficult, but critical, questions of assessing and assigning the economic costs and implications of pollution are discussed in Chapter 5, and the concept of clean technology as a means of avoiding environmental damage at source, rather than reducing environmental damage by end of pipe abatement, is introduced in Part II of the book. Contributions from specialists in industry and academia then discuss the potential impacts and implications of a clean technology approach to specific industrial sectors.

This book should be of interest to anyone concerned with environmental pollution, conservation, legislation and management. It should be invaluable to people in industry, government or research institutions whose remit concerns environmental protection, environmental impact assessment, and environmental monitoring. Additionally, it should be a useful reference for undergraduates and postgraduates who are specialising in an environmentally related discipline, such as environmental biology, chemistry or science, or environmental analysis, policy or management.

R.C.K.
A.J.L.

Contributors

R. Clift	Centre for Environmental Strategy, University of Surrey, Guildford, Surrey, GU2 5XH, UK
D.E. Conlan	ARIC (Atmospheric Research and Information Centre), Department of Environmental and Geographical Sciences, Manchester Metropolitan University, Chester Street, Manchester, M1 5GD, UK
P. Fryer	Department of Chemical Engineering, University of Cambridge, Pembroke Street, Cambridge, CB2 3RA, UK
D. Hammerton	Director of Clyde River Purification Board, Rivers House, Murray Road, East Kilbride, G75 OLA, UK
R.C. Kirkwood	University of Strathclyde, Dept. of Bioscience and Biotechnology, Todd Centre, 31 Taylor Street, Glasgow, G4 ONR, UK
Z.M. Lees	Faculty of Agriculture Microbiology and Plant Pathology, University of Natal, PO Box 375, Pietermaritzburg 3200, South Africa
J.W.S. Longhurst	ARIC (Atmospheric Research and Information Centre), Department of Environmental and Geographical Sciences, Manchester Metropolitan University, Chester Street, Manchester, M1 5GD, UK
A.J. Longley	EPSRC, Clean Technology Unit, Polaris House, North Star Avenue, Swindon, SN2 1ET, UK
R. Mackison	474 Reigate Road, Epsom, Surrey, KT18 5XA, UK
P.S. Owen	ARIC (Atmospheric Research and Information Centre), Department of Environmental and Geographical Sciences, Manchester Metropolitan University, Chester Street, Manchester, M1 5GD, UK
D. Pearce	University College London, Gower Street, London, WC1E 6BT, UK
D.W. Raper	ARIC (Atmospheric Research and Information Centre), Department of Environmental and Geo-

graphical Sciences, Manchester Metropolitan University, Chester Street, Manchester, M1 5GD, UK

E. Senior Faculty of Agriculture Microbiology and Plant Pathology, University of Natal, PO Box 375, Pietermaritzburg 3200, South Africa

C.J. Suckling Department of Pure/Applied Chemistry, University of Strathclyde, Thomas Graham Building, 295 Cathedral Street, Glasgow, G1 1XL, UK

J.W. Twidell The AMSET Centre, School of Engineering and Manufacture, De Montfort University at Leicester, Leicester, LE1 9BH, UK

A.R.F. Watson ARIC (Atmospheric Research and Information Centre), Department of Environmental and Geographical Sciences, Manchester Metropolitan University, Chester Street, Manchester, M1 5GD, UK

Contents

1 Environment and human influence

R.C. KIRKWOOD

1.1 Introduction

Human population growth has resulted in quite massive impacts on natural environments, particularly in regions of industrial and urban development in the Western world. In extreme situations natural habitats (or homes) have been transformed into concrete and tarmacadam labyrinths from which all but a few well-adapted natural species have retreated. In the surrounding countryside the effect of human activity is more subtle but evident in the creation of dams, the development of agriculture, horticulture and forestry and the exploitation of geological (e.g. rock, gravel and sand) and biological resources (e.g. fisheries). The concomitant massive utilisation has resulted in greater fossil fuel combustion (increase in energy), affecting the long-term availability of these non-renewable resources and causing major contamination of the earth's atmosphere and possibly changes in global climate. In addition, the release of organic and inorganic pollutants resulting from a whole range of human activities threatens the future well-being of terrestrial, freshwater and marine environments.

In this Chapter the nature of the major environmental issues will be briefly reviewed and consideration given to strategies for sustainable resource development and maintenance of biodiversity; the concept of clean technology will be briefly introduced.

1.2 The environmental issue

1.2.1 Sustainability and biodiversity

The conservation of our natural heritage is encompassed in the concept of sustainability. Sustainable development can be regarded as the use of ecological habitats (ecosystems) in a manner which satisfies the needs of current generations without compromising the ability of future generations to meet their own requirements. The term natural heritage includes the diversity of flora and fauna, geological and physiographic features; it also encompasses natural beauty and amenity. It includes the land, sea and air, together with the diversity of the flora and fauna. Such biodiversity is found in natural and semi-natural ecosystems but tends to decline with increased human activity as a result of environmental stress due to pollution, disturbance, etc. This is, perhaps, most evident in the case of industrial and urban developments in which natural or semi-natural habitats are lost; the additional impact of human activities results in water and air pollution; the effects on perhaps distant habitats is more insidious.

The capacity of our natural heritage to sustain the pressures of modern human activities may be limited, and the need for sustainable development of natural resources and careful environmental husbandry is becoming appreciated. The requirements of human and other living organisms include unpolluted sources of energy, fuel and pure water. Sustainable utilisation of these renewable resources requires good environmental management. The principles of sustainability must be applied to all proposed developments, to audit land use, manufacturing processes and waste effluent production. The concept of clean technology requires a fundamental appraisal of manufacturing processes which would include an assessment of the impact of the product and by-products.

1.3 The environment, abiotic and biotic components

Ecology has been variously defined as 'the study of plants, microbes and animals in relation to their habitats' or 'the scientific study of the interactions that determine the abundance and distribution of organisms, including the human species'[1]. The success of individual species populations, including humans, depends on the impact of a large number of chemical, physical and biological factors upon natality, mortality and the carrying capacity of the habitat. Natural communities are assemblages of species populations structured by biotic interactions and dependent on the physical/chemical (abiotic) characteristics of the environment; abiotic factors include climatic, topographic, geologic and edaphic (soil) factors. Environmental effects may be subtle and perpetuated by humans.

Climate includes light, temperature, rainfall, humidity, wind, etc., and of these temperature and rainfall conditions are particularly important since they determine the distribution of the world's great vegetation formations (climatic climax communities). Examples include tropical rainforest, deciduous and coniferous forest, whose world distribution reflects climatic conditions. For each abiotic factor, individual species populations have minimum, maximum and optimum values reflecting the tolerance range of the species for that factor. Within a species, variance in tolerance limits exists according to populations of sub-species, ecotypes, biotypes, etc. In addition to variation with latitude, climatic conditions are influenced by topography, altitude, aspect and slope, these being particularly important in relation to habitat microclimate. Geological factors, particularly with regard to rock hardness and type, greatly influence landscape characteristics, and both determine the chemical and physical properties of the soil which overlays the baserock.

The biotic component of the habitat involves the complex infrastructure of plants, animals, and microbes which compose the community. The inter-relationships between individuals of a species may involve (intraspecific) competition for available resources. Interspecific inter-relationships may be complex and include competition, predation, parasitism, mutualism and proto-cooperation. Understanding of the nature of these abiotic and biotic relationships has led to our appreciation of the ecosystem concept.

1.4 The ecosystem, energy flow, biogeochemical cycling and biodiversity

The ecosystem is the fundamental unit of ecology incorporating biotic and abiotic components of the environment. Energy flows through the ecosystem from the plants (autotrophs, primary producers) to the other feeding organisms (heterotrophs). In plants, light energy is transformed during photosynthesis into potential chemical energy which is utilised by the plant in catabolism or is available to other trophic (feeding) levels of the ecosystem (Figure 1.1). Autotroph biomass (weight of plants per unit area) is greater than herbivore biomass > carnivore > detrivore biomass reflecting utilisation of energy, and inefficiencies in its transfer from one trophic level to another. This is reflected in the term biomass pyramid and in the present context is relevant in the biomagnification of persistent chemicals which are transferred from one trophic level to the next in food chains. Associated with this energy flow, and driven by it, are a series of biogeochemical cycles, each of which has a reservoir component (gaseous or sedimentary) and a dynamic cycling component (abiotic → biotic → abiotic) (e.g. Figure 1.2). Soil fertility depends upon such cycling mechanisms but they also represent the mechanism by which biomagnification of heavy metals, persistent pesticides and long half life radioisotopes may occur in food webs!

Biodiversity can be regarded as a measure of ecosystem vigour and stability. In long developed and relatively undisturbed ecosystems, biodiversity may be considerable and complex. It is difficult to predict the long term effect of pollutants on the complex inter-relationships between species populations in ecosystems in which the effects of environmental stress may be incompletely understood. In general, however, recovery of ecosystems from small perturbations is possible; their ability to withstand environmental stress depends, however, upon a number of characteristics, including inertia (ecological buffering capacity), elasticity, amplitude, resilience, hysteresis and malleability.[2]

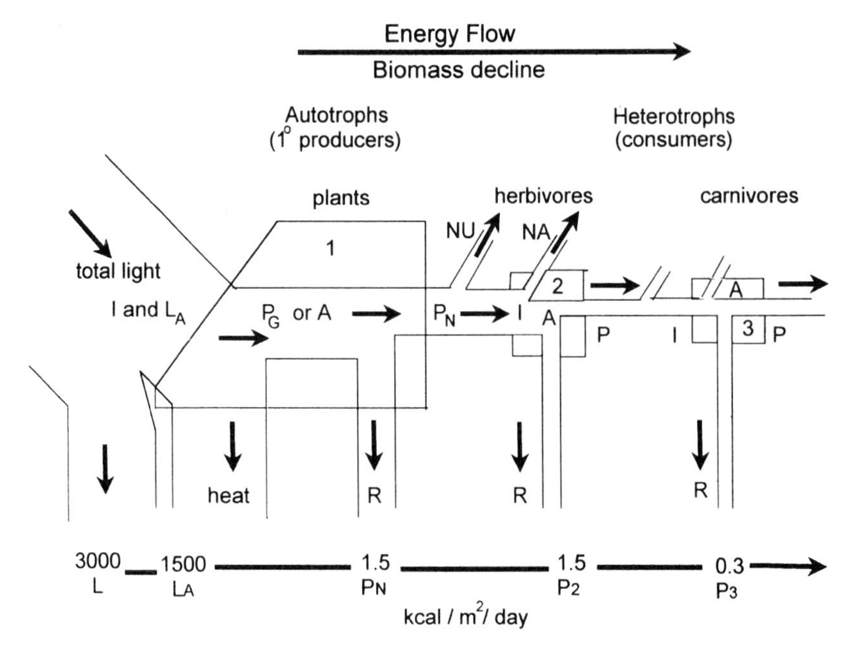

Figure 1.1 A simplified energy flow diagram showing three trophic levels (biomass boxes numbered 1, 2, 3) in a linear food chain. Standard notations for successive energy flows are as follows: I = total energy input: L_A = light absorbed by plant cover: P_G = gross primary production: A = total assimilation: P_N = net primary production; P = secondary (consumer) production; NU = energy not used (stored or exported): NA = energy not assimilated by consumers (egested): R = respiration. Bottom line in the diagram shows the order of magnitude of energy losses expected at major transfer points, starting with a solar input of 3000 kcal per square metre per day (after Odum[1]).

There are a number of guiding principles which relate to the impact of an environmental pollutant. These include its acute and sub-lethal toxicity, teratogenic and mutagenic effects, persistence (half-life), biomagnification in food chains, and synergistic/antagonistic interactions with other chemicals in the environment.

1.5 Human population increase and its influence on the environment

The development of the human population has followed a trend which is typical of any species population, a prolonged 'lag' phase was followed by exponential and logarithmic phases. Currently the population stands at approximately 5 billion people with around 1 billion being added in 15 years. As is the case with any species population, the future trend will depend on the availability of the necessary environmental resources (carrying capacity) of the biosphere. The nature of environmental resistance to further population development must include the availability of clean

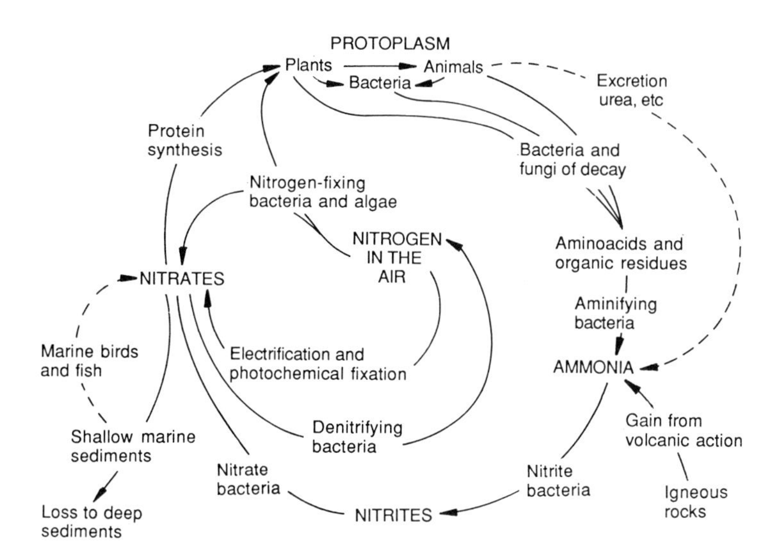

Figure 1.2 The nitrogen biogeochemical cycle. The circulation of nitrogen between organisms and environment is depicted along with micro-organisms which are responsible for key steps (after Odum[1]).

Table 1.1 Main environmental impacts from human activities

Process	Impact
Industrial and urban development	Elimination of natural and semi-natural habitats
Energy production	Particulate material Oxides of sulphur Oxides of nitrogen Carbon gases Partially combusted hydrocarbons Heavy metals
Farming	Soil erosion Eutrophication (nitrates, phosphate, unused feed, silage, effluent, etc.) Pesticide pollution and biomagnification (e.g. persistent compounds)
Industrial production	Organic and inorganic effluents (e.g. PCBs, sulphites and sulphates, heavy metals) pH effects (e.g. acids and alkalis) Eutrophication (e.g. effluents from pulp mills, tanneries, food industries)

air, clean water, unpolluted land and the availability of adequate energy and food supplies.

The growth of the human population has resulted in a range of impacts on natural ecosystems (Table 1.1). The development of primitive agriculture in Europe in prehistorical times resulted in the loss of large tracts of natural deciduous forest; this process is being repeated today in the extensive removal of tropical/sub-tropical rainforest. In new developed countries intensive agriculture depends on new high-yielding cultivars, mechanisation, the use of synthetic fertilisers and pesticides and possibly frequent irrigation. Intensive production of beef, lamb, pork, poultry, milk and eggs from high-yielding domestic stock may be equally demanding in environmental terms.

Thus, the development of human urban and industrial communities has resulted in the complete loss of natural, semi-natural or agroecosystems, together with intensive and sustained demand for energy, water and food and pollution of surrounding and even distant terrestrial and aquatic habitats.

1.6 The problem of environmental pollutants

It has been suggested that there are now approximately 100 000 synthetic chemicals in commercial use with 1000 new compounds being introduced every year.[3] These may find their way into the environment through accidental leakage during manufacture, transport or usage. In addition, some are deliberately applied to the environment in their use as pesticides, antifouling compounds, paints, etc. It was estimated recently that there are more than 1000 basic active pesticide ingredients (a.i.) available in more than 30 000 different commercial preparations.[4] The manufacture, transport and use of these chemicals leads to their introduction into the environment.

1.6.1 Atmospheric pollution

The major environmental problems of today, however, result from the combustion of fossil fuels with the considerable release of particulate material, carbon, nitrogen, sulphur and other gases into the atmosphere. The problems of atmospheric pollution are considered in Chapter 2 and it is perhaps sufficient to mention the three major problems which are attributed to the current high levels of fossil fuel combustion.

The gradual increase of carbon gases, particularly carbon monoxide (CO), carbon dioxide (CO_2) and methane (CH_4) in the atmosphere is believed to be causing a change in global climate due to reflection of heat emitted from the earth's surface; this is generally referred to as global warming or the greenhouse effect. The matter is controversial, though the gradual increase in CO_2 concentration in the atmosphere is beyond dispute. It has been predicted that these changes could lead to a rise in sea-level, increased frequency of gales and storm surges leading to increased danger of coastal flooding and damage to the infrastructure of towns, roads, railways and sea defences.

Another major problem resulting from atmospheric pollution concerns stratospheric ozone depletion. This has been detected by scientists, particularly above the Antarctic, and is believed to be due to the action of carbon gases, particularly chlorofluorocarbons (CFCs) and related organohalides on the ozone layer of the stratosphere. Ozone depletion is believed to permit enhanced ultraviolet (UV) irradiation and the increased hazard of skin cancer in countries with relatively high sunshine intensities.

Acid rain is caused by the reaction of water with deposits of nitrogen and sulphur oxides which result from fossil fuel combustion. Monitoring of water courses has revealed episodic peaks of acidity with falls from perhaps pH 7 to as low as pH 4 in a few hours. In Europe, America and Canada, acid rain has been reported to affect the growth of conifer plantations and also to reduce fish stocks in upland lochs, possibly due to the effect of dissolved aluminium oxide on the functions of the gills.

1.6.2 Water pollution

While these events are of international dimension, the problems of water pollution, though general, tend to be more localised in nature. Loss of water quality may be due to enrichment (eutrophication) with degradable organic materials including sewage or effluents from food or other industries, farms or fish farms. In addition, the leaching of minerals (e.g. nitrogen, phosphates, potash) from agricultural land or from other sources of enrichment can lead to the development of algal blooms which (on death) can cause a biological oxygen demand (BOD). The increased BODs which result from these sources of pollution cause loss of dissolved oxygen (DO) and can lead to death of aerobes including fish. Other problems arise due to the release of effluents from manufacturing industries. These may include the release of heavy metals such as mercury, cadmium, selenium and persistent organic effluents which may accumulate via food webs in terrestrial or aquatic environments causing toxic effects at the upper trophic levels of food webs. The nature of these problems and the legislation and technologies employed to reduce their frequency and impact are discussed in Chapter 3.

1.6.3. Land pollution

Land pollution with persistent chemicals such as heavy metals, asbestos, pesticides, etc., potentially may present considerable problems and the management of hazardous wastes is essential to meet increasingly stringent legislation. The problems arising from landfill with particular reference to environmental and health risks are discussed in Chapter 4.

Increasing awareness of the nature and sources of pollution problems has resulted in increasingly stringent environmental legislation which is implemented by energy and manufacturing industries, by conservation bodies and pollution inspectorates. Adherence to effluent consents is monitored by the analysis of samples taken from air, water or biota. Methods of extraction, clean-up and chemical analysis have become increasingly sensitive and analytical chemists are able to measure some chemical residues at the part per quadrillion level. While approaches to the toxicological testing of new compounds may not have reached the same degree of sensitivity, it is possible to evaluate the effect of acute or sub-lethal doses on test animals under highly controlled laboratory conditions using ecotoxicological tests.

It may be appropriate to take as an example the current minimum requirements for the registration (or re-registration) of a pesticide compound. The legislation relating to the use of pesticides in the UK encompasses a list of relevant Acts, the most important being embodied in the *Food and Environment Protection Act 1985* and *Pesticides Regulations 1986*. Both require the compliance of manufacturer, contractor, farmer and grower, as well as those concerned with food storage and quality. Manufacturers of a new compound must provide data on its activity and mode of action together with information on its toxicology, metabolism, pesticide residue detection in relevant food crops, ecotoxicity and environmental fate (Table 1.2). The increased rigour of the residue studies requires the use of technically sophisticated equipment currently available to analytical chemists which is capable of detecting trace levels of residues. In addition, acceptable toxicological and ecotoxicological properties of the compound are essential for its registration and approval.

1.7 The scope and nature of the problem of toxic chemicals

The accidental introduction of toxic materials into the environment (Table 1.3) may result during the manufacture, transport, storage and use of toxic synthetic chemicals including pesticides.[4] Many of these compounds may be potentially hazardous[5] but cause problems only on accidental release. Pollution due to accidents may incur the severe costs of remediation and compensation. Notable examples include the Bhopal disaster in India in which many people died or were injured as a result of an explosion in a pesticide factory. Law suits of $(US)15 billion were initiated for compensation, the actual settlement being $(US)470 million. In Italy the explosion at a 2,4,5-trichlorophenol factory at Seveso resulted in the release of dioxins into the atmosphere causing direct threat to human health, disruption of community life and indirectly destruction of farm animals; the costs were greater than $(US)150 million. The effects of chemicals on

Table 1.2 Minimum registration requirements for pesticides

	1950	1990
Toxicology	Acute studies 30–90 day rat	Acute studies 30–90 day rat and dog Chronic rat and dog Reproduction Teratogenicity Mutagenicity Carcinogenicity
Metabolism		Rat and dog Plants Farm animals Soil
Residues	A.i.*: 1 mg/kg	A.i. plus metabolites Food crops:0.01 mg/kg Meat:<0.1 mg/kg Milk:0.005 mg/kg
Ecotoxicity	Acute: birds, fish, bees	Acute (for both a.i. and product): birds (2 spp), fish (2 spp), bees, algae, Daphnia Effects on earthworms and other soil fauna, soil microflora, beneficial spp. Repeat dose studies on birds (2 spp), fish, Daphnia. Bird reproduction
Environmental fate		Soil leaching Soil adsorption/desorption Leaf litter decomposition Biodegradation and half life Hydrolysis Photolysis Partition coefficient Volatility

* A.i. = active ingredient.

environmental and human health, however, may be insidious and not measurable in financial terms.

The management of pollution from synthetic chemicals must take account of existing and new compounds and priorities for risk assessment determined in order to ensure minimal impact on human and environmental health. Risk assessments require analysis of the effects and likely exposure of organisms or ecoystems. Evaluation of the critical physicochemical properties of new chemicals should enable prediction of their transport and fate in the environment (see Section 1.9.3) and a range of toxicological or ecotoxicological tests can be carried out to evaluate their potential toxicity (Figure 1.3). The development of such hierarchical systems can be important in risk assessment and decisions as to the level

Table 1.3 A. Categories and sources of toxic chemicals (after Côté[4])

Toxic chemical issues	Toxic chemical sources
Ecotoxicological	Air emissions
contamination of air, water, soil	Liquid effluents
contamination of biota	Solid and hazardous waste
Public health	Spills
food contamination	transport
water contamination	operations
indoor air quality	Pesticide application
ambient air quality	Drug use
Occupational health	Food and water additives
contamination of air	intentional
hazardous chemicals	unintentional

B. Some examples of human health effects caused by chemicals in the workplace (after Côté[4])

Chemical	Effect
Chromium	Dermatoses
Benzene	Effects on blood, leukaemia
Dibromochloropropane	Low sperm counts and sterility
Lead	Sterility, miscarriages, stillbirths, nervous disorders
Vinyl chloride	Angiosarcoma of the liver
Asbestos	Lung cancer
Polychlorinated biphenyls	Chloracne

and control of the chemical in the workplace, external environment, transport, use and disposal.[6]

Generally waste management involves the control of emissions by a range of treatment technologies categorised as primary (e.g. separation and sedimentation), secondary (e.g. aerating lagoons and trickling filters) and tertiary (e.g. reverse osmosis and activated carbon absorption).[7] An effective control programme may involve a combination of these treatment technologies and must be consistent with the relevant control policies and statutes. In the USA, the Toxic Substances Controls Act details the procedure and guidelines for hazard and exposure analysis.[8] In Sweden, the Act on Products Hazardous to Health and the Environment covers a range of hazardous substances including poisons, PCBs, gasoline, cadmium and medical supplies. Conversely in some countries (e.g. UK, France and Germany) broad enabling legislation has been enacted with simplified administrative procedures.

The international nature of the problems caused by toxic chemicals has been emphasised by attempts to dispose of hazardous waste arising from developed countries in less developed areas, including Africa, the Caribbean and South America. This has stimulated international efforts under the auspices of the UN Environment Development Programme (UNEDP) to adopt the Cairo Guidelines and Principles for the Environmentally Sound

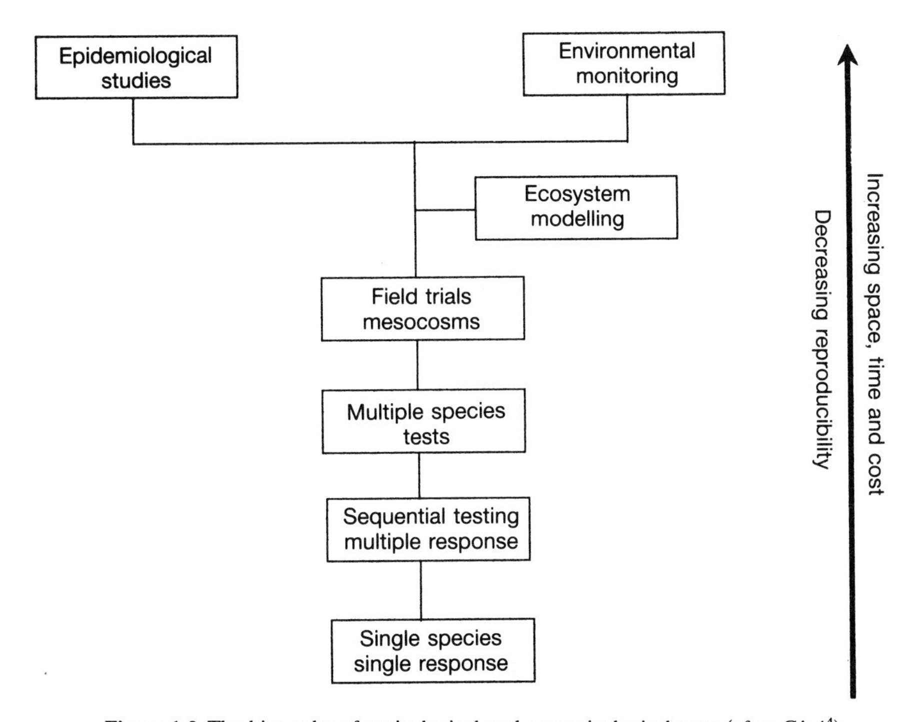

Figure 1.3 The hierarchy of toxicological and ecotoxicological tests (after Côté[4]).

Management of Hazardous Wastes proposed in 1987. The Global Convention on the Control of the Transboundary Movements of Hazardous Wastes has been drawn up and signed (the Berne Convention). Implementation of such conventions is notoriously difficult to achieve and their success may depend on recognition of the need for international co-ordination and consistency in regulatory frameworks. This need to harmonise approaches and information on an international basis is recognised by OECD[9] and other agencies such as the IMO, UNEP, and the World Health Organisation (WHO).

1.8 The strategy of sustainable development

1.8.1 Historical

Historically, the concept of sustainable development of natural resources goes back to the writings of people such as Leopold[10] and Darling[11] who were concerned with human/environment relationships. In the UK the idea that the environment and resource development were inseparable was

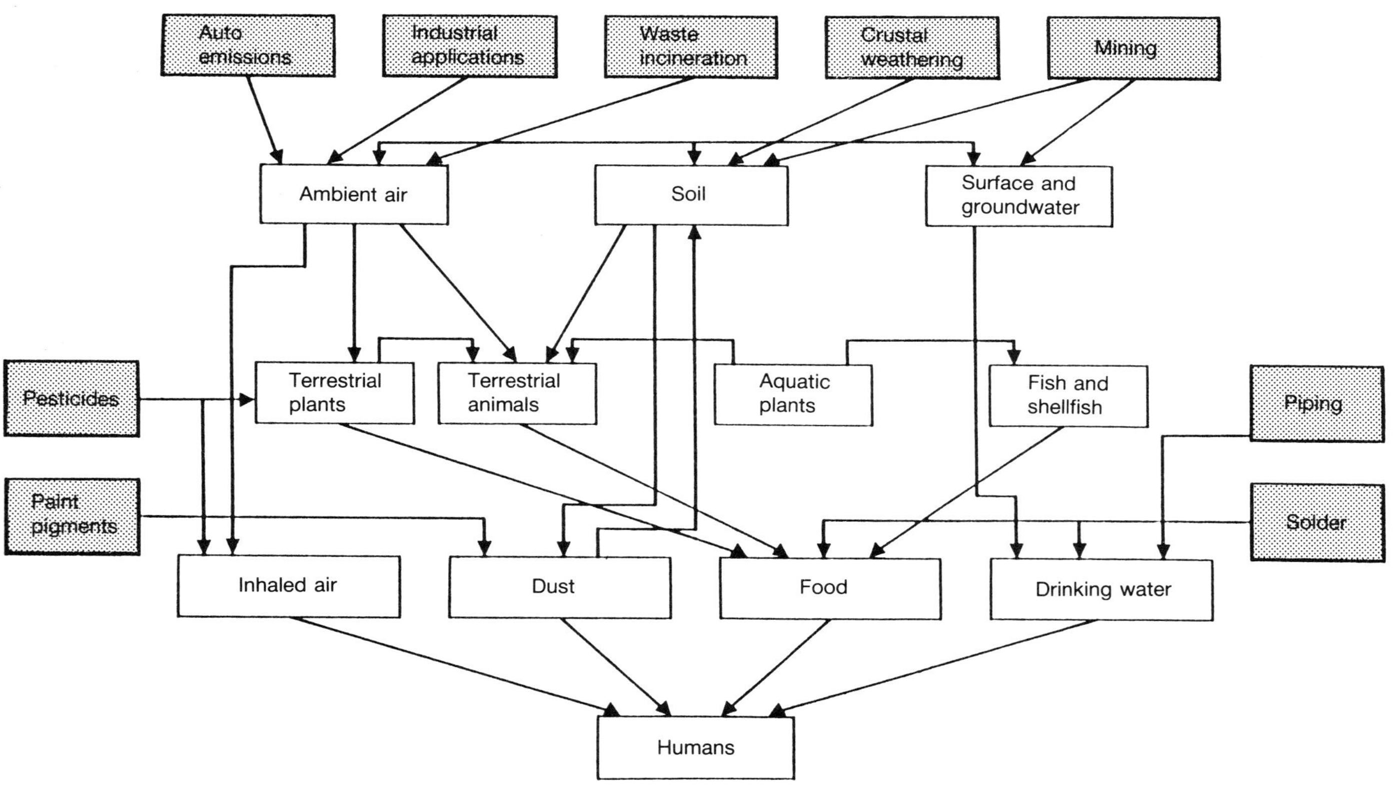

Figure 1.4 Sources and pathways of lead in the environment (after Côté[4]).

reflected in the Parliamentary reports which in the 1940s led to the creation of the Nature Conservancy and the National Parks Commission.

At the international level, the concept of sustainable development has been identified as a means of reconciling the conflicting desires for development and environmental protection in countries at varying stages of economic progress. In 1972, the UN Conference on the Human Environment at Stockholm specifically linked economic development with environmental responsibility. In 1980, the World Conservation Strategy emphasised the need for a new ethic 'embracing plants and animals as well as people'. Three major objectives for conservation in the environment were set out:

1. To maintain essential ecological processes, and life-support systems.
2. To preserve genetic diversity.
3. To *ensure the sustainable use* of species and ecosystems.

In 1983 the Bruntland report (World Commission on Environment and Development) enunciated guiding principles for sustainable development:

1. It should meet the needs of the present without compromising the ability of future generations to meet their own needs.
2. At a minimum it must not endanger the natural systems that support life on earth: the atmosphere, the waters, the soils and the living beings.

The recent UN Conference on Environment and Development (UNCED) held in Rio in 1992 gave further impetus by creation of the UN Commission on Sustainable Development.[12] A comprehensive array of policies and programmes was drawn up under Agenda 21 which provided a framework for action on developing sustainability into the next century. In addition, it called for national strategies for conservation of biodiversity to be integrated into National Sustainable Development Plans. The Rio declaration identified 27 points which placed humans at the centre of concerns for environmental development: 'They are entitled to a healthy and productive life in harmony with nature'. We may presume that 'nature' is entitled to similar considerations.

In fact biodiversity and its conservation have become international themes of considerable importance and a surprisingly large number of conservation programmes are under way, both nationally and internationally.[13] Within the USA, for example, one of the major programmes is the *Systematic Agenda 2000: Charting the Biosphere* which complements other biodiversity initiatives which followed the UNCED meeting in Rio. These initiatives are being prepared by a variety of agencies and organisations including DIVERSITAS, a joint programme of the United Nations Education, Scientific and Cultural Organisation (UNESCO), the International Union of Biological Sciences (IUBS) and the Scientific Committee on Problems of the Environment (SCOPE). These and other non-governmental agencies including the World Resources Institute (WRI), the Worldwide Fund for Nature (WWF), and the World Conservation Union/International Union for the Conservation of Nature and Natural Resources (IUCN) have combined to define a need and provide suggestions of approaches to be followed in measuring and recording diversity. Natural biodiversity has been recognised as our most valuable, threatened, undeveloped living resource and the need for exploration, inventory and conservation is universally recognised.

In 1993, the European Community adopted its Fifth Action Programme on the Environment (*Towards Sustainability*), providing a basis for future Community action on the environment.[14] The term 'sustainable' was used to describe policies and strategies enabling continued economic and social development without detriment to the environment and human resources. It acknowledged that continued human activity and development depended on maintaining the quality of these resources.

In the UK, environmental policy has increasingly recognised the importance of these issues. The White Paper on the Environment, *This Common Inheritance: Britain's Environmental Strategy*[15] and the *Natural Heritage (Scotland) Act 1991* also recognised the importance of the environment in sustaining human existence.

1.8.2 National responsibilities

It is a national responsibility to consider the concept of sustainable development and how it might be applied to fulfil statutory responsibilities for conservation of natural heritage on a national basis. In Scotland, Scottish Natural Heritage (SNH) is the official conservation body, responsible to the Scottish Office Environmental Department (SOEnD). Although a small country, Scotland's natural heritage is particularly diverse and unspoilt, yet its environmental problems are typical of those faced by any developed nation. A recent policy paper, *Sustainable Development and the Natural Heritage*, took a holistic view, encompassing cultural and historical dimensions together with the more obvious factors of wildlife, habitat and landscape.[16]

The SNH rationale for sustainable development was based on a set of fundamental premises and a series of guiding ethical principles. Fundamental premises, based at the global level, were as follows:

1. All human activity is ultimately dependent upon the environment, its natural resources and processes.
2. The human population is growing rapidly.
3. Resource use per capita is high in developed countries and is growing rapidly in developing countries.

4. This high level of human activity is likely to have serious consequences for our quality of life and even for our survival.
5. Our interaction with our environment and its natural processes are so complex that it is often difficult to predict the consequences of our activities.

These premises provide a basis for the theory and practice of sustainable development while a number of ethical principles will act as guidelines in achieving improved sustainability of human activities.

1. *Intergenerational equity* – we should do nothing which puts at risk the natural environment's ability to meet the needs, both material and non-material, of future generations.
2. *International equity* – countries should accept responsibility for the environmental impacts of their own economic activities on other countries; they should avoid 'exporting' their environmental problems.
3. *Societal equity* – one sector of society should not exploit natural resources nor damage the environment at the expense of another.
4. *Inter-species equity* – human society should respect other life-forms; rarely, if ever, are we justified in driving them to extinction for our own purposes.

These guiding principles underlie the SNH policy for sustainable development in a Scottish context. The rationale takes account of the fact that the concept of sustainable development is interpreted differently by economists, ecologists, technologists, farmers, foresters and planners. It also recognises that there are a number of fundamental but unresolved questions on the future of human development.

1. whether economic growth *per se* is sustainable;
2. whether development and conservation are mutually incompatible;
3. whether technology, which has increased human welfare immeasurably in the past, will be able to ensure more sustainable development in future.

No doubt the development of such a commendable rationale is typical of the policy development of many national governments; if it is not, then it should be.

The SNH policy document referred to the important question of the sustainability of economic growth. The Club of Rome report, *Limits to Growth*, gave prominence to the views of many resource economists and environmentalists who believed that economic growth *per se* was inherently unsustainable; they proposed that unless growth was stopped or reversed in a controlled manner, modern society was doomed.[17]

The interest of financiers and industrialists in sustainability was aroused recently by the concepts of the green consumer and green growth. The resulting green developments included the marketing of green products ranging from detergents to environmental unit trusts. There is a view that the initial enthusiasm of environmentalists for such ideas has been tempered by the more cynical acknowledgement that green growth may be more sustainable, but it merely postpones the inevitable collapse which will occur due to excessive human consumption.[16] These and other considerations relating to the economics of clean technology are discussed in Chapter 5.

1.8.3 Sustainable development

The achievement of greater sustainability will require changes in relevant policies, regulatory and planning frameworks to which conservation bodies can contribute, but over which they have no direct control. SNH has developed a set of guidelines for sustainability[16] which, though devised for the Scottish context, have international relevance.

Guideline 1: Non-renewable resources should be used wisely and sparingly at a rate which does not restrict the options of future generations.
Non-renewable resources include fossil fuels, mineral ores, aggregates and peat which can be regarded as a slow-renewing resource; they include also unique geological and landscape features. The rate of depletion should take account of the resource uniqueness, the availability of alternatives, the possibility of substituting alternative resources and the opportunities for re-cycling or re-use of primary materials. Developments in new technology may enable renewable resources to be substituted for non-renewable, e.g. generation of electricity by alternatives to fossil fuel combustion (see Chapter 11).

Guideline 2: Renewable resources should be used within the limits of their capacity for regeneration.
The human exploitation of renewable resources includes the utilisation of forestry, fisheries and agriculture; for each there is a maximum sustainable yield beyond which resource degradation becomes evident. Similarly, use of the atmosphere, rivers, estuaries and the sea as waste depositories depends upon their natural self-cleansing activities. Overloading of their self-cleansing capacity may lead to a long-term decline in the ability of the resource to act as a 'sink' for waste materials, also as a productive resource of clean air, water or food supplies.

Guideline 3: The quality of the whole natural heritage should be maintained and improved.
Where excessive, the exploitation of a resource may trigger a vicious decline in the quality of a renewable resource; for example, run-off from

intensively managed arable land may cause eutrophication of a neighbouring loch. Due to the complexity of such interactions it may be difficult to predict the timescale, nature or extent of their effect. Clearly the development of effective monitoring systems capable of the early and continuing detection of such impacts would be essential in a holistic approach to sustaining our natural heritage.

Guideline 4: In situations of great complexity or uncertainty, the 'precautionary principle' should apply.
The precautionary approach to environmental issues was developed in relation to the German regulatory system and introduced in a UK context by the Royal Commission on Environmental Pollution.[18,19] Subsequently it was adopted by the UK government in *This Common Inheritance: Britain's Environmental Strategy*,[15,20] and was a fundamental principle incorporated in the *Environmental Protection Act 1990*.

The precautionary principle advocates restraint on development if possible environmental damage is reasonably suspected, but not specifically proven. This action is advocated in the following situations where:

1. the impacts could be significant and/or potentially irreversible;
2. the nature and extent of the possible impacts are uncertain;
3. it is impossible to predict or identify the impact of a development due to the complexity of the system interactions.

It is inherent in this approach that the uncertainty is removed by scientific endeavour as soon as possible, thus restoring the scientific basis of decision making.

Guideline 5: There should be an equitable distribution of the costs and benefits of any development.
Socially equitable distribution of costs should prevent a development appearing to be sustainable overall while providing all of the benefits to one sector of society (or generation) and imposing all the costs on another. One mechanism for ensuring the distribution of costs (and benefits) is to require that the polluter or developer pays the full costs involved, including compensation for non-material costs (e.g. energy).

Clearly the future welfare of the human population and its natural heritage equally depends on the adherence of successive generations to the principles encompassed within these guidelines. Enunciation of principles must be accompanied by appropriate legislation and its effective implementation. The concept of clean technology implies an inherent evaluation of human activities particularly in regard to resource utilisation, manufacturing and waste disposal processes; the objectives would be to achieve sustainable development which complies with the ethos of the guidelines above and meets the requirements of national legislation.

Perhaps the greatest need for sustainable development is evident in countries of the Third World and Eastern Europe. In the Third World, for example, non-selective utilisation of rainforest for hardwoods has been carried out at a rate which cannot be sustained long-term and which has resulted in the loss of ecosystems of considerable biodiversity and potential in respect to yield of 'natural products'. The development of selective felling policies combined with replanting should enable a more sustainable development of these valuable rainforests.

In Eastern Europe the problems are different. The relatively low investment in modern technology, perhaps combined with less strict legislation or inadequate implementation, has resulted in widespread contamination of terrestrial and aquatic resources with atmospheric and aquatic pollutants. The return from investment, in terms of sustainability and improved quality of the environment, is likely to be considerably greater in these countries than would occur in more developed nations of the Western world (see Chapter 6).

1.9 Clean technology in relation to major industrial processes

The widely varying types and amounts of organic and inorganic chemicals used in our modern society may produce inherently complex control problems. Factors contributing to their complexity include the following:

1. diversity of their physical, chemical and biochemical properties;
2. interacting multimedia (air, land and water) transport pathways for toxic chemicals;
3. diversity of chemical producers and users;
4. previous disposal or use practices requiring site clean-up;
5. uncontrolled or accidental release of chemicals during production, storage, transport, use and disposal;
6. diversity of regulatory responsibility.

Bishop[7] has discussed the importance of these factors in relation to toxic chemicals and outlined the development of control strategies which can be classified into five functional areas, viz. regulation of production and use; *waste minimisation*; control of release; waste treatment; clean-up of contamination. The concept of clean technology is inherent, particularly to the second area.

Clean technology and clean-up technology are discussed in principle in Chapter 6 and with reference to specific industries including agrochemicals/pharmaceuticals (Chapter 7), plastics (Chapter 8), food (Chapter 9), bulk chemicals (Chapter 10) and energy (Chapter 11). In introducing the topic perhaps it is sufficient to mention that waste minimisation is receiving world-wide emphasis and governmental encouragement. In the USA waste minimisation strategies feature four management approaches (Table 1.4)

Table 1.4 Waste minimization approaches and techniques (after Bishop[7])

Inventory management and improved operations	*Production process changes*
Inventory and trace all raw materials	Substitute non-hazardous for hazardous raw materials
Purchase fewer toxic and more nontoxic production materials	Segregate wastes by type for recovery
Implement employee training and management feedback	Eliminate sources of leaks and spills
Improve material receiving, storage and handling practices	Separate hazardous from non-hazardous wastes
Modification of equipment	Redesign or reformulate end products to be less hazardous
Install equipment that produces minimal or no waste	Optimise reactions and raw material use
Modify equipment to enhance recovery or recycling options	*Recycling and re-use*
Redesign equipment or production lines to produce less waste	Install closed-loop systems
Improve operating efficiency of equipment	Recycle on-site for re-use
Maintain strict preventive maintenance programme	Recycle off-site for re-use
	Exchange wastes

Table 1.5 Control strategies available for the management of chemicals in an industrial context (after Côté[4])

1. Recycle process solutions and chemicals
2. Re-use waste water for secondary uses
3. Recovery for sale or exchange
4. Reduction through waste concentration
5. Waste segregation
6. Increase purity of raw materials
7. End of pipe treatment
8. Land disposal

in addition, waste minimisation audits (Table 1.5) result in step-by-step lists of waste reduction opportunities and ranking guidance for selection of the optimum option for any industry.[4,7] The need for strategic research into clean technology approaches in the manufacture of synthetic chemicals, farming and food production has been recognised in reports commissioned by the UK Agricultural and Food Research Council (AFRC)[a] and Science and Engineering Research Council (SERC)[b] through their Clean Technology Unit. These reports are briefly reviewed below.

1.9.1 Clean synthesis of bulk chemicals

Clean technology includes the discovery and development of cleaner products to substitute for existing polluting ones and the incorporation of

[a] Now Biotechnology and Biological Sciences Research Council (BBSRC).
[b] Now Engineering and Physical Sciences Research Council (EPSRC).

new waste-free processes in the manufacture of existing products. UK companies involved in fine chemical manufacture might reasonably maintain that clean processes are being developed by maximising reaction yields, optimising solvent recovery and minimising energy utilisation. They appreciate the adverse effects of chlorinated solvents and the need to minimise their release into the environment and to find suitable alternatives. Incorporation of clean technology may be influenced by two factors: (1) the capital costs involved and (2) the lack of information on 'clean' alternatives to existing manufacturing processes. In the UK, pressure on industry to adopt clean technology increased with the implementation of the Integrated Pollution Control legislation enacted in the *Environmental Protection Act 1990* to meet EC directives. The timetable for implementation related first to fuel and power, waste disposal and mineral industries (1992), chemical industries (1993), metallurgical and other manufacturing industries (1995).

Biological processes, being 'natural', may enable a cleaner route to the synthesis of chemical products, reducing certain classes of waste (e.g. heavy metals and toxic organic chemicals). Such processes may increase biomass production and thus the problem of biological oxygen demand (BOD). It may be possible, however, to convert waste biological material into a marketable product, and the potential of lignins, lignocelluloses and hemicelluloses for use as a feedstock has been exemplified.

The lack of information on clean alternatives to existing manufacturing processes presents a golden opportunity to researchers in universities and other institutes of higher education to tackle challenges of academic interest and of practical importance. In terms of bulk synthesis a number of areas requiring strategic research effort were identified by Suckling *et al.*,[21] whose recommendations included:

1. optimisation of synthetic reaction conditions involving biocatalysts;
2. identification of new biocatalysts with appropriate environmental properties;
3. the identification of alternative renewable feedstocks with balanced functionality;
4. the integration of separation processes into reactor design.

They commented that basic enabling research relating to a range of products and processes should be complemented by targeted research enabling the incorporation of new technology into a clean process or product. An in-depth discussion of some of these topics is presented in Chapter 10.

The potential for application of clean technology has been acknowledged to be immense and could include:

1. biological control of pests to complement the use of synthetic pesticides;

Table 1.6 The major farming processes (after Legg *et al*[22])

Process	Essential sub-processes
Plant production	Photosynthesis and growth Nutrient utilisation Weed and pest control
Animal production from plants	Preservation of plant material Feed conversion, including grazing animals Protection of animal health and welfare Fish production
Food and industrial feedstocks from plants	Separation of components Storage Processing Production of fine chemicals and pharmaceuticals
Food and industrial feedstocks from animals	Rendering of animal and fish residues Meat, leather and wool processing

2. reduction in the effective dose of pesticides by the use of slow release or improved formulation of pesticides (e.g. the use of adjuvants);
3. minimisation of surfactant emission by the use of controlled release detergents;
4. completely biocompatible surfactants;
5. biodegradable package materials;
6. metal-free photographic films;
7. solvent-free adhesives and gloss paints.

1.9.2 Farming as an engineering process

Farming and the processing of farm products have always been an essential part of human activity. Two major developments have occurred in the twentieth century, however, which have changed agriculture and related activities to such an extent that they may now pose an environmental threat. Intensive farming, in response to the human population explosion, combined with the advent of synthetic chemicals for use as fertilisers and pesticides, has created a series of environmental problems as discussed earlier (Section 1.5).

Legg *et al.*[22] have examined the major farming processes and sub-processes (Table 1.6) and their associated environmental impacts. The growth of green plants involves the synthesis of sugars in light by photosynthesis; ultimately the economic yield of the crop depends on the efficiency of this process. Plants grow most actively when provided with appropriate nutrients, principally N,P,K (Section 1.5) and the fertility resources of the soil can be supplemented by the application of inorganic fertilisers or organic manures (e.g. farmyard manure). The mineral forms of many nutrients may be lost from the crop/soil system to cause major

Table 1.7 Novel approaches to improve the environmental efficiency of crop production (after Legg *et al*[22])

Processes	Environmental problems	New approaches/research
Photosynthesis and growth	Unwanted crop and forest biomass Growth regulatory chemicals	Tailor crop residues for alternative uses Modify genetic control Remote sensing and automatic control systems Algal production systems Alternative crops Bio-fuel production
Nutrient utilisation	Gaseous emission NH_3 and N_2O Nitrate leaching	Close nutrient systems in greenhouses Soil process – N chemistry and interaction between organic wastes and soil structure
	P loss through erosion	Precise application in space and time Hyper-accumulator plants
Weed and pest control	Spray and vapour drift Soil contamination and leaching of pesticides	Reduced chemical usage Development of plant resistance to pests, biological and other non-chemical methods of weed control

contamination of the surrounding environment. For example, transformations of nitrogen compounds to either nitrous oxide or ammonia leads to gaseous emissions which may add to the problems of global warming or acid rain respectively. Nitrates present in high concentrations in the soil solution can leach into rivers or underground water supplies. Legg *et al.*[22] have suggested research approaches which may provide a solution (Table 1.7).

High crop production depends upon the appropriate application of pesticides to control competing weeds and insect pests or disease pathogens. In the UK, current use of pesticides is in the region of 23 000 tonnes of active ingredients per annum, mainly as herbicides (57%) and fungicides (25%); cereal crops may receive at least five pesticide applications.

In addition to environmental concerns arising from intensive crop production, a number of ecological problems have arisen from animal production from plants (Table 1.8). These problems arise from silage production, food conversion by grazing animals, human and animal health treatments and fish production; a number of radically new approaches and necessary research have been suggested. The environmental problems resulting from processes involved in the production of food and industrial feedstocks from plants and animals (Tables 1.9–1.10) have been considered and new approaches suggested which involve biotechnology and genetic engineering.

Table 1.8 Novel approaches to reduce the environmental consequences resulting from animal production from plants (after Legg *et al*[22])

Processes	Environmental problems	New approaches/research
Preservation of plant material	Silage effluent	Biochemical and microbial studies of silage production
Feed conversion including grazing animals: waste production	Methane release Odour, ammonia and nitrous oxide release from wastes	Microbial and biochemical studies of the rumen Microbial and biochemical studies of nutrient status in animal wastes from production to plant uptake Novel waste treatment systems Novel protein sources Optimisation of animal diet
Protection of human and animal health and welfare	Cleaning fluids and antibiotics in wastes Odours from intensive livestock buildings and spread wastes Aerial pollution within livestock buildings, e.g. dust and micro-organisms affecting animal and worker health	Healthier production systems Disease diagnosis Odour prevention Alternative procedures for carcass disposal
Fish production	Contamination of water with pest control agents Unused feed Solid and dissolved wastes	Technologies for production further off-shore Intelligent control for management

1.9.3 Predictive modelling as a tool in clean technology

Environmental exposure to a chemical can be evaluated in mathematical models by combining data for transport and transformation processes with release rates and characteristics. For modelling purposes it is customary to divide the environment into compartments of air, water, soil (or sediment) and biomass; these compartments are divided into sub-compartments which are considered to be homogeneous in chemical phase.

Chemicals can be transported within compartments by homogeneous dispersion processes which do not involve a change in chemical phase.[23] Within or between compartments they may be transported by heterogeneous processes that involve phase changes (e.g. evaporation or sorption); chemical transformation may occur within each compartment. A schematic overview of the major transport and transformation reactions occurring in the environment is shown in Figure 1.5. Within the air and water compartments, homogeneous transport of gaseous and dissolved chemicals is the predominant process (Figure 1.6). Conversely, when a chemical has a high affinity for other environmental compartments (e.g. soil or sediment

Table 1.9 Food and industrial feedstocks from plants (after Legg *et al*[22])

Processes	Environmental problems	New approaches/research
Separation of component	Waste straw, etc., in the field Waste from vegetable production	Genetic engineering to match plant products for their end use Fractionation at source Genetically engineered micro-organisms for cellulose breakdown Rapid, controlled composting
Storage	Pesticide use, residues in food	Non-chemical preservation techniques
Processing	Food processing wastes Wastes from paper production Odour	Controlled anaerobic digestion with variable loading Microbial purification of effluent New process design to minimise waste Rapid controlled composting Odour prevention and removal

Table 1.10 Food and industrial feedstocks from animals (after Legg *et al*[22])

Processes	Environmental problems	New approaches/research
Rendering animal and fish residues	Pathogen control Odour release Bulk disposal	Novel treatment for complete sterilisation Closed processing systems to contain micro-organisms and odour Solid bio-processing to produce animal feed or fertilisers
Meat, leather and wool processing	Fatty and proteinaceous wastes	Separation technologies including bio-technology Pre-separation to avoid complex wastes Rapid controlled composting of solid wastes

particles) heterogeneous transport to these particles and homogeneous transport within these particles may be important.

The nature of transport and degradation processes together with fate analysis are discussed by Wolff and Crossland.[23] The minimum set of physico-chemical properties needed for predicting distribution and for fate analysis includes the octanol/water partition coefficient (log K_{ow}), the vapour pressure and water solubility. Biodegradation in water and soil or the rate of reaction with OH radicals in the atmosphere are also required in

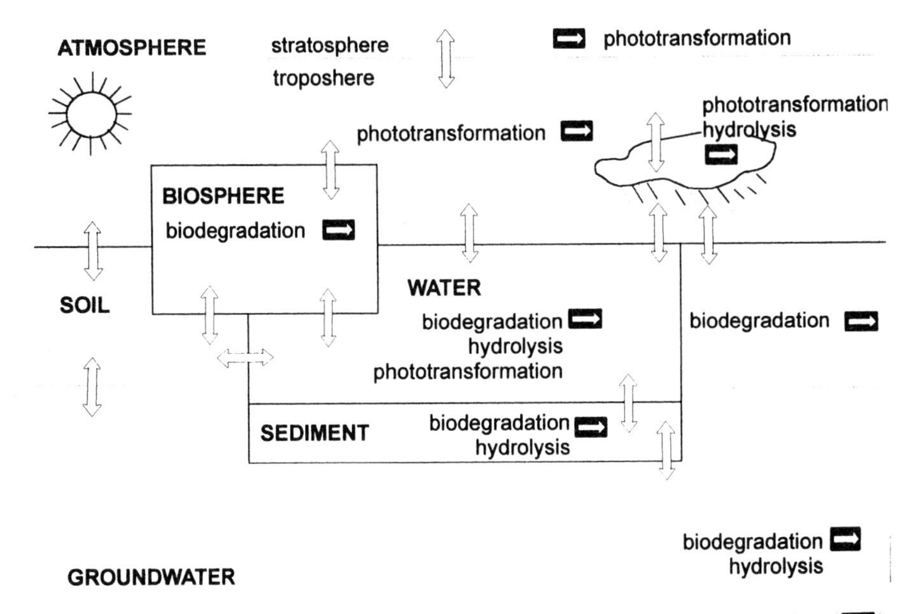

Figure 1.5 Schematic overview of important homogeneous (⇕) and transformation (➡) processes in the environment (after Wolff and Crossland[23]).

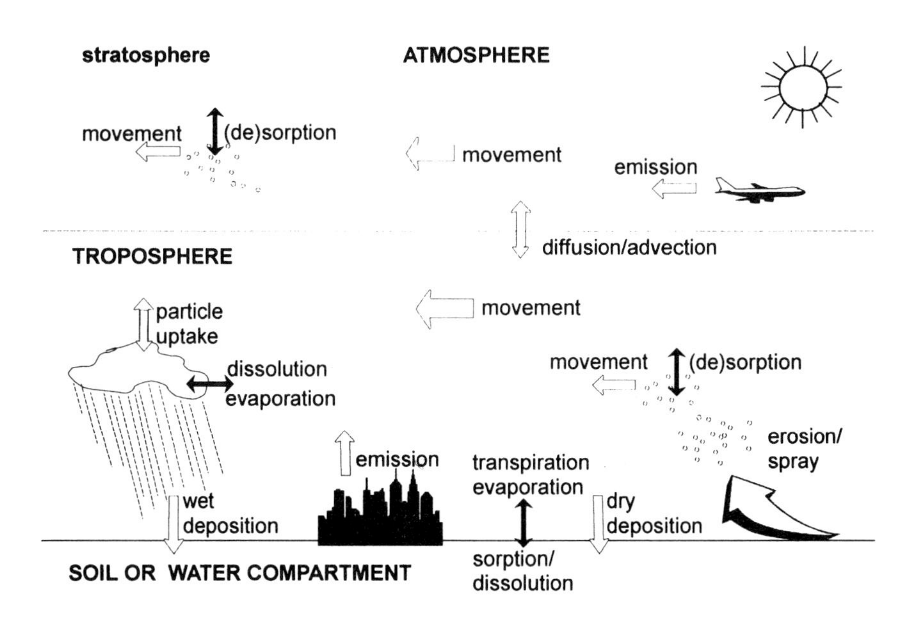

Figure 1.6 Schematic overview of important homogeneous (⇕) and heterogeneous (⬍) transport processes in/into the atmosphere (after Wolff and Crossland[23]).

estimating global persistence. Such mathematical models are reasonably reliable provided that the data inputs are sufficiently accurate; their reliability must be tested, however, in real-world and/or experimental systems.

Assessment of the magnitude of transfer of chemicals between air, water, soil and biota and the resulting concentrations in the respective compartments is a prerequisite for estimating probable exposure, accumulation and toxicity for organisms including humans. Readers are recommended to an excellent review by Scheunert and Klein[24] concerned with predicting the movement of chemicals between environmental compartments (air–water–soil–biota).

The use of modelling approaches in predicting the distribution and fate of pesticides has been carried out by a number of workers including Walker[25]. These studies are particularly relevant in the prediction of pesticide leaching into ground water.[26] The parameters of importance include the rates of adsorption/desorption to the clay–humus colloidal complex in soil/sediment, soil mobility, and the degradation characteristics of the pesticide under aerobic and anaerobic conditions.[27]

1.9.4 Reducing the environmental impact of pesticides

It is widely known that only a small proportion of the applied dose of a pesticide (insecticide, fungicide or herbicide) may reach the target site. In the case of foliage-applied herbicides much of the applied chemical can remain as residues on the leaf surface and may enter the ecosystem by volatilisation, wash-off or abrasion. This can occur because the lipoidal cuticle which reduces water loss from the plant acts as a barrier to the penetration of foliage-applied compounds, particularly those of a polar nature. In fact, the phytotoxicity (and selectivity) of such herbicides depends upon the efficiency of a whole series of inter-related steps including cuticle retention, penetration and tissue absorption[28], translocation to target sites[29] and rates of immobilisation or transformations en route.[30]

Cuticle penetration is dependent on a number of factors including leaf age and waxiness of the cuticle, log K_{ow} of the active ingredient and formulation with appropriate adjuvants including surfactants. At low concentrations (e.g. 0.05%) these adjuvants can have a remarkable effect on the activity of a herbicide, enhancing its absorption and translocation by a factor of perhaps 5–10 fold.[31] Their beneficial action may be due to enhanced surface properties (reduced surface tension, contact angle and increased droplet diameter), predisposition of the cuticle to penetration of the active ingredient; solubilisation of hydrophobic compounds in surfactant micelles may also occur.[28]

The correct choice of adjuvant may be a major factor determining the degree of synergy[32] and the log K_{ow} of the herbicide may be an important factor. For example, a hydrophilic compound (e.g. log $K_{ow} < 1$) appears to require a relatively hydrophilic surfactant while the activity of a lipophilic compound (e.g. log $K_{ow} \sim 5$), however, appears to benefit from a relatively lipophilic surfactant.[33] These findings are consistent with a predictive model developed by Holloway and Stock.[34] A better understanding of the criteria determining selection of a surfactant may enable quite major dose reductions and environmental benefits, though in the case of herbicides detrimental effects on crop/weed selectivity are possible and must be taken into account.

A wide range of soil-applied insecticides, nematicides and herbicides have been detected in ground water[26] and the problem tends to be associated with compounds of intermediate log K_{ow}. There is evidence that these may leach from the soil as the parent compound or as polar metabolites. The development of controlled-release (CR) systems may reduce the volume of pesticide being introduced into the environment. Synthetic microgel polymers within which the active ingredient is incorporated may facilitate the development of seed coatings or granular formulations which, potentially, can provide efficient control of fungal, nematode, insect or weed pests. In the case of granular formulations, placement in a band in the vicinity of the germinating crop seeds may lead to a considerable reduction in the required dose compared with conventional methods. In principle, the active ingredient is ideally located for absorption by the roots or shoots of the developing crop plants and, in the case of xylem-mobile compounds, distribution in the shoot may provide effective control of insect pests or disease pathogens. Such CR microgel formulations may enable the use of relatively water-soluble herbicides for the control of arable (or aquatic) weeds at reduced dose rates for an extended period, with reduced danger of soil leaching and environmental pollution.

This approach appears to have potential as a means of providing economic localised treatment with minimal environmental side-effects.[35] In addition, the development of patentable novel formulation technologies such as controlled release may enable extension of the patent life of existing useful compounds despite increasingly rigorous environmental legislation.

1.9.5 Biotechnological approaches towards improving environmental acceptability of novel pesticides

The development of selective chemicals for pest control has been a fascinating success story which has unfolded during the last four decades (see Chapter 7). Most, if not all, of the currently used pesticides have arisen through the systematic screening of novel chemicals, identification of active 'lead' compounds, synthesis and formulation of the most active analogues. While this approach has been eminently successful there is a view that, ideally, novel compounds should arise from rational design based on understanding of the molecular architecture of specific enzyme receptor sites. Potentially such target site specificity could ensure maximum species selectivity and minimum environmental side effects (on non-target species).

The rational design of herbicidal compounds requires understanding of the receptor proteins at the potential target sites including photosynthesis,[36] carotenoid,[37] amino acid,[38] and lipid biosynthesis,[39] as well as pathways in intermediary metabolism.[40] The biosynthetic routes for synthesis of amino acids in plants are particularly attractive targets for toxicologically safe herbicides and for pesticides aimed at plant pathogens.[38] Thus, as an adjunct to empirical screening, the derivation of new compounds may involve direct testing for potential inhibitors based on gene cloning, and over expression and purification of bacterial enzymes associated with biosynthetic pathways. It has been speculated that as our knowledge of plant metabolism and plant biotechnology increases so the mechanisms of herbicidal inhibition will be discovered[40] and the nature of the receptor sites elucidated.

1.10 Concluding remarks

In concluding this introductory Chapter, it may be appropriate to draw attention to the recent report by the Centre for Exploitation of Science and Technology (CEST) which provided a systematic review of key environmental problems and potential environmental opportunities for UK industry.[41] The opportunities to provide technologies and services which could contribute to solution of these problems is enormous. In the next ten years it is estimated that over £140 billion could be spent in the UK on the solution of environmental problems; the corresponding projections to Europe and the USA are anticipated to be £860 and £1060 billion respectively. The major areas of potential lie in the reduction of greenhouse gas emissions, improvement of water quality and waste management, and it is suggested that between 1991 and 2000, relevant expenditure may be in the region of around £50, 24 and 18 billion respectively (Figure 1.7)

Legislation is the most effective mechanism by which environmental pressure is applied to industry and it is expected that legislation pressure will increase in future. This trend is being led by a small number of environmentally-advanced countries and the CEST report suggests that companies should be encouraged to learn about policy developments in these countries. A long-term view should be taken of regulatory trends and

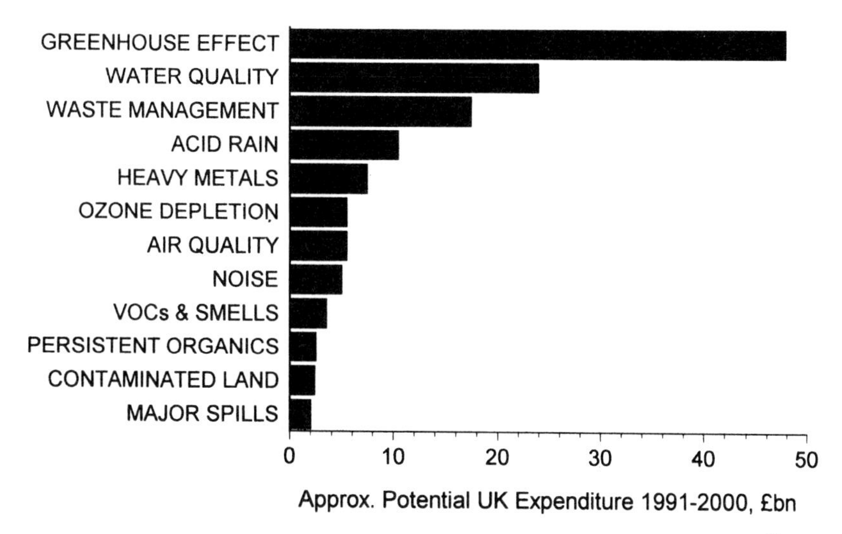

Figure 1.7 Potential UK expenditure on environmental problems (after Good[41]).

there should be improved dialogue on environmental policy between industries and government.

In the longer term, improvements in the environmental performance of industry must depend upon the generation of new knowledge, particularly with regard to the science underlying manufacturing processes. Institutes of higher education have a major role to play in relation to the generation of new knowledge; the opportunities for progress through multidisciplinary approaches are infinite providing the funding is available. It is essential that environmental solutions shift from 'end of pipe solutions' to being inherent in every facet of manufacturing and production processes.

In this Chapter the reader has been introduced to some of the current environmental problems and the philosophy of sustainability; these environmental concerns are explored more fully in Chapters 2–5. In the second part of the book the concept of clean technology is discussed (Chapter 6), and in the succeeding chapters the reader may be able to judge the progress which individual industries have made in approaching environmental problems. It is to be hoped that some of the successes and potential opportunities for development of clean technology will be evident. The ultimate objective of this approach must be to ensure that the stewardship of our natural heritage is consistent with the ethos of sustainability. Exploitation and development of natural resources must be such that the needs of succeeding generations of humans and wildlife are sustained and that the environmental impact of human activities is minimised by rigorous environmental audit combined with ingenious application of the concept of clean technology.

References

1. Odum, E.P. (1971) *Fundamentals of Ecology*, 3rd edn., W.B. Saunders Co. London.
2. Sheehan, P.J. (1991) Ecotoxicological considerations. In: *Controlling Chemical Hazards* (eds. R.P. Côté and P.G. Wells), Unwin Hyman, London, pp. 79–118.
3. Connell, D.W. (1987) Ecotoxicology – A framework for investigations of hazardous chemicals in the environment, *Ambio*, **16**, 47–50.
4. Côté, R.P. (1991) The nature and scope of the toxic chemicals issue. In: *Controlling Chemical Hazards* (eds. R.P. Côté and P.G. Wells), Unwin Hyman, London, pp. 1–18.
5. GESAMP (Joint Group of Experts on the Scientific Aspects of Marine Pollution) (1989) The Evaluation of the Hazards of Harmful Substances Carried by Ships: Revision of GESAMP Reports and Studies, No 17, Reports and Studies, No 35 (London: International Maritime Organisation).
6. Daniels, S.L. and Park, C.N. (1991) Integrating effect and exposure information: an industrial viewpoint. In: *Controlling Chemical Hazards* (ed. R.P. Côté and P.G. Wells), Unwin Hyman, London, pp. 119–143.
7. Bishop, D.F. (1991) Control strategies and technologies. In: *Controlling Chemical Hazards* (ed. R.P. Côté and P.G. Wells), Unwin Hyman, London, pp. 144–174.
8. EPA (Environmental Protection Agency) (1984) *Proposed Guidelines for Exposure Assessment*, Federal Register, Vol. 49, No 227, November (Washington DC), 46304–12.
9. OECD (1984) *The OECD Chemicals Programme*, OECD, Paris.
10. Leopold, A. (1949) *A Sand County Almanac*, OUP, New York.
11. Darling, F.F. (1955) *West Highland Survey: An Essay in Human Ecology*, HMSO, London.
12. Anon (1993) *The Earth Summit: The United Nations Conference on Environment and Development* (UNCED), Graham and Trotman.
13. Colwell, R.R. (1993) Biodiversity. An international challenge for environmental biotechnology. *AGRO-food-INDUSTRY HI-TECH.*, **4**(6), 4–5.
14. Anon (1993) Commission of the European Communities: *Towards Sustainability – A European Programme of Policy and Action in Relation to the Environment and Sustainable Development*, (Cmnd 92), 23 Final, Brussels.
15. Anon (1990) *This Common Inheritance: Britain's Environmental Strategy* (Cmnd 1200), HMSO, London.
16. Anon (1993) *Sustainable Development and the Natural Heritage*: The SNH Approach. Policy Paper, Publications Section, Scottish Natural Heritage, Battleby, Redgorton, Perth, PH1 3EW.
17. Meadows, D.H., Meadows, D.L., Randers, J. and Behrens, W.W. (1972) *The Limits to Growth – A Report for the Club of Rome*, Pan, London and Sydney.
18. Anon (1976) Royal Commission on Environmental Pollution: Fifth Report, *Air Pollution Control an Integrated Approach* (Cmnd 6371), HMSO, London.
19. Anon (1988) Royal Commission on Environmental Pollution: Twelfth Report, *Best Practicable Environmental Option*, (Cmnd 310), HMSO, London.
20. Anon (1992) This Common Inheritance: The Second Year Report on the UK's Environmental Strategy (Cmnd 2068), HMSO, London.
21. Suckling, C.J., Halling, P.J., Kirkwood, R.C. and Bell, G. (1992) *Clean Synthesis of Effect Chemicals*, AFRC–SERC Clean Technology Unit, Swindon, Wilts, SN2 1ET.
22. Legg, B.J., Cumby, T.R., Day, W., Miller, P.C.H. and Phillips, V.R. (1992) *Farming as an Engineering Process*, AFRC–SERC Clean Technology Unit, Swindon, Wilts, SN2 1ET.
23. Wolff, C.J.M and Crossland, N.O. (1991) The environmental fate of organic chemicals. In: *Controlling Chemical Hazards* (ed. R.P. Côte and P.G. Wells), Unwin Hyman, London, pp. 19–46.
24. Scheunert, I. and Klein, W. (1985) Predicting the movement of chemicals between environmental compartments (air–water–soil–biota). In: *Appraisal of Tests to Predict the Environmental Behaviour of Chemicals* (ed. P. Sheehan, F. Korte, W. Klein and Ph. Bourdeau), J. Wiley and Sons Ltd.
25. Walker, A. (1991) *Pesticides in Soils and Water: Current Perspectives*, BCPC. Monograph No 47, BCPC, Farnham, Surrey, GU9 7PH.

26. Boehm, N. and Anderson, T.P. (1990) Soil-applied controlled-release products and suSCon® technology. In: *Controlled Delivery of Crop Protection Agents*, (ed. R.M. Wilkins), Taylor and Francis, London, pp. 125–148.
27. Leake, C.R. (1991) Fate of soil-applied herbicides: factors influencing delivery of active ingredients to target sites. In: *Target Sites for Herbicide Action* (ed. R.C. Kirkwood), Plenum Press, New York, pp. 189–218.
28. Kirkwood, R.C. (1991) Pathways and mechanisms of uptake of foliage-applied herbicides with particular reference to the role of surfactants. In: *Target Sites for Herbicide Action* (ed. R.C., Kirkwood), Plenum Press, New York, pp. 220–244.
29. Bromilow, R.H. and Chamberlain K. (1991) Pathways and mechanisms of transport of herbicides in plants. In: *Target Sites for Herbicide Action* (ed. R.C. Kirkwood), Plenum Press, New York, pp. 245–284.
30. Owen, J.W. (1991) Herbicide metabolism as a basis for selectivity. In: *Target Sites for Herbicide Action* (ed. R.C. Kirkwood), Plenum Press, New York, pp. 285–314.
31. Gaskin, R.E. and Kirkwood, R.C. (1989) The effect of certain non-ionic surfactants on the uptake and translocation of herbicides in bracken (*Pteridium aquilinum* L. Kuhn). In: *Adjuvants and Agrochemicals*, Vol 1. *Mode of Action and Physiological Activity* (eds. P.N.P. Chow, C.A. Grant, A.M. Hinshalwood and E. Simunsson), CRC Press Inc., Boca Raton, FL., pp. 129–139.
32. Hatzios, K.K. (1991) Modifiers of herbicide action at target sites. In: *Target Sites for Herbicide Action* (ed. R.C. Kirkwood), Plenum Press, New York, pp. 169–188.
33. Kirkwood, R.C., Knight, H., McKay, I. and Chandrasena, J.P.N.R. (1992) The effect of a range of nonylphenol surfactants on cuticle penetration, absorption and translocation of water-soluble and non-water soluble herbicides. In: *Adjuvants and Agrichemicals* (ed. C.L. Foy), CRC Press Inc., Boca Raton, FL., pp. 117–126.
34. Holloway, P.J. and Stock, D. (1990) Factors affecting the activation of foliar uptake of agrochemicals by surfactants. In: *Industrial Application of Surfactants II*, (ed. D. Karsa), Special Publication No 77, Royal Society of Chemistry, Cambridge, UK, pp. 303–307.
35. Wilkins, R.M. (ed.) (1990) *Controlled Delivery of Crop-Protection Agents*, Taylor and Francis, London.
36. Dodge, A.D. (1991) Photosynthesis. In: *Target Sites for Herbicide Action* (ed. R.C. Kirkwood), Plenum Press, New York, pp. 1–28.
37. Bramley, P.M. (1991) Carotenoid biosynthesis. In: *Target Sites for Herbicide Action* (ed. R.C. Kirkwood), Plenum Press, New York, pp. 95–123.
38. Mousdale, D.M. and Coggins, J.R. (1991) Amino acid synthesis. In: *Target Sites for Herbicide Action* (ed. R.C. Kirkwood), Plenum Press, New York, pp. 29–56.
39. Harwood, J.L. (1991) Lipid synthesis. In *Target Sites for Herbicide Action* (ed. R.C. Kirkwood), Plenum Press, New York, pp. 57–94.
40. Pallett, K.E. (1991) Other primary target sites for herbicides. In *Target Sites for Herbicide Action* (ed. R.C. Kirkwood), Plenum Press, New York, pp. 124–168.
41. Good, B. (1991) *Industry and the Environment: A Strategic Overview*, Centre for Exploitation of Science and Technology (CEST), 5 Berners Road, London, N1 0PW.

2 Atmospheric pollution: components, mechanisms, control and remediation

J.W.S LONGHURST, P.S. OWEN, D.E. CONLAN, A.F.R. WATSON and D.W. RAPER

2.1 Introduction

This chapter considers the environmental problems posed by the release to atmosphere of sulphur, nitrogen and carbon species. The structure of the atmosphere is briefly described and the key air pollutants are reviewed in terms of sources, concentrations and impacts. Relevant air quality standards and guidelines set by the European Community and the World Health Organisation are presented. Emissions of the major pollutants are described at a range of spatial scales and particular attention is given to the emission profile of the UK as a typical example of emissions from an advanced industrial state. The transport, reaction pathways and deposition of pollutants in the troposphere is discussed with particular reference to the role of photochemistry.

Case studies of four contemporary air pollution problems complete the chapter: toxic organic air pollutants; acid deposition; stratospheric ozone depletion; and the enhanced greenhouse effect. In each of these examples a holistic approach is adopted to consider the nature and scale of the problems posed and the available control and remediation options.

2.2 The structure of the atmosphere

The earth's atmosphere is remarkably thin in comparison with the dimensions of the earth. Half of the mass of the atmosphere lies below an approximate mean height of 5.5 km (less than 0.001% of the earth's radius) above sea level[1] and about 99% of the mass of the atmosphere lies within the lowest 30 km above sea level. The four most important constituent gases in the atmosphere by concentration are nitrogen, oxygen, argon and carbon dioxide. Table 2.1 shows the major constituents of the unpolluted lower atmosphere.[2]

The earth's atmosphere is divided into four main regions according to the temperature variations found therein. The first of these regions, the troposphere, lies between the earth's surface and about 10 km altitude.

Table 2.1 Typical composition of unpolluted air in the lower atmosphere[2]

Gas	Concentration (%)	Weight (%)
Nitrogen (N_2)	78.084	75.527
Oxygen (O_2)	20.946	23.143
Argon (Ar)	0.934	1.282
Carbon dioxide (CO_2)	340 ppm	0.0456
Neon (Ne)	18.18 ppm	1.25×10^{-3}
Helium (He)	5.24 ppm	7.24×10^{-5}
Methane (CH_4)	1.5 ppm	7.75×10^{-5}
Krypton (Kr)	1.14 ppm	3.30×10^{-4}
Hydrogen (H_2)	0.50 ppm	3.48×10^{-4}
Nitrous oxide (N_2O)	0.30 ppm	7.6×10^{-5}
Xenon (Xe)	0.087 ppm	3.9×10^{-5}
Ozone (O_3)	0.02 ppm	6×10^{-6}
Nitrogen dioxide (NO_2)	0.001 ppm	6×10^{-7}

This region is subdivided into two sections: the boundary layer and the free troposphere. The boundary layer is characterised by high turbulence and rapid mixing, and is usually confined to between 500 and 2000 m above the earth's surface. The boundary layer can be thought of as a mixing layer for particulates and trace gases expelled from the earth's surface. For many gases and particles, the increased turbulence of this region allows for deposition back onto the earth's surface. However, if species remain airborne for more than a few days they may be transported into the free troposphere where they are able to travel long distances, carried by atmospheric circulation. The free troposphere is less turbulent than the boundary layer and is characterised by relatively strong vertical mixing; molecules can traverse the entire depth of the troposphere over the course of a few days. Within the troposphere, temperature decreases with altitude at a rate of about 7 K km^{-1}. This lapse rate varies widely from area to area, but never exceeds 10 K km^{-1} except near ground level. The weather systems are contained within the troposphere, with high winds and the highest cirrus clouds being found at the upper margin of the troposphere known as the tropopause. The tropopause is the boundary at the top of the troposphere where the temperature stops decreasing with altitude. The altitude of the tropopause varies greatly with time and geography but is at its highest at the equator (about 20 km above the earth's surface) and lowest at the poles (approximately 10 km above the earth's surface).

The stratosphere lies immediately above the tropopause and is a zone of increasing temperature with increasing altitude until a maximum value of 273 K is reached at 50 km above the equator and mid-latitude regions.[3] If radiation were absorbed and then emitted evenly by the earth's surface it would be expected that there would be a steady decrease in temperature with increasing altitude throughout the atmosphere, as occurs in the troposphere. However, at the tropopause abrupt changes can be seen in the concentrations of some of the more important atmospheric trace constituents; water vapour levels drop sharply while ozone concentrations can increase by an order of magnitude within a few kilometres. It is this sudden increase in the level of ozone in the atmosphere which is indirectly responsible for the temperature increase within the stratosphere. At about 50 km altitude, a maximum in temperature is found. At this point, further increases in altitude are accompanied by falling temperatures. This temperature maximum defines the position of the stratopause in the earth's atmosphere. The pressure at the stratopause is approximately 1 mb, compared to a pressure of 1013 mb at the earth's surface. This indicates that the troposphere and stratosphere together make up approximately 99.9% of the total mass of the atmosphere.

2.3 Key air pollution species

The important air pollutant species are those with emissions which, when considered at the local, national or international scale, may exceed the standards which have been set to protect human health and the environment as a whole. These air pollutants are generally thought to include nitrogen dioxide (NO_2) and its reaction products, sulphur dioxide (SO_2) and its reaction products, ozone (O_3), particulate matter, lead (Pb), volatile organic compounds (VOCs) and carbon monoxide (CO).

2.3.1 Nitrogen oxides

Nitrogen oxides or NO_x is a collective term used to refer to two species of oxides of nitrogen, nitric oxide (NO) and nitrogen dioxide (NO_2). Most anthropogenically emitted nitrogen dioxide is derived from NO. Once NO is emitted into the atmosphere the majority of it is rapidly oxidised to NO_2, through reaction with O_3, or other oxidants. Nitrogen dioxide is generally considered to be more important than NO in terms of human health effects and to reflect this the EC directive values (see Section 2.4) are expressed in concentrations of NO_2. Nitrogen dioxide is a reddish brown gas, a strong oxidant and soluble in water.[4] The main sources of nitrogen dioxide are both natural and anthropogenic. The emission of NO_x during combustion is temperature dependent with more NO_x being emitted from higher temperature combustion processes. In 1991, typical 98 percentile values in the UK varied from 51 to 103 ppb.[5]

2.3.2 Sulphur dioxide

Sulphur dioxide is a colourless gas which reacts on the surface of various airborne solid particles and is water soluble. It is the principal pollutant associated with acid deposition. Sulphur dioxide can react with OH to form HSO_3 which ultimately forms sulphuric acid (H_2SO_4). Sulphuric acid is generally present as an acid aerosol which is often associated with other pollutants. Sulphur dioxide is produced by the combustion of sulphur components which are a natural constituent of fossil fuels, particularly coal and oil. As a consequence, the main sources of SO_2 include domestic fuel combustion, industrial processes and power stations. Small quantities are also emitted from diesel engined vehicles.

High concentrations of SO_2 can give rise to human health effects which are respiratory in nature and can aggravate bronchitis.[4] Other environmental effects of this gas include its toxicity to plants either by direct gaseous toxicity or by acid deposition. Corrosion to buildings, both stonework and metals, is increased in the presence of SO_2.[4] In 1991–92 typical 98 percentile values in the UK varied from 28 to 235 $\mu g\ m^{-3}$.[5]

2.3.3 *Ozone*

Ozone is the tri-atomic form of molecular oxygen. It is a powerful oxidising agent and therefore highly reactive. In the stratosphere, ozone inhibits harmful ultraviolet radiation from reaching the earth's surface. However, in the lowest region of the atmosphere, the troposphere, ozone can be deleterious to human health and vegetation. The majority of the O_3 in the troposphere is formed by the action of sunlight on nitrogen dioxide. The rate at which O_3 is produced is determined by a number of factors including the concentrations of NO, NO_2 and O_3 and the intensity of sunlight. Highest concentrations of O_3 usually occur on hot summer's days as a result of the dependence of the ozone formation reaction on sunlight intensity. Ozone concentrations increase downwind of urban centres as NO (emitted from vehicle activity in urban environments) can destroy O_3. The major human health effects of O_3 include eye, nose and throat irritation, chest discomfort, coughs and headaches.[4] At high concentrations it can cause the alveoli of the lungs to fill up with fluid and impair the absorption of oxygen into the blood. Typical average UK concentrations are around 30 ppb.[5]

2.3.4 *Carbon monoxide*

Carbon monoxide is produced by the incomplete combustion of fossil fuels or organic material. It has a strong affinity for haemoglobin, the oxygen carrying substance in the blood. When oxygen is displaced by CO it can progressively lead to oxygen starvation and severe, acute, health effects. Approximately 90% of UK emissions of CO are emitted from motor vehicles and hence the highest concentrations are generally close to busy roads and in enclosed spaces such as multi-storey car parks. In 1991 maximum 8-hour CO concentrations were in the range 0.9–14.6 ppm at selected UK sites.[5]

2.3.5 *Lead*

This is the most common of the heavy metals emitted to the atmosphere. Emissions arise primarily from motor vehicles using leaded petrol. However, with the widespread introduction and use of unleaded petroleum, atmospheric emissions of lead have significantly declined. Lead is a cumulative poison and long-term exposure can be detrimental to the central nervous system. In 1991 typical UK lead values were in the range 13–1390 ng m^{-3}.[5]

2.3.6 *Particulate matter*

This is a general term covering a complex mixture of organic and inorganic substances present in the atmosphere as both liquids and solids. Particulate matter is removed from the atmosphere by dry and wet deposition processes. Some of the observed effects of short-term exposure to high levels of particulates (often as black smoke in combination with sulphur dioxide) include acute respiratory morbidity, and decrements in lung function. Exposure to particulates may give rise to feelings of discomfort and annoyance. In addition, soiling of exposed surfaces and visibility impairment can give rise to nuisance problems where dust generating activities are undertaken. In 1991–92, typical 98 percentile black smoke concentrations in the UK varied from 24 to 162 $\mu g\ m^{-3}$.[5]

2.3.7 *Volatile organic compounds (VOCs)*

VOCs comprise a wide range of chemical compounds including hydrocarbons (alkanes, alkenes and aromatics), oxygenated hydrocarbons (alcohols, ketones, acids, and ethers) and halogenated species. Methane is one of the major VOCs; however, its environmental impact is primarily its contribution to global warming. The other non-methane VOCs impact more locally via generation of tropospheric ozone and photochemical smogs, therefore most commentators consider VOCs separately from methane. Typical UK annual average concentrations for selected VOCs are as follows; propane 0.86–6.98 ppb, benzene 0.34–1.98 ppb and toluene 0.5–258 ppb.[5]

2.4 Air quality standards and guidelines

Concentrations of air pollutants in the ambient environment of European Union states are regulated by statutory air quality limit values. These

Table 2.2 Nitrogen dioxide. EC Directive – based on calendar year

Guideline	Description	Criteria	Value ($\mu g\ m^{-3}$)
EC Directive	Limit value	98% of hourly means	200
	Guide value	98% of hourly means	135
	Guide value	50% of hourly values	50

Table 2.3 Sulphur dioxide. EC Directive, limit values

Reference period	Smoke ($\mu g\ m^{-3}$)	Limit values for sulphur dioxide ($\mu g\ m^{-3}$)
Year (median of daily values)	68	if smoke ≤ 34:120 if smoke > 34:80
Winter (median of daily values, October–March)	111	if smoke ≤ 51:180 if smoke > 51:130
Year (peak) (98 percentile of daily values)	213	if smoke ≤128:350 if smoke >128:250

Table 2.4 Sulphur dioxide. EC Directive, guide values

Reference period	Smoke ($\mu g\ m^{-3}$)	Guide values for sulphur dioxide ($\mu g\ m^{-3}$)
Year (arithmetic mean of daily values)	34–51	40 to 60
24 hours (daily mean value)	85–128	100 to 150

values have been set by EC directives for the pollutants nitrogen dioxide (directive 85/203/EEC) and sulphur dioxide/smoke (directive 80/779/EEC). The limit values within these directives are incorporated into member state law and are designed for the protection of human health. In addition, these directives set guide values which are at a lower concentration than the limit value, for the protection of the environment as a whole. The guide values are not statutory. The SO_2 and smoke directive links both these pollutants together such that when the smoke concentration exceeds a trigger value a lower limit for SO_2 is enforced. No ambient standard is set for VOCs. EC Directives also cover lead (directive 82/884/EEC) and tropospheric ozone (COM (92) 236 final). EC directive standards and guidelines are shown in Tables 2.2 to 2.8.

The World Health Organisation set air quality guidelines for Europe in 1987.[6] These cover, amongst other pollutants, NO_2, SO_2 and O_3 with guideline values set for the protection of human health. Guidelines are

Table 2.5 Black smoke. EC Directive – based on pollution year

Guideline	Description	Criteria	Value ($\mu g\ m^{-3}$)
EC Directive	Limit value	Pollution year (median of daily values)	68
	Limit value	Winter (median of daily values) (October to March)	111
	Limit value	Pollution year (mean of daily values)	213
	Guide value	Pollution year (mean of daily values)	34–51
	Guide value	24 hours (daily mean value)	85–128

Table 2.6 Suspended particulates. EC Directive limit values for suspended particulates as measured by the gravimetric method – based on calendar year

Reference period	Limit value ($\mu g\ m^{-3}$)
Year	150 (arithmetic mean of daily mean values)
Year (made up of units of measuring periods of 24 hours)	300 (95 percentile of all daily mean values in year)

Table 2.7 Ozone. Proposed EC Directive values

Guideline	Description	Criteria	Value ($\mu g\ m^{-3}$)
EC Directive	Population information threshold	1 hour mean	180
	Population warning threshold	1 hour	360
	Health protection threshold	8 hour mean	110
	Vegetation protection threshold	1 hour	200
	Vegetation protection threshold	24 hours	65

Table 2.8 Lead. EC Directive – based on calendar year

Guideline	Description	Criteria	Value ($\mu g\ m^{-3}$)
EC Directive	Limit value	Annual average	2

Table 2.9 World Health Organisation air quality guideline values for individual substances based on effects other than cancer or odour[6]

Substance	Time weighted average	Averaging period
Cadmium	1–5 ng m^{-3} 10–20 ng m^{-3}	1 year rural 1 year urban
Carbon disulphide	100 $\mu g\ m^{-3}$	24 hours
Carbon monoxide	100 mg m^{-3} 60 mg m^{-3} 30 mg m^{-3} 10 mg m^{-3}	15 min 30 min 1 hour 8 hours
1,2-Dichloroethane	0.7 mg m^{-3}	24 hours
Dichloromethane	3 mg m^{-3}	24 hours
Formaldehyde	0.1 mg m^{-3}	30 min
Hydrogen sulphide	150 $\mu g\ m^{-3}$	24 hours
Lead	0.5–1.0 $\mu g\ m^{-3}$	1 year
Manganese	1 $\mu g\ m^{-3}$	1 year
Mercury	1 $\mu g\ m^{-3}$ (indoors)	1 year
Nitrogen dioxide	400 $\mu g\ m^{-3}$ 150 $\mu g\ m^{-3}$	1 hour 24 hours
Ozone	150–200 $\mu g\ m^{-3}$ 100–120 $\mu g\ m^{-3}$	1 hour 8 hours
Styrene	800 $\mu g\ m^{-3}$	24 hours
Sulphur dioxide	500 $\mu g\ m^{-3}$ 350 $\mu g\ m^{-3}$	10 min 1 hour
Tetrachloroethylene	5 mg m^{-3}	24 hours
Toluene	8 mg m^{-3}	24 hours
Trichloroethylene	1 mg m^{-3}	24 hours
Vanadium	1 $\mu g\ m^{-3}$	24 hours

available for a number of VOCs (e.g. toluene) although for others such as benzene no safe level can be recommended. These guidelines are advisory and do not have any statutory status. All WHO guidelines are shown in Table 2.9.

In common with other governments, the UK Department of the Environment has introduced its own air quality classification system. The classes include 'very good', 'good', 'poor' and 'very poor' (Table 2.10). Currently only nitrogen dioxide, sulphur dioxide and ozone concentrations can be assessed against this classification system. These air quality criteria have been adopted on a national scale to provide the public with easily accessible and understandable information on air quality on a standard region basis.

Table 2.10 Department of the Environment Air Quality Classification (ppb)

Class	Nitrogen dioxide	Sulphur dioxide	Ozone
Very good	< 50	< 60	< 50
Good	50–99	60–124	50–89
Poor	100–299	125–399	90–179
Very poor	>300	>400	>180

2.5 Emissions of air pollutants

Species considered as air pollutants usually have both natural and anthropogenic sources of emission. In general terms anthropogenic sources dominate in industrialised regions. Anthropogenic sources include power generation, refineries, industry, transport and commercial/domestic sources.[7] Natural sources include biogenic emissions from terrestrial, tidal and oceanic areas and non-biogenic emissions from natural combustion, geothermal activity, lightning, airborne soil particles and water aerosols.[7]

2.5.1 Sulphur dioxide

Moller[8] estimates the natural and anthropogenic emissions of SO_2 to be of the same order, with recent estimates of natural emissions being in the range 100–1000 MT SO_2 yr^{-1}, and anthropogenic emissions 120–160 MT SO_2 yr^{-1}.[9] Such estimates must be treated with caution as they extrapolate from a small number of measurements to the global scale.

However, whilst there is similarity in order of size there are distinct spatial differences between natural and anthropogenic sources of emission. Biogenic sources are concentrated in tropical areas.[8] Anthropogenic emission sources are predominantly located in the industrialised northern hemisphere, covering less than 10% of the earth's surface, where some 68% of anthropogenic emissions occur.[7,10] More than 90% of sulphur in the northern hemispheric atmosphere is of anthropogenic origin and thus on the regional scale, man-made emissions dominate.[7] Recently emissions of marine-derived sulphur species have attracted considerable scientific interest. Such emissions may be of local significance in spring and summer.[11]

Whilst estimates of global anthropogenic emissions are available[8,9] estimates of emissions are most detailed in industrialised areas where fuel combustion and other data are used to estimate emissions. For example, European emissions in the late 1980s are estimated to be of the order of 42 million tonnes expressed as SO_2.[7] Emissions for 1990 are shown in Table 2.11[10]. On a national scale where good estimates of fuel consumption and its sulphur content are available, and information on population density

Table 2.11 1990 Emissions of sulphur and nitrogen as reported by EMEP[10]

Country	Sulphur (000 tonnes as S)	Nitrogen (000 tonnes as NO_2)
Austria	45	222
Belgium	222	334
Bulgaria	1010	376
Czechoslovakia	1222	987
Denmark	90	283
Finland	130	290
France	630	1750
FRG	470	2600
GDR	2400	630
Greece	250	746
Hungary	505	238
Ireland	84	135
Italy	1090	1761
Netherlands	104	552
Norway	27	233
Poland	1605	1280
Portugal	102	122
Romania	900	390
Soviet Union	4168	4248
Spain	1158	839
Sweden	85	404
Switzerland	31	184
Turkey	177	175
United Kingdom	1887	2727
Yugoslavia	739	420
Others	36	40
Total	19 167	21 968

and other factors is reasonably reliable, emission calculations are more precise but still have an uncertainty of 10–20%[9] (see Section 2.5.5).

Major source areas of anthropogenic emission include the northern part of England, the Low Countries, and a band of countries stretching east from the former Federal Republic of Germany into Eastern Europe.[7] A large area of high emission density occurs in the eastern United States where 75% of all US emissions are estimated to arise east of the Mississippi river.[7]

Various workers have reviewed temporal trends for the emission of sulphur in Europe and the USA.[7] Emissions at the turn of the century were dominated by coal burning for heating and industry but as a consequence of the post-war economic boom in the 1950s emissions sharply increased as a result of power station emissions and increased oil consumption.[9] By 1979 Europe accounted for 44% and North America 24% of global anthropogenic emissions.[7] Since that date utilisation of flue gas desulphurisation and increased energy efficiency have resulted in decreases in emissions in many countries. Despite the reductions recorded in the northern hemisphere, global emissions continue to increase.[11]

2.5.2 *Oxides of nitrogen (NO_x)*

Almost all oxidised nitrogen is released as nitric oxide (NO). The emission strengths of oxides of nitrogen are less certain than for SO_2 and the total emission is estimated by Irwin[9] to be in the range 25–99 MT N yr^{-1}. As with SO_2 there are both anthropogenic and natural sources of emission for NO and NO_2. The principal natural sources are biomass burning, lightning, microbial activity, biological processes, ammonia oxidation and stratospheric input.[9] Natural sources may account for some 33% of global emissions[9] although some authors[12] suggest a rough equivalence between natural and anthropogenic emissions. Anthropogenic sources include high temperature combustion of fossil fuels in power stations and motor transport with the latter source now being the most important. Estimation of anthropogenic emissions is complicated by the formation of NO_x from nitrogen present in the fuel and from thermal oxidation of atmospheric nitrogen.[13] Hence emissions are dependent upon combustion conditions and fuel properties. Major NO_x source areas are similar to those for SO_2. In North America and Europe anthropogenic emissions may account for between 75% and 93% of all emissions of NO_x.[14] For example, annual European anthropogenic emissions in the late 1980s are estimated to be of the order of 21.5 million tonnes expressed as NO_2.[7] 1990 emissions are shown in Table 2.11.[10]

2.5.3 *Ammonia*

Ammonia is the most important neutralising compound in the atmosphere. Its major sources include animal wastes, fertiliser applications, and industrial emissions. Smaller contributions arise from human respiration, landfill, traffic and uncultivated soil.[15] The European spatial pattern of ammonia emissions is quite different from that of SO_2 or NO_2 as agricultural source areas, rather than industrial areas, dominate the emission profile.[9] European emissions of ammonia are estimated at 6.95 MT N yr^{-1}.[11] Temporal analysis of emissions indicates that in recent years emissions have increased as a consequence of intensive animal husbandry practices, with emissions from livestock having increased by 50% between 1950 and 1980.[16]

2.5.4 *Volatile organic compounds (VOCs)*

VOCs are emitted from transport sources, industrial processes and from solvent evaporation. Certain VOCs are important in the formation of acidic species in the atmosphere[9] through photochemical generation of oxidising species. Natural sources are also important, particularly forestry emissions. Regional scale emission estimates are available for the USA, Europe and Western Europe[7] where the major sources are road transport and industrial solvents. As with other estimates of emission strengths, national data are considered more reliable.

2.5.5 *Estimating emission strengths*

Most industrial nations attempt to estimate the emissions of various pollutants from within their boundaries. The reasons for doing so include compliance monitoring with international legislation. Various bodies such as the European Union and the United Nations Economic Commission for Europe encourage their member states to undertake the estimation procedure according to broadly similar rules. Hence, the results of the procedure allow the comparison of individual national totals between nations.

The case of the United Kingdom is fairly typical amongst European nations and the procedures undertaken in the UK are described here as an example of the way in which emissions are estimated.[5,17] Typical species included in national emission inventories are shown in Table 2.12.

As emissions of air pollutants arise from a wide range of sources often with little detailed information on temporal variation, or from a specific individual source,[17] it is necessary to compile emission estimates from statistical information. For many pollutants, stationary fossil fuel combustion processes are an important source of emissions and these emissions can be estimated by applying an appropriate emission factor to statistics on annual fuel consumption.[17] Vehicle emissions from petrol engines are estimated using speed-related emission factors and information about road usage and speed distribution.[5,17] Estimates of emission from DERV-fuelled vehicles are made using emission factors applied to fuel consumption statistics.[17] Certain pollutants such as VOCs have emissions sources which do not

Table 2.12 Typical pollutants included in an atmospheric emissions inventory[17]

Pollutant
Black smoke
Sulphur dioxide
Oxides of nitrogen
Carbon monoxide
Methane
Carbon dioxide
Ammonia
Nitrous oxide
Non-methane volatile organic compounds
Lead

Table 2.13 Percentage contribution of major sources to total UK emissions[17]

Emission Source	SO_2	NO_2	Black smoke	CO	VOCs	CO_2
Power stations	72	29	6	1	1	34
Industry	19	9	14	1	50	26
Road transport	2	51	46	90	40	19
Domestic	3	2	33	4	2	14
Total	96	91	99	96	93	93

involve combustion. These include natural gas leakage in the national distribution network, petrol evaporation, petrol refining, industrial processes and solvent evaporation.[17]

For most pollutants contained in the UK emission inventory, information is available according to emission source, by end user, and by type of fuel. In many cases a lengthy time series of data exists.[5]

2.5.5.1 Sulphur dioxide emissions in the UK. The 1991 emission of SO_2 is estimated at 3 565 000 tonnes.[5] Power generation is the most important source as can be seen in Table 2.13. The EC large combustion plants directive 88/609 requires a reduction in total SO_2 emissions from existing combustion installations with a capacity greater than 50 megawatts thermal of 20% by 1993, 40% by 1998 and 60% by 2003, taking 1980 emissions as the baseline. By 1991 emissions from large combustion plants were below the 1993 EU target. Estimated total emissions for each pollutant vary in their accuracy with emissions of SO_2 being most accurate as they are estimated from coal and fuel oil consumption by power stations and other fuel users.[17] The totals are likely to be within an accuracy of ± 10–15% of the true figure.[5,17]

2.5.5.2 Black smoke emissions in the UK. Between 1970 and 1980, emissions of black smoke nearly halved mainly because of reduced domestic coal consumption. Since 1980 emissions from domestic coal consumption have continued to fall but these have been offset by the emissions from the increasing use of diesel vehicles.[17] The 1991 UK emission is estimated at 498 000 tonnes (see Table 2.13).[5] The accuracy of the estimate is likely to be in the range ± 20–25%.[17]

2.5.5.3 Nitrogen dioxide emissions in the UK. The 1991 emission estimate for the UK was 2 747 000 tonnes[5] (see Table 2.13). Major sources include power generation and transport. The EC large combustion plants directive 88/609 requires a reduction in total NO_x emissions from existing combustion installations with a capacity greater than 50 megawatts thermal of 15% by 1993 and 30% by 1998 taking 1980 emissions as the baseline.

Emissions from large combustion plants in 1991 in the UK were 18% below the 1980 emission baseline. Figures for NO_x emissions are less accurate than for SO_2 because they are based on relatively few measurements of emission factors and depend on combustion conditions such as driving conditions which can vary.[17] In addition, many of the data used have been derived from small samples which contain margins of error such as vehicle speed distributions.[17] It is estimated that NO_x emissions are accurate to ± 30%.[17] Catalytic convertors on cars are being used to help comply with EU directive 91/441/EEC which imposes limits on the NO_x content of exhaust emissions from new cars.

2.5.5.4 Carbon monoxide emissions in the UK. In 1991 the total UK emissions from all sources were estimated at 6 735 000 tonnes[5] (see Table 2.13). Transport is the dominant source and motor spirit is the dominant fuel type. CO emissions are also based on emission factors derived from relatively few measurements of emissions[17] from different types of boiler, which can vary with combustion conditions. It is estimated that CO emissions are accurate to ± 50%.[17]

2.5.5.5 Lead emissions in the UK. The main sources of lead in air are from lead in petrol, lead in coal and from metal working[17] (see Table 2.13). Emissions from vehicles represent the most important source. In 1992, 1800 tonnes of lead were emitted from combustion of leaded petrol.[5] The maximum UK emission was in 1976 when 8000 tonnes of lead were emitted from vehicles.[17] Reductions in the permitted amount of lead in petroleum from 0.45 g l^{-1} in 1981 to 0.15 g l^{-1} in 1985 reduced the emission of lead substantially.[15] Increases in the sales of unleaded petroleum have further reduced lead emissions. Currently more than 50% of UK petrol sales are unleaded.[5] Lead emissions from motor vehicles are likely to be accurate to ± 10–15% and are largely dependent on uncertainties in the actual lead content of fuel and retention of lead in the vehicle.[17]

2.5.5.6 Volatile organic compounds emissions in the UK. Emission estimate trends are available for emission sources, by end user and by fuel type. Processes and solvents are the largest source, road transport the largest end user source category.[5] Of the fuel use categories motor spirit is the most important. The 1991 UK emission was 2 678 000 tonnes[5] (see Table 2.13). Non-methane VOC emissions are subject to a wide margin of error since data relating to individual industrial processes and solvent use are incomplete.[17] UK emissions are probably accurate to ±50%.[17]

2.5.5.7 Carbon dioxide emissions in the UK. The UK contributes about 2.2% to global man-made emissions of CO_2 which are currently about 7400 MT yr^{-1} (expressed as carbon).[5] In 1991, UK emissions totalled

159 million tonnes[5] (see Table 2.13). Fossil fuel combustion is the major source of CO_2 emissions and fuel consumption statistics, combined with emission factors for each source and type of fuel, are used to estimate CO_2 emissions. These factors are based on the carbon content of fuels and an assumption is made that a proportion of the carbon is retained in the ash of solid fuels. The accuracy of the emission total is thought to be in the range ±5–10%.[17]

2.5.5.8 Methane emissions in the UK. Methane (CH_4) concentrations in the atmosphere have been rising at a rate of 1% per annum in the last few decades although there has been a slow-down in the rate of increase in recent years.[5,17] The annual global release of CH_4 to the atmosphere is about 500 million tonnes. The 1991 emission from the UK was estimated to be 3 555 000 tonnes.[5] Important sources of methane include decomposition of wastes in landfill sites, releases from coal mines and from offshore oil and gas activities. The uncertainty attached to the emission total is considered to be high.[17]

2.5.5.9 Nitrous oxide emissions in the UK. There are thought to be two main sources of N_2O emissions. The first results from the production of adipic acid used in the production of nylon.[17] The second is from denitrification processes in soils.[17] Emissions from soils appear to depend on a range of factors including soil type, soil moisture, fertiliser type and timing of application, land use and climatic factors such as wind speed, air temperature, rainfall and humidity. The 1991 UK emission is estimated to be 175 000 tonnes (see Table 2.14). The uncertainty attached to the emission total is considered to be high.[17]

2.5.5.10 Ammonia emissions in the UK. The total ammonia emission from the UK is very uncertain due to the wide range of factors which can influence emissions from sources such as fertilised soils, animal waste products and humans.[16] Table 2.15 shows the estimated percentage share of major sources to the UK annual emission total.

Table 2.14 Source contribution to UK nitrous oxide emissions[17]

Emission source	Contribution (%)
Adipic acid manufacture	56
Emissions from soil	41
Other	3
Total	100

Table 2.15 Source contributions to UK ammonia emissions[17]

Emission source	Contribution (%)
Animals	67
Fertilisers	22
Humans	4
Other	7
Total	100

2.6 Transport, reaction and deposition of air pollutants

Upon combustion of fuel, either during industrial processing, transport, or from natural activities, emission to the atmosphere occurs. The subsequent mixing of the emitted pollutant with the receiving air is largely determined by the height of emission and the prevailing weather pattern. High level emissions, such as those from power stations, leave the chimney at high velocity, often with considerable thermal buoyancy. Turbulence in the atmosphere generated by convection and the friction between wind and surface features acts to disperse an emission in both the vertical and horizontal planes[14] with vertical dispersion being the principal mechanism by which emissions are diluted.[18] Mixing with oxidants, present in the atmosphere from other sources, occurs and such interaction of emissions and oxidants may give rise to a variety of reaction products.

2.6.1 Removal and reactions of sulphur and nitrogen

Various removal mechanisms act upon emitted pollutants. In the case of sulphur dioxide a proportion will be removed continuously by dry deposition, except during rainfall. Oxidation processes, particularly photo-oxidation, will give rise to sulphuric acid, aerosol or sulphate, by reactions in the gas phase, in the liquid phase or on the surface of solids.[18] Removal from a plume may occur through dry (adsorption directly to a surface or uptake by plants), wet (rain, sleet, snow, hail) or occult (impaction of cloud or droplets onto vegetation or other surfaces) deposition. Where acidic material is involved then the term acid deposition is used to encompass all of these removal processes.

The dry deposition process is affected, *inter alia*, by the characteristics of the airborne substances, meteorology–surface interactions and surface properties.[7] Dry deposition can be estimated from the product of ground level concentration and the deposition velocity of the species in question.[18] This latter concept is defined as the flux to a surface per unit concentration in air. Dry deposition is thus assumed to be proportional to concentration.[18] Measurements have shown that deposition velocities vary from surface to surface, seasonally and diurnally.

Removal by dry deposition is more important for plumes emanating from low level sources, but such emissions do become well mixed in the boundary layer within a few hours and the major part of the emission is transported downwind in the same manner as for high level sources.[18] The concentration at long range from a low level source is, on average, only 10–20% less than that from a high level emission of the same initial concentration.[18]

However, if a plume is subjected to rainfall then removal of sulphur dioxide will occur, with the amount removed depending, *inter alia*, on the amount of sulphur dioxide converted to sulphate. Rates of removal as wet deposition are greater than for dry deposition with the concentration of sulphur in rainwater being determined by the solubility of sulphur dioxide and the acidity of rainwater. The process is self limiting; as sulphur dioxide dissolves it increases the acidity of rainwater thus limiting further dissolution. As the sulphate molecule is a cloud condensation nucleus, wet deposition is a significant removal mechanism.

Gas phase oxidation of sulphur dioxide by hydroxyl radicals to sulphur trioxide and then to sulphuric acid or ammonium sulphate is important; however, the rate of conversion is slower.[18] Aqueous phase reactions involving hydrogen peroxide and ozone are important, but their relative importance depends upon their concentration and the timescale of removal. Oxidation by hydrogen peroxide is rapid and is limited by the availability of the oxidant. Oxidation by ozone is slower due to its low solubility and the reaction is inversely related to acidity.[18] The extent to which this reaction controls the rate of removal of sulphur dioxide is determined by the amount of sulphur dioxide present, duration of rainfall, and the presence of alkaline species such as ammonia which affect the acidity of water.[19]

The main nitrogen species emitted from a combustion process is nitric oxide. This is not removed from the atmosphere very effectively as dry deposition, as it is quickly oxidised to nitrogen dioxide, nitrate aerosol or nitric acid. Little aqueous phase oxidation is likely due to the low solubilities of nitric oxide and nitrogen dioxide. Nitric acid (see Section 2.6.2.3) is volatile and will exist in the gas phase in the atmosphere, unlike sulphuric acid which will be present in the fine particle range $<0.2\ \mu m$.[19] Consequently, sulphuric acid particles may act as condensation nuclei or cause visibility degradation.

Both sulphur and nitrogen species will react with alkaline materials in the atmosphere such as calcium, magnesium and ammonia to form, for example, sulphates or nitrates; nitric acid, however, can volatilise from the latter.[19]

2.6.2 Photochemistry

2.6.2.1 Formation and reaction of tropospheric ozone. The formation of tropospheric ozone is dependent upon whether peroxy (RO_2) radicals formed in the atmosphere react with HO_2 or NO. If they react with HO_2 then the route of ozone formation is aborted. However, if the peroxy radicals react with NO, then the newly-formed methoxy radical may react with molecular oxygen (see reactions (2.13) and (2.14)), to form HO_2. The HO_2 radical can then go on to react with NO, producing NO_2 (see reaction (2.12)). Photolysis of NO_2 at λ = 423 nm[20] may then take place

$$NO_2 + h\nu \rightarrow NO + O \qquad (2.1)$$

The oxygen atom produced then undergoes a three-body reaction with molecular oxygen

$$O + O_2 + M \rightarrow O_3 + M \qquad (2.2)$$

However, the HO_2 produced could also react with ozone in a relatively slow reaction to regenerate the OH radical

$$HO_2 + O_3 \rightarrow OH + 2O_2 \qquad (2.3)$$

Ozone is the most important precursor to OH in the troposphere and an understanding of the tropospheric ozone budget facilitates understanding of tropospheric processes. For example, whether the oxidation of methane will ultimately become a net producer or destroyer of ozone depends on the [NO]/[O_3] ratio.[21] In remote unpolluted areas, where NO_x concentrations are relatively low, the ratio will become small enough for ozone loss to become the dominant process. Industrial areas on the other hand, where pollution is high, will have relatively high NO concentrations and so oxidation of methane, in this instance, will be a producer of ozone. This explains why levels of tropospheric ozone are always highest in industrial and densely populated areas.

2.6.2.2 The hydroxyl radical. The hydroxyl radical is the most important oxidising species present in the troposphere and its production is dependent upon the existence of tropospheric ozone. Some of the ozone present in the troposphere is produced *in situ*, mainly in the more polluted areas where NO_x is present, the rest being obtained from downward transportation of stratospheric ozone. Photodissociation of ozone by ultraviolet radiation, at wavelengths between λ = 290 and 310 nm, can then take place producing the precursor of OH, $O(^1D)$, which then reacts with water

$$O_3 + h\nu \rightarrow O(^1D) + O_2 \qquad (2.4)$$

$$O(^1D) + H_2O \rightarrow 2OH \qquad (2.5)$$

Since the initial reaction (2.4) is photochemically driven the domination of tropospheric chemistry by OH is limited to daylight hours. The average concentration of OH is approximately 10^6 molecule cm^{-3}.[22] This low concentration of OH is balanced by its high reactivity with respect to a wide range of atmospheric species, to the extent that most trace constituents that have an appreciable lifetime in the troposphere will undergo oxidation by OH. Two of the major reactions of the OH radical are with CO and CH_4

$$OH + CO \rightarrow H + CO_2 \quad (2.6)$$

$$OH + CH_4 \rightarrow CH_3 + H_2O \quad (2.7)$$

Both processes produce an active species which is capable of adding to molecular oxygen to produce a peroxy radical. The reactions are both three-body processes as stabilisation of the newly formed product is required

$$H + O_2 + M \rightarrow HO_2 + M \quad (2.8)$$

$$CH_3 + O_2 + M \rightarrow CH_3O_2 + M \quad (2.9)$$

In unpolluted air, such as is found in remote rural areas, where NO_x concentrations are low, the main pathway of the peroxy radicals is further reaction with HO_2 and other peroxides

$$HO_2 + HO_2 \rightarrow H_2O_2 + O_2 \quad (2.10)$$

$$CH_3O_2 + HO_2 \rightarrow CH_3OOH + O_2 \quad (2.11)$$

One fate of the peroxides is that they can dissolve in cloud droplets and then be removed from the troposphere in the form of rain. In this way, reactions (2.10) and (2.11) can be thought of as loss processes.

If significant concentrations of NO_x are present in the troposphere, such as are formed over industrial or densely populated areas, a different fate awaits the peroxy radicals. HO_2 and CH_3O_2 react rapidly with NO

$$HO_2 + NO \rightarrow OH + NO_2 \quad (2.12)$$

$$CH_3O_2 + NO \rightarrow CH_3O + NO_2 \quad (2.13)$$

Reaction (2.12) regenerates the hydroxyl radical, while (2.13) produces a methoxy radical which can further react with O_2 to produce formaldehyde

$$CH_3O + O_2 \rightarrow HCHO + HO_2 \quad (2.14)$$

However, formaldehyde is photolysed at $\lambda < 338$ nm, leading to the regeneration of radicals

$$HCHO + h\nu \rightarrow H + HCO \quad (2.15)$$

The two radical fragments, H and HCO, re-enter the HO_x chain via reaction with molecular oxygen forming HO_2. The HO_2 goes on to react with NO as in reaction (2.12) to regenerate the OH radical. This sequence of reactions can be envisaged for other hydrocarbons and is important to the formation of tropospheric ozone.

The rapid removal and production of the OH radical through reactions such as those outlined leads to a steady-state concentration of OH in the daytime troposphere. The OH radical can thus be thought of as an efficient scavenger of both biogenic and anthropogenic trace constituents of the troposphere.

2.6.2.3 The nitrate radical. The nitrate radical (NO_3) is formed in the earth's atmosphere via the reaction between nitrogen dioxide and ozone

$$NO_2 + O_3 \rightarrow NO_3 + O_2 \qquad (2.16)$$

Although the NO_3 radical has been known for over 100 years, its importance in night-time tropospheric chemistry oxidation processes has only become apparent over the last couple of decades. Even though NO_3 is less reactive towards certain classes of organic compounds in comparison with OH, concentrations of NO_3 in the troposphere are found to be as high as 10^9 molecule cm^{-3},[22] i.e. three orders of magnitude larger than concentrations of OH. Hence the oxidising potentials of the two dominant species are comparable. In fact for some species, such as dimethyl sulphide ($(CH_3)_2S$) and some naturally occurring terpenes, the reaction with NO_3 may even dominate over the daytime reaction with OH.

A source of NO_3 is the dissociation of N_2O_5

$$N_2O_5 + M \rightarrow NO_3 + NO_2 + M \qquad (2.17)$$

but as N_2O_5 mainly is formed by the reaction (2.17), it can be seen that N_2O_5 formation is crucially dependent on the formation of NO_3 in reaction (2.16). The decomposition of N_2O_5 is strongly temperature dependent, redissociating at higher temperatures to yield NO_3 and NO_2. N_2O_5 is the anhydride of nitric acid (HNO_3); one fate of the anhydride is to react heterogeneously with water molecules to produce the acid and hence contribute to the acidification of the atmosphere.

NO_3 is rapidly photolysed at all altitudes. The products of this photodissociation are both NO_2 + O (ozone producing) and NO + O_2 (ozone destroying). The wavelength dependence of the products of photolysis is uncertain.[23] Fortunately, the product distribution is of little consequence to the overall chemical balance of the atmosphere. As a result of this rapid photolysis, and as a consequence of the main precursors of NO_3 (O_3 and NO_2) being mainly found in high concentrations over polluted areas, the nitrate radical has its greatest effect in polluted night-time atmospheres. However, as OH is a photochemically generated radical, the two species complement each other and urban atmospheres are found to exhibit a diurnal variation between the two different oxidation environments.

The NO_3 radical is generally thought to be relatively unreactive towards closed-shell species, inorganic compounds and saturated hydrocarbons. The situation is very different, however, for open-shell molecules (such as NO) and unsaturated hydrocarbons. With NO, the following reaction takes place

$$NO + NO_3 \rightarrow NO_2 + NO_2 \qquad (2.18)$$

This is a rapid reaction and is used in the laboratory as a gas-phase titration reaction. Other such open-shell species are peroxy radicals and these species will be discussed later.

With organic substrates there are two main kinds of initial step that may occur: hydrogen abstraction or addition to the unsaturated bond

$$NO_3 + RH \rightarrow HNO_3 + R \qquad (2.19)$$

$$NO_3 + {>}C{=}C{<} \rightarrow {>}C(ONO_2)C{<} \qquad (2.20)$$

The species RH (reaction (2.19)) can be any organic compound with a removable hydrogen atom, so that the species R formed is an organic radical. In this situation the radical can subsequently add oxygen to form RO_2. In the case where RH is formaldehyde, the HCO radical will ultimately form the HO_2 radical (reaction (2.15)), as a result of further reaction with molecular oxygen. For higher aldehydes, and their chlorinated derivatives, the radicals formed (e.g. the acyl radical) go on to react with O_2 to form the acylperoxy radical ($RCO.O_2$) and are therefore potential sources of peroxyacetyl nitrates (see Section 2.7.4). It is interesting to note that nitric acid is a direct product of the abstraction, contributing to atmospheric acidification.

The addition of the NO_3 radical to unsaturated bonds is the other main reaction pathway. The initial adduct formed in this reaction can go on to eliminate NO_2 to form the corresponding epoxide. There is, however, another fate for the adduct; as it is also a radical it can add oxygen in the presence of air and a variety of products could result. In the example of propene the adduct reacts with O_2 in air as follows

$$CH_3CHC(ONO_2)H_2 + O_2 \rightarrow CH_3C(O_2)HC(ONO_2)H_2 \qquad (2.21)$$

Under simulated atmospheric conditions, the ultimate oxidation products observed include HCHO, CH_3CHO, 1,2-propanediol dinitrate (PDDN), nitroxyperoxypropyl nitrate (NPPN), and α-(nitroxy)acetone. Since formaldehyde is a product, it can, again, regenerate the OH molecule which then enters into a night-time chain oxidation of organic compounds.

The secondary pollutants (i.e. pollutants which were not emitted but were formed in the atmosphere) already mentioned, e.g. PDDN, PANs,

NPPN and α-(nitroxy)acetone, are all known to be toxic and noxious species. The species α-(nitroxy)acetone has been shown to be a mutagen.

The above illustrates the importance of the NO_3 radical in the overall oxidation of hydrocarbons, both from natural and anthropogenic sources. Primary organic pollutants can be oxidised and removed during the night, toxic and otherwise noxious compounds, such as peroxyacetyl nitrate, may then be formed and can lead to such phenomena as photochemical smog. Nitrate products can act as temporary reservoirs of NO_x in the atmosphere, and these compounds can also be transported to remote regions, far away from where they were originally formed.

2.6.3 Particulate matter

Particles of an aerodynamic diameter greater than 2.5 μm are described as coarse particles and those of less than 2.5 μm are fine particles.[4] Coarse materials are those usually derived from the earth's crust such as soil dust, sea spray or from fugitive dust releases from roads and industrial activities. Fine particles comprise secondarily formed aerosols, combustion particles and recondensed organic and metallic vapours. The acid component of particulate matter will generally occur as fine particles. Coarse particles are mainly formed by mechanical attrition. They tend to be large and have high settling speeds which results in relatively short atmospheric lifetimes. Fine particles can, in the range 0.2–2 μm, have a much longer atmospheric lifetime, often some 7–30 days, whilst the finest particles, <0.2 μm, have a very short lifetime as they regularly coagulate into larger particles.[4]

A variety of terminology is used to describe particulate matter. This includes method of sampling (e.g. total suspended particulates (TSP)), site of deposition in humans (e.g. inhalable thoracic particles) or physical characteristics (e.g. PM_{10}, which refers to an aerodynamic diameter of less than 10 μm).[4] The general term 'suspended particulate matter' (SPM) is used to embrace all airborne particles. Whilst there is no agreed definition, particles of less than 2.5 μm are often described as respirable. The phrase 'total suspended particulates' (TSP) describes the gravimetrically determined mass loading of airborne particles where particles are collected upon a filter for weighing. TSP is usually used in connection with US high volume air samplers.[4]

Particles are also distinguished by their origin. This classification uses the term 'primary particulate' to refer to those particles directly emitted to the atmosphere whilst 'secondary particulate' is used to describe those formed by reactions with other pollutants.[4] In urban environments most secondary particulate matter occurs as sulphates and nitrates formed in reactions involving SO_2 and NO_x. Reported concentrations of particulate matter vary according to the sampling technique. In the UK urban areas, typical annual mean concentrations of 10–40 μg m^{-3} using the black smoke

Table 2.16 Main components of urban airborne particulate matter[24]

Soil-derived minerals
Elemental carbon from combustion
Organic compounds from combustion
Ammonium salts from acid neutralisation
Sodium and magnesium chloride from sea spray
Calcium sulphates from construction activities and soil/rock derived material
Sulphates from SO_2 oxidation
Nitrates from NO_x oxidation

method and 50–150 $\mu g\ m^{-3}$ using the gravimetric method have been reported.[4] Peak values are in the range 100–250 $\mu g\ m^{-3}$ (black smoke) or 200–400 $\mu g\ m^{-3}$ (gravimetric method).[4] Corresponding rural values will vary from 0–10 $\mu g\ m^{-3}$ when measured by the black smoke method.[4] The differences between urban and rural areas are explained in terms of the difference in sources of particulate emission. Whilst particulate matter has a number of sources such as mineral working, stockpiling, power plant, industrial activities, vehicular traffic, coal burning and incineration, most of these are preferentially concentrated in urban environments. In rural areas wind-generated dust from agricultural or mineral extraction sources will be locally important. Natural sources of dust emission include volcanic activity and dust storms which will generally be remote from urban industrial regions but on isolated occasions may contribute a small amount of dustfall when large scale meteorological conditions allow.

An analysis of the chemical composition of SPM from the Leeds, UK, area[24] has indicated that urban airborne particulate matter comprises eight main components as shown in Table 2.16. The largest component, comprising approximately 40% of the total mass of particles, was carbonaceous matter.[24] Approximately 20% of the total mass of particles was determined to be insoluble mineral matter derived from wind-blown dust. Sulphates represented the third largest category of particulate matter, with the remainder being split equally between chlorides, ammonium and a group comprising non-chloride sea salts and dust.[24]

2.7 Toxic organic pollutants

The environmental impact of organic pollutants is both global, owing to their contribution to the enhanced greenhouse effect, and regional (particularly pronounced in urban–industrial regions) via their contribution to ozone production in the troposphere. However, organic pollutants are also an increasingly important component of air quality because of concerns over their toxicity.[4,6] In addition, many of these compounds are odorous and give rise to amenity problems.

Fuels and their combustion products also contain species which have considerable implications for human health such as benzene or benzo(a)-pyrene. The WHO estimate a 4 in 1 million risk of leukaemia on exposure to a concentration of 1 $\mu g\ m^{-3}$ of benzene.[6] The relationship between exposure to atmospheric organics and cancers is still a subject for both debate and research. Existing WHO guidelines for a range of organic species are shown in Table 2.9.

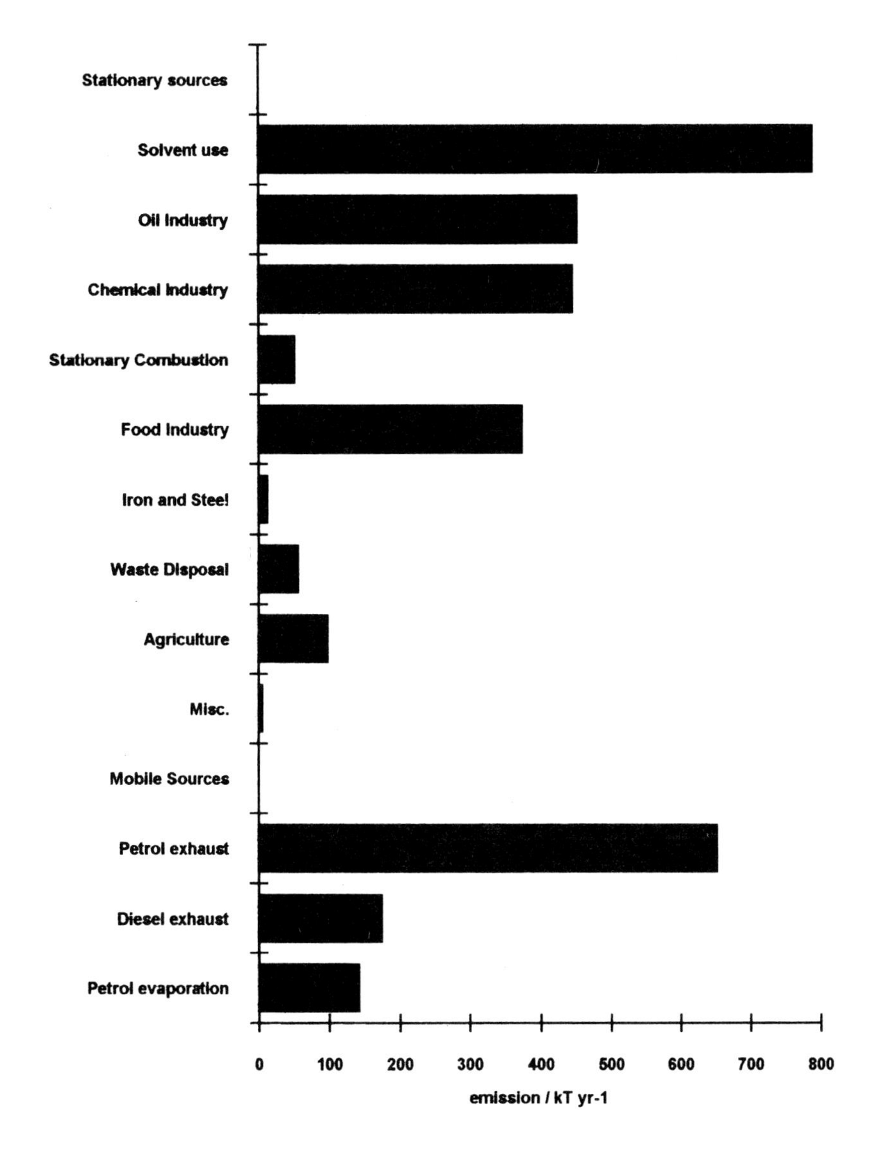

Figure 2.1 Emission estimates for (non-methane) VOC sources in the UK (1991), after Marlowe.[27]

The majority of atmospheric organic pollutants are a class known as volatile organic compounds (VOCs). Examples include benzene, toluene (the most prevalent non-methane hydrocarbon in the troposphere), aldehydes and alkenes. VOCs are emitted from a variety of sources, with solvent use and transport as major areas (see Figure 2.1), hence urban sites tend to have a higher concentration of VOCs than do rural.[4]

2.7.1 *Photochemical oxidation*

The major environmental impact of non-methane VOCs arises from their potential for generating smogs and ozone via photochemical oxidation.[4,21,25] Specific reaction mechanisms for ozone generation are still the subject of research; however, a generalised reaction scheme for alkane hydrocarbons is shown in Box 1.

Box 1

Ozone generation via an alkane reaction pathway[4,21]

The sequence of reactions to generate ozone via VOCs is initiated by the photolysis of ozone ($\lambda \leq 310$ nm) to generate electronically excited oxygen atoms O^*:

$$O_3 + h\nu \rightarrow O^* + O_2$$

(The initial source of tropospheric ozone is via the photolysis of NO_2 – see Section 2.6.2.)

- The majority of the excited O^* are 'quenched' through reaction with N_2 or O_2 to form ground state oxygen atoms. However, a small fraction are energetically capable of reactions with water vapour to form *two* hydroxyl radicals

$$O^* + H_2O \rightarrow 2OH\bullet$$

- Reaction with OH• is almost exclusively the loss process for all tropospheric hydrocarbons, RH. Reaction with the OH• radical causes H atom subtraction from the C–H bond to form an alkyl radical, R, which can subsequently initiate a sequence of oxidation reactions:

$$RH + OH\bullet \rightarrow R + H_2O$$

$$R + O_2 \rightarrow RO_2$$

$$RO_2 + NO \rightarrow RO + NO_2$$

- NO_2 is photolysed by light of $\lambda \leq 435$ nm regenerating NO, which subsequently produces ground state oxygen atoms which in turn form ozone.

$$NO_2 + h\nu \rightarrow NO + O$$

$$O + O_2 + M \rightarrow O_3 + M$$

M is a chemical species (usually N_2 or O_2) which dissipates the excess energy of the newly formed ozone molecule via collision.

- This simple reaction scheme illustrates that the photolysis of a single ozone molecule can lead to the formation of two ozone molecules via oxidation of 2 RH molecules.

Different reaction pathways can be followed depending upon atmospheric conditions and VOC type. Aldehydes (RCHO), especially formaldehyde, are believed to have a dramatic effect on the conversion rates of

NO–hydrocarbon–air mixtures to NO_2, and are also instrumental in the formation of peroxyacetyl nitrate (PAN), another important component of photochemical smog (see Section 2.7.4). The potential of a VOC to generate ozone is quantified via its photochemical ozone creation potential (POCP), tables of which can be found in the literature.[4]

2.7.2 VOC emissions

In the United States the Environmental Protection Agency (EPA) recognised the need to produce a VOC national emissions inventory and concentrations database. This was first published in 1983 with subsequent revisions as the sources of data increased. In 1988 Shah and Singh[26] produced a more detailed database which gave details of daily indoor and outdoor concentrations for a number of VOCs ranging from aromatic hydrocarbons to CFCs for use in terms of risk analysis and environmental management. In the UK, the Quality of Urban Air Review Group[4] went further by producing a speciated VOC emission inventory which also outlines the prime sources of pollutants. Marlowe[27] has reviewed the sources and industrial processes emitting VOCs in the UK.

Efforts have been made to target airborne VOCs to emission sources and to validate the emission inventories, within urban and rural areas.[28,29] Wind trajectory analysis, factor analysis, and chemical mass balance statistical models for data interpretation have been used for a number of point and area sources. Results showed that the majority of urban area sources, such as vehicle exhausts, evaporation of petroleum products, and solvent emissions by commercial and industrial end users produced the majority of VOCs, although emissions from large point sources only appear to increase concentrations within 1 km of the point source.

2.7.3 Other toxic organic pollutants

2.7.3.1 Polycyclic aromatic hydrocarbon compounds (PAHs).[4] PAHs are a class of organic pollutants formed from multinumbered aromatic ring compounds, examples of which include naphthalene, benzo(a)pyrene and anthracene. As is common in heterocyclic organic chemistry, N, O, S heteroatoms can also be found in the ring structure. Major sources of PAHs include the pyrolysis of fossil fuels, and mobile transport sources as shown in Table 2.17. PAHs are found in low concentrations (typical urban values are in the range 0.1–35 ng m^{-3}) in the vapour phase and on particulates (0.1–3.3 ng m^{-3}). The health concerns associated with PAHs arise from the fact that this group of materials are noted carcinogens.

2.7.3.2 Dioxins and polychlorinated biphenyls (PCBs). Like PAHs, dioxins and PCBs are formed as by-products of organic compound

Table 2.17 Estimated PAH emissions to air from diffuse sources in Sweden and the UK for 1985 (T yr^{-1})[4]

Source	Sweden	UK
Anthracite	–	29.4
Bituminous coal	5.8	91.0
Wood	325.0	2.5
Gasoline	3.2	17.1
Motor diesel oil	1.6	7.9
Coke production	3.6	27.8
Total	339.0	176.0

pyrolysis, more specifically chlorinated organic material.[4] Major sources include incinerators, domestic and industrial coal combustion, power generation and vehicle emissions; however, potential sources (notably pesticides) have been limited since the 1970s owing to their widely reported links with mutagenic and fetotoxic responses.[30]

The annual average PCB concentrations are reported to be just under 1.5 ng m^{-3} while urban site measurements of dioxins average 5 pg m^{-3}.[4,30] These concentrations are observed to peak in summer months and to fall during winter. This suggests that one of the major sources of these materials may be evaporative, favoured by warm weather.[4]

2.7.4 Peroxyacetyl nitrate

Peroxyacetyl nitrate (PAN) is an important contributor to photochemical smog. Photochemical or 'Los Angeles' smog, so called because the phenomenon was first identified in that city (but the phenomenon is not limited to there), is a consequence of a number of specific situations: a high density of motor cars confined to a limited geographical area (i.e. the Los Angeles basin) coupled to the existence of stagnant air and the formation of a low-lying inversion layer. The essential ingredients for the smog are petrol fumes and the array of derived substances resulting from incomplete combustion of the gasoline. The derived substances include alkenes, aldehydes, ketones and higher hydrocarbons. The physical effects of photochemical smog include a brown haze over the horizon by about midday and eye irritation. The latter is a consequence of the formation of PAN (as well as ozone), well known as a powerful lachrymator, in the local atmosphere.

As well as being identified in highly polluted areas, PAN has now been found to be present in the clean air of remote oceanic regions.[31] Concentrations of PAN varying between 10 and 400 parts per trillion by volume (pptv) have been observed at altitudes up to 8 km over the Pacific ocean.

Peroxyacetyl nitrate exists in equilibrium with NO_2

$$CH_3C(O)OONO_2 \rightleftharpoons CH_3C(O)OO + NO_2 \quad (2.22)$$

and its thermal decomposition is highly temperature dependent: at lower temperatures PAN is stable. Above the boundary layer of the atmosphere the temperatures are sufficiently low for PAN to become a long-range transporter of NO_x.

2.7.5 *Control of toxic organic emissions*

2.7.5.1 Legislative. As monitoring of VOCs is largely in its infancy, legislative mechanisms are still being defined in many areas. The most detailed legislation regarding solvent emissions has come from the United States. In October 1986 the United States Congress passed the *Emergency Planning and Community Right-to-Know Act 1986*, otherwise known as 'Title III of SARA' (*Superfund Amendments and Reauthorization Act 1986*), a section of which includes the Toxics Release Inventory (TRI). The TRI, while being concerned with all environmental media, requires that information regarding emissions into the environment is made available to the public. Nine of the twenty-five chemicals with the largest TRI releases come into the class of VOCs. The public availability and release of TRI data has put significant pressure on companies to change their practices. Further acts such as the *Pollution Prevention Act 1990* and the *Clean Air Act Amendments* (*1990*)[25] have stated that 'facilities shall demonstrate a 90% reduction . . . which will be determined with respect to emissions in a base year not earlier than 1987' in order for the facilities to avoid regulatory requirements. TRI will be used to monitor these regulations.[32]

In the UK the majority of legislation has centred upon reducing VOC emissions from transport sources (Directive 70 220 EEC), even though most emissions inventories suggest that solvent use is the primary source of VOC pollution. This most recent EC emission standards (Directive 91/441/EC) requires all newly produced vehicles (diesel and petrol engines) used in the UK to meet stringent emission standards from the beginning of 1993. More recently, under the auspices of the United Nations Economic Commission for Europe a number of nations have agreed a programme to limit emissions as part of the VOC protocol.

2.7.5.2 Control technology and alternative fuels. The main technological emission control techniques for VOCs primarily apply to transport sources.[25] First is the use of three way catalytic converters to control VOC emission from exhausts. This converts hydrocarbons to the less harmful CO_2 at an efficiency of some 85–95%.[10] Catalytic control may also be used in conjunction with 'lean burn' technology in the internal combustion engine. Although lean burn technology increases hydrocarbon emissions by increasing the fuel–air ratios within the engine, the use of catalysts and the fact that other gaseous pollutant emissions are reduced, coupled with improved fuel economy mean that this particular technology combination will find increased applications. Diesel engines, because of their lean burn characteristics, already meet the new 1993 EC directives without the need for a catalyst, although it is probable that some form of catalytic control may be required in the future if legislative control requires yet tougher standards.

The second approach to control is to consider the use of alternative fuel sources with lower VOC emissions rates, such as liquefied petroleum gas, in place of petrol or diesel. However, use of such fuels is in its infancy and they have the disadvantage that some methane is lost to the atmosphere with potentially damaging contributions to the enhanced greenhouse effect.

2.7.6 Analytical techniques

Legislation and the imposition of control technologies require measurement in order to assess strategies and police emissions. There are two key measurement techniques for airborne toxic organic compounds.

2.7.6.1 Optical techniques. Long path absorption techniques such as differential optical absorption spectroscopy (DOAS) can be used. The technique may be useful with aromatic organic mixtures such as BTX (benzene, toluene, and xylene). Optical methods for VOCs (such as DOAS) scan in the near UV region (250–300 nm) and can provide average concentrations over several hundred metres with short reporting periods (typically in the range 1 min to 1 h). Detection limits of sub-ppb over a 10 km light path have been reported. The main drawback to this method is the problem with deconvolution of overlapping interfering spectra, notably from SO_2 and O_3. Consequently this method still requires development work to provide suitable measurement protocols.

2.7.6.2 Gas chromatography (GC).[34,35,36] With ambient levels of VOCs being relatively low, preconcentration of the sample is usually required for GC analysis. This is usually achieved by use of solid absorbent in diffusive monitors. Choice of absorbent depends on a number of factors such as quantitative collection efficiency, recovery of trapped vapour, low background contribution from the sorbent and little affinity for water, a major interferent. Perhaps the most commonly reported sorbents for organic vapours are Tenax (2,6-diphenyl-*p*-phenylene oxide), a porous polymer, and activated charcoal. The trapped vapour is thermally desorbed from the sorbent onto a cryogenic trap whereupon it is injected onto the chromatographic column. Careful choice of a chromatographic column (e.g. CP Sil 5 CB) and detector system can ensure that the method is sensitive and species specific.

The commonly used detectors are the flame ionisation detector (FID), which has also been adapted for use commercially within continuous monitoring analysers, and the electron capture device (ECD). The ECD has the advantage of lower levels of detection whilst the FID is a more general purpose detector for VOCs with the advantage of being insensitive to water vapour. Alternatively, if identification of organic material is required, a mass spectrometer can be used as detector. Detection limits of low ppbv can be achieved using GC methodology.

2.8 Acid deposition

2.8.1 The scale and importance of acid deposition

The environmental implications of acid deposition and acidification remain the subject of extensive international and national research programmes. In these programmes the local, regional and global sources and effects of sulphur dioxide, oxidised and reduced nitrogen species and volatile organic compounds are addressed. The complete relation between cause and effect is by no means fully understood, due in part to the extremely complicated interactions between pollutants within atmospheric, terrestrial and freshwater environments. The result of such interactions may be either a direct or an indirect effect upon a susceptible target.[7,37,38] However, the broad relation between sulphur and nitrogen emissions and environmental change, particularly with regard to freshwater environments, is understood well enough to enable national and supra-national control policies to be formulated. Examples include the European Community Large Combustion Plant Directive and the United Nations Economic Commission for Europe's (UNECE) protocols on emissions on sulphur dioxide or nitrogen oxides and their transboundary fluxes.[7,37,38].

The scale and importance of the contemporary acid deposition phenomena tend to suggest that acid deposition is a relatively recent environmental problem. It was, however, first described in 1852 by R.A. Smith working in the city of Manchester and it was Smith who neologised the term acid rain.[7] At this time, sulphur was clearly the main pollutant and the effects of acid deposition were clearly an urban and near urban phenomenon. Environmental effects attributable to acid deposition either described or hypothesised by Smith included damage to building materials, metalwork, and textiles; vegetation effects were also described.[39] The southern Pennines of England received acid precipitation as a direct consequence of acid emissions in Manchester and other cities and have continued to do so in excess of 200 years (see Section 2.8.4.2). The combination of dry and wet deposition of sulphur compounds profoundly affected the *Sphagnum* moss communities, leading to species reduction, habitat change, moorland drying and subsequent erosion.[7]

Other historical measurements of precipitation composition are described by Cowling.[40] While the distribution of pollutants from industrial and urban areas was generally no more than regional in scale, longer range transport of pollutants to Norway from the UK is considered to have occurred as early as 1881.[7]

Whilst Cowling[40] provides an extensive review of the historical interest in precipitation chemistry, only a fragmented data set of limited quality is available[9,11] for the period up until the renaissance of interest in the subject of acid deposition in the late 1960s and early 1970s,[41] when effects upon freshwater were observed. From this date research into the causes and effects has been driven by a political impetus as, individually and co-operatively, nations responded to the challenges posed by acid deposition.[7]

One of the most important of the next phase of investigations was the Norwegian SNSF project which began in 1972.[42] Scientific and political interest in the United States was stimulated by the discussion of Likens *et al.*[43] of the regional distribution of acid precipitation and its significance for terrestrial and aquatic ecosystems in North America. In Europe, as a consequence of the Swedish case study,[41] the Organisation for Economic Cooperation and Development began its review of the long range transport of air pollution in 1973.[44] Under the auspices of the United Nations the European Monitoring and Evaluation Programme (EMEP) began in 1977 and shortly afterwards the National Atmospheric Deposition Programme became operational in the United States. The National Acid Precipitation Assessment Programme was established by Congress under the *Acid Precipitation Act* (P.L. 96–294, Title vii) of 1980.[45] In 1982 Sweden convened an international conference on acid rain and in 1983 the joint UK, Swedish and Norwegian Surface Water Acidification Programme was proposed. Contemporary scientific understanding of the cause and effect of acid deposition was reviewed at the International Symposium on Acidic Precipitation held at Muskoka, Canada in 1985. More recently UNECE has established the International Cooperative Programme to assess the effects of atmospheric pollutants on materials, and on forests, lakes and rivers.[7] Progress in scientific understanding of acid deposition and its consequences was reviewed at the quinquennial Acid Deposition Conference held in Glasgow in 1990.

2.8.2 Monitoring programmes and patterns of deposition

Measurement of precipitation chemistry is routinely carried out by a number of agencies on a regional or continental scale including, for example, the European Air Chemistry Network (EACN), the European Monitoring and Evaluation Programme (EMEP) and the World Meteorological Organisation.[7] Patterns of precipitation chemistry show considerable year-to-year variation. Intra-network comparisons are validated by common operating protocols, identical collectors and inter-laboratory comparison exercises. Absence of these controls on quality militates against inter-network comparisons.

In the current phase of EMEP some 92 monitoring stations measure precipitation and acid deposition precursors according to a rigorous protocol that ensures comparability of data.[7,38] Gaseous sulphur dioxide, nitrogen dioxide, ozone, nitric acid and ammonia, particulate sulphate, nitrate and ammonium, precipitation amount, concentration of nine ions and conductivity are all regularly determined. A general pattern found within the EMEP area is of maximum concentrations of aerosol sulphate and non-marine sulphate in precipitation in central-eastern Europe, with a smaller maximum of aerosol sulphate in western Europe.

The spatial patterns of pH (the logarithmic expression of hydrogen ion [acidity] concentration) of precipitation across Europe indicates that most of the UK (excluding Northern Ireland, north-west Scotland and south-west Cornwall), southern Scandinavia, northern France, northern Austria, Switzerland and northeastern Europe is bound by the pH 4.5 isoline. The most acidic precipitation occurs in central Europe bound by the pH 4.1 isoline, encompassing northern Belgium, Germany, the Czech Republic, the Slovak Republic and Poland.[7,38] These patterns cannot be considered particularly robust as they are strongly influenced by changes in emissions and in meteorological conditions.

A comparison of North American and European precipitation chemistry measurements concluded that sulphate and nitrate concentrations and wet deposition rates are greater in central Europe but levels are comparable for wet deposited sulphur in North America, east of the Mississippi, and southern Scandinavia at about 20–40 kg of sulphate per hectare per annum.[46]

Monitoring of acid deposition on a national basis began in the UK at the beginning of 1986.[7,9,11] Data from the monitoring networks, established by Warren Spring Laboratory, on behalf of the Department of the Environment, indicate that acidity increases from west to east with a maximum between the Humber and the Wash. Maximum concentrations of non-marine sulphate, nitrate and ammonium occurred in the east, particularly south of the Wash. The greatest deposition occurred in areas of highest rainfall, particularly parts of Highland Scotland, north Wales, Cumbria and the Pennines, where acid deposition is comparable to areas of southern Scandinavia. The largest concentrations of non-marine sulphate occur in low rainfall easterly air flows and the highest deposition occurs in high rainfall westerly air flows.

This broad pattern of an increasing acidity concentration gradient from west to east in the UK is consistent with the hydrogen ion pattern described by EMEP. The concentration gradients of other ions within the UK are also consistent with European scale patterns.

The concentration of ions in precipitation shows considerable variation with season, with location of measurement, with type of precipitation and even within precipitation events.[11] Temporal changes in the importance of nitrate in precipitation have been identified and this changing pattern is consistent with observed decreases in sulphur emissions and increases in oxides of nitrogen emission.[7,9,11]

Distinct temporal patterns have been observed in emissions of acid deposition precursors and the composition of their reaction products in precipitation and deposition chemistry. In the UK and Europe, emissions of precursor gases are at a maximum in winter and maximum concentrations of sulphate and nitrate are generally observed in precipitation in spring and early summer.[11] As many different ions exhibit distinct seasonal patterns despite differing sources of origin, it tends to indicate a common meteorological explanation such as wind direction. Much of the deposition experienced in certain areas of Western Europe is episodic with a large proportion of the annual deposition occurring on a relatively small number of days.[11] Such events are associated with air masses whose trajectories indicate passage over major emission source areas. Obviously, such deposition has significant implications for receptor environments. In certain areas of the world a considerable proportion of annual precipitation occurs as snowfall. Rain and snow have different mechanisms of pollutant removal and their composition can be quite different. The concentration of nitrate is higher in snow than rain with lower concentrations of sulphate found in snow samples than in rain. The high concentrations in snow are also due to dry deposition onto snow lying on the ground. Hence, when snowmelt occurs a large flush of these ions is released to watercourses with the meltwater.

2.8.3 Effects of acid deposition on freshwater systems

Concern about the effects of acid deposition on aquatic ecosystems arose for two reasons: its potential impact on aquatic biota and on human health.[7] Acidification of surface waters has serious implications for aquatic biota and species diversity. In particular it has a major effect on fish populations.[47] Elevated concentration of aluminium in surface waters associated with low pH is the main toxic agent. Aluminium solubility is largely determined by pH and various hydroxy-aluminium complexes are formed depending on the level of acidity. In laboratory tests, survival of brown trout fry has been shown to be reduced in a concentration of aluminium of 250 $\mu g\ l^{-1}$, especially if the associated calcium concentration is low. Salmonids are particularly affected and it is thought that mortality rates substantially increase during episodic acid events.[7]

Human health can also be potentially affected by acidification of aquatic systems. Concern has been expressed over the leaching of toxic metals (e.g. lead, cadmium, mercury, aluminium and copper) from catchments and from the corrosion of water supply pipes.[7] In particular there has been recent concern over levels of aluminium in drinking water.[10]

Up to the early 1970s many researchers were of the opinion that acid deposition was the main factor responsible for the acid status of surface waters. In particular the disappearance of fish from lakes in the southernmost counties of Norway and Sweden was assumed to be associated with the increase in European emissions of sulphur. This led to the conclusion that acidity in precipitation was the cause of declining fish stocks. However, it is now thought that both acid deposition and natural processes contribute to the acidification of surface waters.

2.8.3.1 Vulnerability of surface water to acidification. Acid deposition falling directly on to the surface of a free flowing stream is unlikely to contribute to acidification. On the other hand, acid deposition falling directly on to standing water can potentially contribute to acidification. However, this largely depends, *inter alia*, upon the alkalinity of the catchment and the ratio of the drainage area to lake area.[47] Catchment studies have shown that the vulnerability of both lake and stream waters to acidification by atmospheric deposition depends on the biogeochemistry and hydrology of the whole catchment.[7] Precipitation follows various pathways through a terrestrial ecosystem before it reaches a stream or a lake and factors such as vegetation, land use and soil type can contribute to, or resist, acidification.[47]

2.8.3.2 Vegetation and land use. Forest cover modifies the quality of intercepted bulk deposition (both wet and dry deposition) as it moves through the forest canopy. Trees, in particular coniferous trees, are very efficient scavengers of dry deposited gases and aerosols including oxidised and reduced nitrogen species, sulphur dioxide and sulphate. Therefore, the chemistry of throughfall is altered due to the leaching of these deposits and of excreted metabolites on the leaf surface. Soils under coniferous forests can become particularly acidic due to ion exchange processes and leaching.[7] Changed land use practices can also influence the susceptibility of surface water to acidification. For example, timber harvesting appears to decrease, and reforestation appears to increase, surface water acidification.

2.8.3.3. Bedrock geology. Areas underlain by granites and related igneous rocks and non-calcareous sandstones offer little buffering because of their slow weathering properties and catchments located in such geologies generally have low alkalinities (< 200 $\mu g\ l^{-1}$) and are regarded as susceptible to acidification.[7]

2.8.3.4. Soil type. Except when precipitation falls directly on to the surface water in a catchment or on to exposed rock, precipitation either flows across, or percolates through, soils before entering streams or lakes.[7] The movement of precipitation through soil horizons changes its chemistry and ultimately affects surface water chemistry. Throughflow, occurring in upper soil horizons which are naturally acidic, has the potential to wash out internally generated acidity, although most of the acid input from this process to a lake or stream probably occurs during episodic events.

2.8.3.5. Episodic acidification. Significant short-term pulses or flushes of acid in small upland catchments can occur on an hourly timescale and relate to hydrological changes in the catchment. These short-term acid flushes have been linked to detrimental impacts on fisheries in Scandinavia and northeastern USA.[7]

Many studies – for example, from the Adirondack Mountains of New York and Hubbard Brook, USA; Birkenes Creek, southern Norway; northeast Scotland, mid Wales and Cumbria, UK – have shown the importance of hydrological pathways in episodic surface water acidification.[7]

Short-term elevated H^+ and Al^{3+} concentrations in surface waters are attributable to heavy rainfall, snowmelt or a combination of the two.[48] Precipitation deposited as snow has been shown to affect surface water quality during snow melts. It has been observed that during and after snowmelt the concentration of ionic species in stream water differed from that at base flow in 37 Norwegian catchments. It is hypothesised that because the initial wave of meltwater is enriched with ions such as H^+ and SO_4^{2-} it could be the cause of acid flushes.[48]

2.8.3.6 Evidence of acidification. There is considerable difficulty in establishing trends of acidification in streams and rivers. However, historic pH levels for lakes and reservoirs can be reconstructed from records of fossil diatom assemblages present in lake sediments. Diatoms are microscopic algae which float freely in lake water. They have hard skeletons made of silica which are characteristic of each species and are relatively resistant to decay. Different diatom species have different pH preferences. Dead diatoms sink to the bottom of the lake where the remains accumulate as sediment. Thus, if a core of sediment is taken the stratification of the diatoms will detail the acidification history of the water. If sediment layers can be dated then acidification can be associated with known historical events. Diatom analysis is used widely in Europe and North America.[49] In the UK, Battarbee and his colleagues have systematically used diatom analysis to provide evidence of acidification in a number of regions including Galloway, Scotland, Cumbria, England and in parts of Wales. Lake sediments also contain a record of land use change (pollen analysis), air pollution (trace metal and soot analysis) and other changes in the aquatic ecosystem.[47]

In almost all of the waters examined in Scotland, England and Wales, analysis has shown stable diatom floras up to 1850. From this date, it has been suggested that atmospheric pollution from industrial sources has been responsible for subsequent acidification and diatom changes. Further evidence for the acidification of these waters is given by increases in both heavy metals and soot contained in sediment cores.

Data from the six lakes studied in Galloway, Scotland,[50] were consistent with a decline in pH of between 0.5 and 1.2, within the last 150 years. The onset of recent acidification varied from lake to lake from about 1850 to 1925 although, in general, acidification in the nineteenth century was slow and an acceleration occurred in the twentieth century between 1930 and 1950.

The effects of acid deposition on freshwater fish are particularly pronounced in Norway and Sweden. Southern Norway contains some 33 000 km^2 of acidified freshwater systems which are unable to support a normal ecological community.[38]

Brown trout and Atlantic salmon are the most affected species and populations of Arctic charr, perch and roach are all at risk in certain areas.[38] The decline of fish population in acidic waters in southern Norway started as early as the 1920s but the most rapid losses occurred during the decade 1960–1970 and by 1978 the population of Atlantic salmon had disappeared from southern Norway.[38] In the same area, more than half the brown trout population had disappeared by 1985. In Sweden it is estimated that 18 000 lakes and 100 000 km of running water have pH values where damage to freshwater organisms can be expected.[38] Episodic fish kills, due to rapid changes in water quality as a result of acid surges, have severely affected salmon rivers on the west coast of Norway. Most affected have been smolts of Atlantic salmon and spawning migrating salmon on return to acidified home waters.

2.8.4 The effects of acid deposition on vegetation

Sulphuric acid, nitric acid and ammonium ions are the main agents of damage to the terrestrial systems rather than acidity *per se*. The roles of ammonia and ammonium ions in particular have only relatively recently been recognised. Ammonia is the major neutralising compound in the atmosphere. Extensive work has been undertaken in the Netherlands, where ambient concentrations of ammonia are high, to identify how high depositions effect the natural vegetation, in particular heathland systems.

Acid deposition has been reported to effect vegetation in various types of ecosystem including agricultural crops, heathlands, wetlands and forests.[7,38] Concern over the effects of acid deposition and air pollution upon vegetation is of more recent origin than that for freshwater.

Rather than attempt a comprehensive treatment of the effects of such pollutants on all terrestrial ecosystems, attention is briefly focused upon two systems: forests and moorlands.

2.8.4.1 Effects upon forest systems. Forest systems have always been subjected to environmental stress; harsh climate, poor nutrition and pathogens have always affected forest productivity. However, since the 1970s an unprecedented decline of forest ecosystems has occurred in central Europe.[7,38] Forest decline has also occurred in North America. Species affected include silver fir, Norway spruce, beech, Scots pine, larch and oak, with old spruce, fir and oak the most affected. Symptoms include yellowing of needles, loss of older needles, crown thinning, growth reductions and growth alterations.[38]

The 1990 forest damage survey[51] reports damage in all of the 25 participating countries of Europe and North America. Older trees are often more heavily defoliated but in some countries very high defoliation is observed on newer stands. In several countries, forests at higher elevations were more defoliated than those at lower elevations. Several thousand hectares of forests on exposed mountain tops in Germany are the most severely affected or even dying. Similar conditions prevail in Bulgaria, the Czech and Slovak Republics and Poland.[38]

A number of hypotheses have been suggested to explain the observed decline in forest health in many European nations. Proposed explanations for forest decline include: soil acidification and aluminium toxicity; needle reddening disease; the combined effects of ozone, acid deposition and nutrient deficiencies at high altitude; and increased anthropogenic nitrogen deposition.[7] Whilst forest decline continues to be an important issue in the ecological effects of air pollution and acid deposition, it is highly unlikely that one decisive 'factor' will be identified which universally explains forest decline. In more recent years, site-specific studies have become more important and many valuable data have been acquired. These studies have re-emphasised the importance of considering local stress factors, the balance of which vary according to location.[7]

2.8.4.2 Effects of acid deposition upon moorlands. Peatlands are important ecosystems both nationally in the UK and in a global context. They cover approximately 8.6% of the total land area of the British Isles[52] and about 3% of the earth's total land surface.[53] They form an important catchment area for reservoirs and are of ecological and recreational value.[54]

Peatlands are themselves particularly sensitive to acid deposition for various reasons. These include their ombrotrophic nature in that they receive all their elements from the atmosphere and consequently any change in the atmospheric supply will effect the plants. Plants in moorland communities are adapted to the supply of elements being low together with a high annual precipitation amount. Also many of these peatlands are dominated by bryophytes, and in particular *Sphagnum* species. Such plants lack a protective cuticle and are only one cell thick; therefore they are directly and continuously exposed to the atmospheric supply of elements both from wet and dry deposition.

Marked vegetation changes have been recorded within the past two centuries in the peatlands of the southern Pennines of England.[55] The most important change is probably the loss of *Sphagnum*, a major component of most northern peatlands. *Sphagnum* was dominant over large areas in the fourteenth, fifteenth and sixteenth centuries, but began to decline in the 1700s.[56] The complete disappearance of the *Sphagnum* cover is strongly correlated with a layer of black soot in the peat profile, at approximately 5–10 cm from the peat surface, which coincides with the period of the Industrial Revolution in the eighteenth and nineteenth centuries.[57–59] The dominance of bryophytes before the Industrial Revolution has been documented by Farey,[60] and Moss[61] recognised 13 species of *Sphagnum* occurring in the southern Pennines, only two of which were then described as rare. However, in 1964 only five species were listed as growing in the area and only *S. recurvum* was described as common.[58] *Calluna* too has declined, largely as a result of sheep grazing, so that the present day moorland is dominated by *Eriophorum vaginatum* L., *E. angustifolium* L., *Empetrum nigrum* L. and *Vaccinium myrtillus* L.

The present-day rarity of *Sphagnum* in the southern Pennines has been related to sulphur and nitrogen pollutants. These moorlands have been subject to high concentrations of atmospheric pollutants for over two centuries.[62,63] *Sphagnum* species appear to be particularly susceptible to atmospheric pollution as discussed above. Indeed, experiments have shown that the growth of a number of *Sphagnum* species is severely restricted when fumigated with high concentrations of SO_2 (131 $\mu g\ m^{-3}$).[64,65] Increased nitrogen deposition has also been shown to restrict the growth of *Sphagnum*.[66] Ferguson and Lee[68] report failure of *Sphagnum* after an 18-month reintroduction experiment to Holme Moss, southern Pennines, though 'control' plants grew successfully at a relatively unpolluted site in North Wales.

More recent work has shown that unlike *Sphagnum* from unpolluted sites, *Sphagnum* in the southern Pennines has an effective mechanism which prevents the toxic effects of aqueous sulphur dioxide and bisulphite.[68,69] Fe (III) and Mn (II), metals which are high in concentration in the southern Pennine peats due to recycling of past industrial emissions, act as catalysts for the oxidation of these sulphur ions – a mechanism which may have been important in the survival of local colonies of *Sphagnum* in the past when sulphur dioxide concentrations were significantly higher than at the present day.

Sphagnum species do occur in the southern Pennines today, but the populations are restricted and largely isolated. *S. papillosum* grows on at least two plateau sites, Cowms Moor and Hey Ridge Moor, in the southern Pennines.[70] *S. cuspidatum* Hoffm. is found very locally in bog pools on the northwest moors, while *S. recurvum*, *S. auriculatum var. auriculatum* Schimp., *S. palustre* L. and *S. squarrosum* Crome, plants with wider ecological tolerances, can be found in flush sites.

However, Sphagna, especially ombrotrophic species, still remain a rarity in the southern Pennines. One such site, Cowms Moor, has the unusual feature, in a southern Pennine context, of *Sphagnum* growing there today in considerable abundance, *S. papillosum* returned to sites at three altitudes on Cowms Moor in approximately the year 1900. At the lowest altitude site on Cowms Moor there is evidence that *Sphagnum* never disappeared at or after the time of the Industrial Revolution. It remains unclear why small patches of *Sphagnum* were not killed by past high levels of pollution.

Present day acid deposition, both in terms of nitrogen and sulphur deposition, was shown to be greater at the highest altitude on Cowms Moor (505 m asl). This was found to be related to the amount of higher precipitation at the highest altitude and to the importance of occult deposition in upland areas. Indeed at these sites, especially at the highest altitude, concentrations of solutes in cloudwater were shown to be several times higher than those in rainfall. Occult bulk quotients were as high as 9.5.[71] *Sphagnum* growth on the highest altitude site on Cowms Moor was found to be reduced compared to that recorded at the lower altitude sites. Experimental evidence suggests that the greater acid deposition rate at the highest site, at least partially, is the cause of the restricted *Sphagnum* growth.

2.8.4.3 Critical loads and levels. The concept of a critical load is now utilised to determine if acid deposition is affecting vegetation. A critical load is defined as a quantitative estimate of an exposure to one or more pollutants below which significant harmful effects on specified sensitive elements of the environment do not occur according to present knowledge.[37] A critical level has also been defined: this is the concentration(s) of pollutant(s) in the atmosphere above which direct adverse effects on receptors such as plants, ecosystems or materials, may occur according to present knowledge.[37] A critical load is intended to act as a threshold deposition which ecosystems can tolerate without damage occurring. Maps of critical loads for various receptors have been published for Europe by several groups.[72,73] In theory the acid deposition fields can be compared to these critical loads to devise exceedence maps for the country in question. This has been achieved in the UK[74] but some debate within the scientific community as to the usefulness of such an approach has occurred. For example, what criteria should be used to determine if an ecosystem is undergoing damage? Criteria may include, for example, loss of species diversity, visible and/or invisible effects to plants, or changes to the soil system such as an increase in nitrate concentrations. However, generally it has been agreed in the UK that the aim of the critical loads approach should be to return waters and soils to their pre-industrial status.

2.8.5 Other effects

2.8.5.1 Materials. Buildings, materials and cultural property are subject to weathering decay under the action of meteorological factors such as precipitation, wind and solar radiation. This is a natural process which acts in the absence of human involvement. Air pollutants including acid deposition and its precursors can, however, accelerate the ageing of materials and buildings and cause secondary effects such as increased expenditure for protective measures, increased cleaning and potentially a diminished utility for the design purpose.[37]

Material damage can occur either as atmospheric corrosion or as corrosion in soil/water systems. Atmospheric corrosion is mainly, but not exclusively, due to dry deposition of pollutants and may affect a range of important technical materials. Corrosion resulting from soil and water acidification is a threat to technical and economically valuable materials such as water pipes, cables, building and telecommunication networks.[37,38]

There is some evidence to support the view that reducing air pollution levels to those that are below the critical level for biological systems will provide a satisfactory reduction in the rate of material degradation.[37]

2.8.5.2 Health. Potential health effects of acid deposition can be divided into direct effects of precursors and indirect effects such as mobilisation of, and increased risk of exposure to, heavy metals in drinking water and foodstuffs.[37] Direct effects have been extensively studied and reviewed by international and national organisations. Concentrations at which SO_2 and NO_x affect human health have been established and guidelines produced (see Section 2.4).

Human health can also be potentially affected by acidification of aquatic systems. Concern has been expressed over the mobilisation of metals in domestic water supply systems, caused by the leaching of toxic metals (e.g. lead, cadmium, mercury, aluminium and copper) from catchments and from corrosion of water supply pipes. In particular there has been recent concern expressed over levels of aluminium in drinking water and research has indicated that aluminium may be implicated in the development of Alzheimer's disease.[37]

Swimming in lake waters acidified to pH 4.6 has not been found to adversely affect human eyes or skin.[38]

2.8.6 *Technology for control of acid deposition precursors*

Control is needed both at power plants and other sources. The benefits of energy-efficiency measures, which undeniably can reduce energy demand and hence emissions, are self evident and are not discussed further here.

2.8.6.1 Stationary sources. There are a variety of different approaches to dealing with emissions from large combustion plants, ranging from fuel treatment prior to combustion (e.g. coal cleaning or blending), through to flue gas treatment after combustion (e.g. flue gas desulphurisation (FGD) and selective catalytic reduction).[37,38] The technologies available for implementing control measures are numerous and diverse. Each one however, has both its advantages and limitations, related to cost, removal efficiency and operational experience. The nature and amount of waste and by-product resulting from the control technologies play a major part in influencing a potential operator's decision over which system is the most appropriate.

Traditionally sulphur dioxide has been removed from a large combustion plant by FGD. The choice has primarily been between a limestone–gypsum method and the regenerable Wellman–Lord method.[75–77] More recently a variety of nitrogen oxide emission control technologies has been introduced.[76] These are divided into two categories: combustion modification and post-combustion exhaust gas treatment. Combustion modification involves conversion of the burner unit to reduce peak flame temperatures and hence formation of thermal oxides of nitrogen. Post-combustion control involves catalytic destruction of nitrogen oxides. Both methods have found application in Europe, North America and Japan although the former is less expensive and less efficient.[76,37]

More recently attention has turned to alternative combustion processes in preference to FGD based control systems. Particularly favoured amongst such systems are the combined cycle gas turbine (CCGT) and various fluidised bed combustion systems. These technologies offer higher efficiencies than conventional power generation systems and in the case of CCGT the fuel is inherently low in sulphur. Fluidised bed systems can reduce sulphur *in situ* by incorporation of lime or other sorbents in the fluidised bed.[37]

2.8.6.2 Mobile source control. Motor vehicle exhaust is a significant and growing source of emission of acid deposition precursor gases. Catalytic converters are recognised as the only commercially available control technology able to significantly reduce this air pollution burden, at relatively low expense, without unduly affecting vehicle performance and fuel consumption.[37,38] Emissions of oxides of nitrogen and volatile organic compounds can be reduced by about 80–90% through the use of catalysts. In North America, catalysts have been in use since the mid 1970s and whilst certain European nations (Sweden, Norway, the former F.R. Germany, the Netherlands and Austria) encouraged their uptake from the mid 1980s, it is only from model year 1993 that catalytic convertors have become mandatory on petrol engined vehicles sold in the European Community.

A variety of other techniques are available to control emissions from mobile sources. These include alternative fuels (such as electric vehicles / hybrid vehicles / natural gas fuels / fuels derived from vegetable sources like rape-seed), substitution of diesel for petrol and new engine designs. The problem with all of these options and with catalytic controls is that they only reduce the individual vehicle contribution. If the total number of vehicles continues to increase then, at somc point in thc not too distant future (perhaps 2007–2014 in the UK), the benefits of the control technology will be overtaken by the increase in total vehicle numbers and emissions will return to their early 1990s total.

The impetus to control emissions from stationary and mobile sources has been international agreements such as the EC large combustion plants directive or the UNECE protocols on transboundary fluxes of sulphur and nitrogen.

2.8.7 International regulation

The United Nations Economic Commission for Europe (UNECE) is the major international forum for the measurement and abatement of acid deposition. The UNECE Convention on Long Range Transboundary Air Pollution was adopted in 1979 and entered into force in 1983 as the first multilateral treaty to protect the atmospheric environment.[10,37,38] Thirty-two parties have ratified this convention which lays down principles and provides a framework for co-operation. Four protocols to the convention provide the instruments by which emission reductions can be achieved and base line monitoring carried out. These are:

1. The protocol on the long term financing of the co-operative Programme for Monitoring and Evaluation of the Long Range Transmission of Air Pollutants in Europe.
2. The protocol on the reduction of sulphur emissions or their transboundary fluxes by at least 30% adopted in Helsinki, Finland, in July 1985 (the so-called 30% protocol).

Table 2.18 Proposed sulphur emission reductions from UN Economic Commission for Europe member states signatory to the Geneva Convention on Long Range Transboundary Air Pollution

Emission reductions proposed from a 1980 emission total[10]		
State	Percentage reduction	Deadline (year)
Sweden	80	2000
Netherlands	80	2000
Finland	80	2003
Austria	70	1995
Former FRG	65	1993
Luxembourg	58	1990
Switzerland	57	1995
France	50	1990
Canada	50	1994
Norway	50	1994
Belgium	50	1995
Denmark	50	1995
Liechtenstein	30	1993
Italy	30	1993
Former Czechoslovakia	30	1993
Former East Germany	30	1993
Former USSR	30	1993
Bulgaria	30	1993
Ukraine	30	1993
White Russia	30	1993

The EU, UK, Greece, Iceland, Ireland, Poland, Portugal, Romania, San Marino, Spain, Turkey, USA, Yugoslavia and the Vatican City State are signatories to the Geneva Convention but have not committed themselves to a percentage reduction within a specified timeframe.

3. The protocol concerning the control of emissions of nitrogen oxides or their transboundary fluxes, signed in Sofia, Bulgaria, in November 1988.
4. The protocol concerning the control of emissions of volatile organic compounds or their transboundary fluxes, signed in Geneva, Switzerland, in 1991.

The 30% protocol has been ratified by 16 nations and entered into force in September 1987. This protocol, the so-called '30% club', aims for a 30% reduction in 1980 sulphur emissions by 1993.[37,10] This protocol is being renegotiated in the early 1990s. Table 2.18 shows the sulphur emission reductions proposed by members of the UNECE. The protocol concerning nitrogen oxides was signed by 25 nations and came into effect in 1991. It requires the signatories to ensure that after 1994 emissions do not exceed the 1987 level.[37] Whilst not initially specifying percentage reductions on emissions it does require a second step to be undertaken which is that emission reduction measures must be taken which are consistent with the concepts of critical load and level.[37] Negotiations concerning the second phase of the NO_x protocol began in 1990–91 with renegotiation due in 1995. Emission reduction measures are required to be in place by the end of 1996.[10]

The VOC protocol is designed to limit the formation of tropospheric ozone and other photochemical oxidants. Parties to the protocol must reduce emissions of VOCs by 30% by 1999 taking a specified year between 1984 and 1990 as a baseline.[10]

Until recently emission reductions proposed have been essentially political agreements relying upon social and economic acceptance of the reduction rather than the capacity of environmental systems to receive deposition. More recently the concepts of a critical load and a critical level (see Section 2.8.4.3 for the definitions of a critical load and level) has been introduced into the political process regulating emissions.

Through the auspices of the UNECE, critical loads and levels have now been defined for a range of pollutants and sensitive environments. Work will continue through the UNECE to refine the accuracy and precision of the critical load and critical level concept. Current and critical loads and levels are mapped under the guidances of the UNECE and critical values now form major inputs to abatement strategy models. International agreements of this type are required to effectively control the emission of acid deposition precursors and their subsequent environmental effects.

2.9 Stratospheric ozone depletion

2.9.1 Formation of stratospheric ozone

Oxygen is the second largest component of the atmosphere, and mixes with nitrogen in its elemental state rather than in the form of dissolved nitrate ions. Oxygen also coexists with hydrogen, methane and nitrous oxide without reacting in the way that the laws of chemical thermodynamics would dictate. This thermodynamic disequilibrium is fuelled by biological processes which are vital to the production of the oxidisable components of the atmosphere and these processes are also responsible for the formation of oxygen, the main oxidant of the atmosphere.

Two sources of atmospheric oxygen are the dissociation of water and the interaction of CO_2 with H_2O which occurs during photosynthesis

$$2H_2O \rightarrow 2H_2 + O_2 \qquad (2.23)$$

$$nH_2O + nCO_2 \rightarrow \{CH_2O\}_n + nO_2 \qquad (2.24)$$

The absorption of solar radiation is vital to both of the reactions shown above. Reaction (2.23) requires ultraviolet, and reaction (2.24) visible radiation. The absence of oxygen in the atmospheres of the earth's closest neighbours, Venus and Mars, coupled with the lack of life on these planets, indicates that the most probable major producer of the oxygen found in the earth's atmosphere is the photosynthesis reaction (2.24).

The first stages of single-cell organism development require an oxygen-free (reducing) environment. This type of environment existed on earth over 4 billion years ago. As the primitive forms of plant life multiplied and evolved, they began to release minute amounts of oxygen through the photosynthesis reaction [reaction (2.24)]. The build up of oxygen in the atmosphere led to the formation of the ozone layer in the stratosphere. The layer filters out incoming radiation in the ultraviolet part of the spectrum. Thus, with the development of the ozone layer came the development of more advanced life-forms and hence the release of even more oxygen into the atmosphere. So the interdependence of life on atmospheric composition and atmospheric composition on life established itself.

Ozone is formed in the stratosphere (10–50 km) and mesosphere (50–80 km) by photochemically induced reactions of oxygen:[78]

$$O_2 + h\nu \rightarrow O(^1D) + O(^1D) \quad (2.25)$$

$$O(^1D) + O_2 + M \rightarrow O_3 + M \quad (2.02)$$

The maximum ozone concentration in the stratosphere occurs at an altitude of about 25 km. Ozone absorbs ultraviolet solar radiation in the region $200 < \lambda < 300$ nm, whereas no other major atmospheric species absorbs over this particular wavelength range. At levels below 70 km, virtually all of the absorbed energy is converted into the form of kinetic energy within the ozone molecules, i.e. into increasing the temperature of the stratosphere.

The strong concentration gradients observed just above the tropopause are a reflection of the fact that there is relatively little mixing between the two lower regions of the atmosphere. In contrast to the situation in the troposphere, it is found that particles may remain in the stratosphere for a period of many years after they are produced. As a result of this slow vertical mixing and, accordingly, long residence times, the stratosphere can act as a reservoir for certain species of importance in atmospheric pollution, especially long-lived species such as CFCs.

2.9.2 Stratosphere–troposphere exchange

An example of the interaction of the two lower regions of the atmosphere is the occurrence of stratosphere–troposphere exchange. The average annual exchange of stratospheric air to the troposphere is between 22% and 44% of the stratosphere's total mass. The exchange of tropospheric air, in terms of percentage, is much less at about 4.5–9%.[79] Trace components from the troposphere are transported up to the stratosphere mainly in the tropics as a result of the powerful convection currents that are generated by solar heating. As tropospheric air rises, the moisture contained within it condenses out as it passes through the cold tropopause (temperatures at this point are approximately 210 K). As a result, stratospheric air is very dry in comparison with tropospheric air. This dry air is easily detected when it returns to the troposphere in mid to high latitudes, and it provides a significant source of tropospheric ozone. In fact, the incursion of the air is marked by exceedingly high ozone levels, up to 200 ppb compared to a normal background level of about 30 ppb.

Transport of stratospheric air downwards into the higher latitude troposphere is aided mainly by the tropospheric folding associated with low-pressure weather systems. The physical process of intrusion of stratospheric air will not carry on down to ground level, but tropospheric turbulence will eventually transport this air downwards. A study of the amount of stratospheric ozone to reach ground level over a period of about two months indicated that between 25% and 93% of the ozone detected at various times originated in the stratosphere.[79]

An important aspect of the interaction of the two regions is the transportation of active chlorine from the lower regions of the troposphere into the stratosphere where it can deplete ozone. The existence of the ozone hole is a direct consequence of the presence of chlorinated compounds (e.g. CFCs) in the atmosphere. The reason for the popularity of CFCs is the same as the reason for the concern over CFC releases resulting in ozone layer destruction, i.e. their chemical inertness. As a result of this inertness they have very long lifetimes within the troposphere; this allows them to be transported to the equator from the heavily industrialized regions of the world. At the equator they are transported up to higher altitudes, i.e. into the stratosphere. In the stratosphere the chlorine atoms are released from their parent CFC by photochemical decomposition, thus enabling them to destroy ozone.

Ninety-five percent of all the CFCs in existence were produced in the developed world. However, due to the direction of wind patterns around the globe and the unusually isolated and cold conditions at the South Pole, atmospheric concentrations of chlorine are generally higher over the Antarctic than over the Northern Hemisphere where the bulk of the CFC pollutants were produced. As a result of the enormous popularity of the chlorofluorocarbons, there is six times the concentration of chlorine in the atmosphere today that there was in the 1900s. Similarly, the concentration of fluorine in the lower atmosphere in the 1930s was negligible. Nowadays fluorine is ubiquitous throughout the polluted troposphere and each of its primary sources (CF_3Cl, $CBrF_2Cl$, C_2F_6, C_2F_5Cl, etc.) is of anthropogenic nature.

By the late 1980s and early 1990s, significant decreases in stratospheric ozone in both northern and southern hemispheres at middle and high latitudes were being observed and consequently large increases in surface ultra-violet radiation were recorded in Antarctica during periods of low ozone.[80]

In both the spring and winter of 1992, the total ozone column over Britain and Europe was reduced by 10–20% compared to ozone levels at any other time of the year. In the United States of America, NASA has noted that it is increasingly likely in the coming years that substantial Arctic ozone losses will occur during particularly cold, protracted winters and mid-latitude ozone trends will continue to decline.[81] NASA also reported in February 1992 that there were unprecedented levels of chlorine pollution over the northern hemisphere. The reactive chlorine levels over cities including London, UK, Moscow, Russia, and Amsterdam, the Netherlands, on 11 January 1992 were comparable to those observed within the Antarctic ozone hole.[81]

There is no longer any scientific doubt that the observed ozone losses in the stratosphere can be largely attributed to man-made substances such as CFCs, HCFCs, carbon tetrachloride, methyl bromide and methyl chloroform. These compounds can release chlorine or bromine which react with O_3 to produce chlorine or bromine monoxide and an oxygen molecule, which results in a chain destruction of ozone. Each year ozone losses over the Antarctic average 50% and predicted losses of ozone over the northern hemisphere could reach 30% in spring by the end of the century.

Even with a phase out-date of 1 January 1996 for all offending source compounds, a return to 2 ppbv levels of stratospheric chlorine (the level of [Cl] in 1976) is not predicted until some time between the middle and the end of the next century, while chlorine concentration is expected to peak in the stratosphere at over 4 ppbv around the year 2000.[80]

The link between ozone depletion and increasing UV-B radiation at the earth's surface is now well established,[82] as are damaging effects of an increase in UV–B radiation on our ecosystem. A sustained 10% loss of ozone would lead to an increase in non-melanoma skin cancers by 26%.[80]

It has already been reported, for example, that plankton production was reduced by 6–12% under the Antarctic ozone hole in the spring of 1990, this being equivalent to a 2–4% annual reduction in production.[83] As plankton are at the base of the food chain, this is clearly threatening to fisheries, marine wildlife and, with time, the entire marine ecosystem. Since more than 50% of the world's biomass is located in aquatic ecosystems, even small reductions in productivity could have serious implications.

2.9.3 International control programmes: the Montreal Protocol

The discovery of the severity of the ozone hole over Antarctica accentuated the need for urgent international measures to control the global use of

CFCs. On the basis of the Vienna Convention, the Montreal Protocol on Substances that Deplete the Ozone Layer was negotiated and signed by 24 countries in September 1987. The protocol called for the parties to freeze consumption of CFC-11, 12, 113, 114 and 115 at 1986 levels by 1 July 1989. A goal was set to a reduction of 50% relative to 1986 levels in CFC production and consumption by 1999. Following the revision of the Protocol in June 1990, at which an agreement was reached that both the production and consumption of all CFCs should be phased out by the year 2000, the European Union agreed to a new European Regulation (594/91) bringing forward CFC phase-out to July 1997.

In order to deal with the special difficulties experienced by developing countries, it was agreed that they would be given 10 years' grace, so long as their use of CFCs remained under 0.3 kg per person per year. The global average is about 0.2 kg, but the figure for the industrialised world is about 1 kg per person per year.

The Montreal Protocol was revised once again in Copenhagen, Denmark, in November 1992. In Copenhagen, Ministers and government officials from 74 countries agreed to a revision of the phase-out deadlines of controlled substances ahead of the original schedule agreed in Montreal. The following guidelines were established:

1. Complete phase-out of ozone-depleting CFCs by 1 January 1996, four years ahead of schedule; 75% will be phased-out by 1994.
2. Acceleration of the elimination of halons (bromine containing compounds) by six years to 1994.
3. 100% phase-out of methyl chloroform by 1996, nine years ahead of schedule, with a 50% reduction by 1994.
4. 100% phase-out of carbon tetrachloride by 1996, 85% reduction by 1994.
5. Phase-out of HCFCs by 2030, with a 35% reduction by 2004, 65% by 2010, a 90% reduction by 2015 and a 99.5% reduction by 2020.
6. 100% phase-out of HBFCs by 1996.
7. A multilateral ozone fund was created to help developing countries phase out controlled substances.
8. Six technologies were adopted for destroying the controlled substances.

The next proposed review of the situation will be in 1996.

2.10 The greenhouse effect

2.10.1 Carbon dioxide (CO_2)

Whenever a gas that is a weak absorber in the visible and a strong absorber in the infrared is a constituent of a planetary atmosphere, it contributes towards raising the surface temperature of that planet. CO_2 is one such gas and also, to a lesser extent, are H_2O and O_3. This warming has been commonly referred to as the 'greenhouse effect' and the gases that produce it are collectively known as greenhouse gases.

Even though CO_2 concentrations in the atmosphere are very small in comparison with nitrogen or oxygen, its existence in the atmosphere has a significant effect on the surface temperature of the earth. Although CO_2 is a naturally occurring species, the contribution of man-made CO_2 (from combustion processes) in the atmosphere is increasing. In 1990 the atmospheric concentration of carbon dioxide was 353 ppmv, about 25% greater than the pre-industrial (1750–1800) value of about 280 ppmv, and higher than at any time in the last 160 000 years.[84] Since 1957, the CO_2 concentration in the air has been measured at Mauna Loa, Hawaii. The record shows a very clear increase of about 0.5% per year (1.8 ppmv) due to anthropogenic emissions.[84] The atmospheric increase in CO_2 during the past decade corresponds to just under half of the total emissions during the same period with the remainder being taken up by the oceans and land. The time taken for atmospheric carbon dioxide to adjust to changes in sources or sinks is of the order of 50–200 years, determined mainly by the slow exchange of carbon between surface waters and deep layers of the oceans. Consequently, CO_2 emitted into the atmosphere today will influence the atmospheric concentrations of CO_2 for centuries into the future.[3] Models have been used to estimate that even if anthropogenic emissions of CO_2 could be kept constant at present day rates, atmospheric carbon dioxide would increase to 415–480 ppmv by the year 2050 and to 460–560 ppmv by the year 2100. In order to stabilise concentrations at present day levels, the Inter Governmental Panel on Climate Change (IPCC) estimate that an immediate reduction in global anthropogenic emissions by over 60% would be necessary.[84]

2.10.2 The earth's natural greenhouse

Long-wave ultraviolet (UV) radiation (λ = 300–400 nm) is only weakly absorbed by the earth's atmospheric gases, hence the incoming UV radiation from the sun penetrates through to the planet's surface easily. Radiation emitted from the earth, on the other hand, occurs at much longer wavelengths and is of much lower intensity than that emitted from the sun. These long-wave infrared (IR) emissions from the earth corres-pond to a wavelength region at which the greenhouse gases absorb strongly, resulting in the outgoing radiation being trapped by these gases. This additional long-wave radiation warms up the lower regions of the atmosphere and perturbs the radiation balance between the earth and its gaseous envelope.

2.10.3 *Impacts of the enhanced greenhouse effect*

Concern about increasing CO_2, and other greenhouse gas, concentrations in the lower part of the atmosphere is centred around research that shows, through complex computer models of the global climate system, that mean surface temperatures will rise by approximately 3°C for a doubling in CO_2 concentration.[84] Positive feedbacks, such as decreased albedo (the overall reflectivity of a planet) as a result of shrinkage of the polar snow caps, can amplify small temperature changes. Such rises in temperature could, in turn, lead to more water vapour being added to the atmosphere. As water vapour is also an important greenhouse gas, this positive feedback could result in further increases in temperature. Measures are now beginning to be implemented in an attempt to curb the dramatic increase in CO_2 concentrations over the last two centuries.

2.10.4 *International control programmes: the Convention on Climate Change*

The Convention on Climate Change, which met in Geneva in 1992, decreed that all developed countries should attempt to halt CO_2 increases and bring about a stabilisation of CO_2 emissions to 1990 levels by the year 2000. The convention was adopted by the Inter-Governmental Negotiating Committee (INC) on 9 May 1992. It was signed, after 14 months of negotiations, in June 1992 at the Rio Earth Summit by 162 governments in total. The overall objective of the treaty was identified as the 'protection of the climate system as a global ecological resource'.

The convention requires all countries that ratify the Treaty to produce a national programme containing measures that will limit the amount of total greenhouse gases produced and introduce methods for protecting CO_2 sinks (e.g. forests and vegetation). Whilst developed countries must aim to reduce their emissions of CO_2 by the year 2000, developing countries will be helped by the developed countries' technology and funding to draw up their programmes and reduce their emissions.

The UK's programme will aim to save 10 million tonnes of CO_2 by 2000 (6% of current levels).[20] New targets for non-renewable and renewable energy capacities will be set. The UK Government will promote energy efficiency and will include the phased implementation of value added tax on domestic energy usage. Energy efficiency standards will gradually be introduced for household appliances such as washing machines and dish washers. More public programmes for advice and information will be implemented.

All parties will also cooperate into further research on climate change and its effects on ecosystems. The proposals put forward in the convention will be regularly reviewed, at the Conference of the Parties (COP), the first of which will be in 1995. By August 1993, over 30 countries had ratified this Treaty.[85] Fifty signatures are required for the treaty to become a legally binding document.

2.11 Conclusions

This chapter has reviewed the major air pollution problems facing society in the late twentieth century. The problems identified are essentially related to the method of economic development practised by industrial nations. Structural adjustments to this development pattern are required which move towards modes recognised as being economically and environmentally sustainable (i.e. sustainable development patterns). This is necessary if these problems are to be dealt with in the industrialised world, avoided in the developing world and the benefits of development spread more equitably between and within societies. To achieve sustainable development will require the integration of economic and environmental decision-making processes. A desirable and likely outcome from such a shift in decision making will be the adoption of clean technologies, which require smaller inputs of energy and materials to produce the same level of output with a reduction in the emission of air and other pollutants.

References

1. Wallace, W.H. and Hobbs, P.L. (1977) *Atmospheric Science (An Introductory Survey)*, Academic Press, New York.
2. McCormac, B.M. (1971) *Atmospheric Pollution*, D. Reidel, Dordrecht.
3. Heicklen, J. (1976) *Atmospheric Chemistry*, Academic Press, New York.
4. Quality of Urban Air Review Group (1993) *Urban Air Quality in the U.K.* Department of the Environment, London.
5. Department of the Environment (1993) *Digest of Environmental Protection and Water Statistics*. No 15, 1992. HMSO, London.
6. World Health Organisation (1987) *Air Quality Guidelines for Europe*. WHO Regional Publications, European Series No. 23, Copenhagen.
7. Longhurst, J.W.S., Raper, D., Lee, D., Heath, B., Conlan, D.E. and King, H. (1993) Acid deposition: a select review 1852–1990. Part 1. *Fuel*, **72** (9), 1261–1280.
8. Moller, D. (1984) Estimation of the global man-made sulphur emission. *Atmospheric Environment*, **18**, 29–39.
9. Irwin, J.G. (1989) Acid rain: emissions and deposition, *Archive of Environmental Contamination and Toxicology*, **18**, 95–107.
10. Hanneberg, P. (ed.) (1993) *Acidification and Air Pollution: A Brief Guide*. Swedish Environmental Protection Agency, Solna.
11. Irwin, J.G., Campbell, G., Cape, J.N., Clark, P.A., Davies, T.D., Derwent, R.G., Fisher, B.E.A., Fowler, D., Kallend, T., Longhurst, J.W.S., Martin, A.M., Smith, F.B., Warrilow, D.A., Lee, D.S. and Metcalfe, S. (1990) *Acid Deposition in the United Kingdom 1986–1988*, A third report of the UK RGAR, Warren Spring Laboratory, Stevenage.
12. Ehhalt, D.H. and Drummond, J.W. (1992) The tropospheric cycle of nitrogen oxides. In:

Chemistry of the Unpolluted and Polluted Troposphere (eds Georgii, H.W. and Jaeschke, W.), D. Reidel, Dordrecht.

13. Longhurst, J.W.S. (1989) Oxides of nitrogen in the Greater Manchester conurbation, *Environmentalist*, **9** (4), 253–260.
14. Buckley-Golder, D. (1984) *Acidity in the Environment*, Energy Technology Research Unit Report 23, Harwell.
15. Buijsman, E., Mass, H.F.M. and Asman, W.A.H. (1987) Anthropogenic ammonia emissions in Europe, *Atmospheric Environment*, **21**, 1009–1022.
16. ApSimon, H.M., Kruse, M. and Bell, J.N.B. (1987) Ammonia emissions and their role in acid deposition, *Atmospheric Environment*, **21**, 1939–1946.
17. Warren Spring Laboratory (1993) National Atmospheric Emissions Inventory. WSL, Stevenage.
18. Crane, A.J. and Cocks, A.T. (1989) The transport, transformation and deposition of airborne emissions from power stations. In: *Acid Deposition: Sources, Effects and Controls* (ed. Longhurst, J.W.S.), British Library, London and Technical Communications, Letchworth, pp 1–13.
19. Legge, A.H. and Crowther, R.A. (1987) *Acid Deposition and the Environment: A Literature Overview*, Acid Deposition Research Programme, report 11, University of Calgary.
20. DeMore, W.B., Sanders, S.P., Golden, D.M., Molina, M.J., Hampson, R.F., Kurylo, M.S., Howard, C.J. and Ravishankara, A.R. (1990) *Chemical Kinetics and Photochemistry Data For Use In Stratospheric Modelling Evaluation*, No 9, JPL, 90–91, Pasadena.
21. Wayne, R.P. (1991) *Chemistry of Atmospheres*, 2nd edn, Oxford University Press.
22. Finlayson-Pitts, B.J. and Pitts Jr., J.N. (1986) *Atmospheric Chemistry*, John Wiley and Sons, New York.
23. Penkett, S.A. (1983) Atmospheric chemistry – PAN in the natural atmosphere, *Nature*, **302**, 293–294.
24. Clarke, A.G., Wilson, M.J. and Zeki, E.M. (1984) A comparison of urban and rural aerosol composition using dichotomous samplers, *Atmospheric Environment*, **18**, 1767–1775.
25. Roth, P.M., Ziman, S.D., and Fine J.D. (1993) Tropospheric Ozone. In: *Keeping Pace with Science and Engineering: Case Studies in Environmental Regulation* (ed. Uman, M.F.), National Academy Press, Washington DC, pp. 39–91.
26. Shah, J.J. and Singh, H.B. (1988) Distribution of volatile organic chemicals in outdoor and indoor air, *Environmental Science and Technology*, **22** (12), 1381–1388.
27. Marlowe, I.T. (ed.) (1992) *Emissions of Volatile Organic Compounds in the United Kingdom: A Review of the Emission Factors by Species and Process*, WSL Report LR 882 (PA), Warren Spring Laboratory, Stevenage, UK.
28. Sweet, C.W. and Vermette S.J (1992) Toxic volatile organic compounds in urban air in Illinois, *Environmental Science and Technology*, **26** (1), 165–173.
29. Scheff, P.A. and Wadden, R.A. (1993) Receptor modelling of volatile organic compounds. 1. Emission inventory and validation, *Environmental Science and Technology*, **27** (4), 617–625.
30. Moore, J.A., Kimbrough, R.D. and Gough M. (1993) The dioxin TCDD: a selective study of science and policy interaction. In: *Keeping Pace with Science and Engineering: Case Studies in Environmental Regulation* (ed. Uman, M.F.), National Academy Press, Washington DC, pp 221–243
31. Singh, H.B., Salas, L.J. and Viezee, W. (1986) The global distribution of peroxyacetyl nitrate, *Nature*, **321**, 588–591.
32. Newburg-Rinn, S.D. (1992) Right to know and pollution prevention legislation: opportunities and challenges for the chemist. In: *Pollution Prevention in Industrial Processes* (eds Breen, J.J. and Dellarco, M.J.), ACS Symposium Series No 508, American Chemical Society, Washington DC.
33. Morley, R.H. (1986) Remote monitoring techniques. In: *Handbook of Air Pollution Analysis*, 2nd edn (eds. Harrison, R.M. and Perry, R.), Chapman & Hall, London.
34. McIntyre, A.E. and Lester, J.N. (1986) Hydrocarbons and carbon monoxide. In:

Handbook of Air Pollution Analysis, 2nd edn (eds. Harrison, R.M. and Perry, R.), Chapman & Hall, London.

35. Quintan, J.M., Uribe, B. and Lopez Arbeloa, J.F. (1992) Sorbents for active sampling. In: *Clean Air at Work* (eds Brown, R.H., Curtis, M., Saunders, K.J. and Vandenriessches, S.), Royal Society of Chemistry, Cambridge UK, pp. 123–134.
36. Kelly, T.J., Callahan, P.J., Plell, J. and Evans G.F. (1993) Method development and field measurements for polar volatile organic compounds in ambient air, *Environmental Science and Technology*, **27**, 1146–1153.
37. Longhurst, J.W.S., Raper, D.W., Lee, D.S., Heath, B.A., Conlan, D.E. and King, H. (1993) Acid deposition: a select review 1852–1990. Part 2, *Fuel*, **72** (10), 1363–1380.
38. Longhurst J.W.S. (1990) Acid deposition, in *Longman World Guide to Environmental Issues and Organisations* (ed. Buckley, P.) Longman, London, pp 3–23.
39. Smith, R.A. (1872) *Air and Rain. The Beginnings of a Chemical Climatology*. Longmans Green & Co., London.
40. Cowling, E.B., (1982) Acid precipitation in historical perspective, *Environmental Science and Technology*, **16** (2), 110A–123A.
41. Bolin, B. (1972) *Sweden's Case Study for the United Nations Conference on the Human Environment, 1972. Air pollution across national boundaries. The impact on the environment of sulphur in air and precipitation*, Norstedt & Sons, Stockholm.
42. Overrein, L.N., Seip, H.M. and Tollan, A. (1981) *Acid Precipitation–Effects on Forest and Fish*, Final report of the SNSF Project, 1972–1980, Report FR 19/80, Oslo.
43. Likens, G.E., Bormann, F.H. and Johnson, N.M. (1972) Acid rain, *Environment*, **14**, 33–40.
44. Organisation for Economic Cooperation and Development (1977) *The OECD Programme on Long-range Transport of Air Pollutants*, OECD, Paris.
45. National Acid Precipitation Assessment Programme (1983) *1982 Annual Report to the President and Congress*, NAPAP, Washington.
46. Research and Monitoring Coordinating Committee (1986) *Atmospheric Sciences. Assessment of the state of knowledge on the long range transport of air pollutants and acid deposition, Part 2*, Environment Canada, Ottawa.
47. Acid Waters Review Group (1988) *Acidity in United Kingdom Fresh Waters*, AWRG, HMSO, London.
48. Dillon, P.J. (1983) Chemical alterations of surface waters by acid deposition in Canada. In *Ecological Effects of Acid Deposition*, National Swedish Environmental Protection Board, Report PM 1636, Solna, pp. 275–287.
49. Brown, D.J.A. (1989) Freshwater acidification and fisheries decline. In *Acid Deposition: Sources, Effects and Controls* (ed. Longhurst, J.W.S.), British Library, London and Technical Communications, Letchworth, pp. 117–142.
50. Battarbee, R.W., Flower, R.J., Stevenson, A.C. and Rippey, B. (1985) Lake acidification in Galloway: a palaeoecological test of competing hypotheses, *Nature*, **314**, 350–352.
51. International Co-operative Programme on Assessment and Monitoring of Air Pollution Effects on Forests (1992) *Forest Damage and Air Pollution. Report of the 1990 Forest Damage Survey in Europe*, BFH, Hamburg.
52. Taylor, J.A. (1983) The peatlands of Great Britain and Ireland. In: *Mires: Swamp, Bog, Fen and Moor*, B. Regional Studies (ed. Gore, A.J.P.). Elsevier Scientific Publishing Co., Amsterdam, pp. 1–46.
53. Clymo, R.S. (1987) Interactions of *Sphagnum* with water and air. In: *Effects of Atmospheric Pollutants on Forests, Wetlands and Agricultural Ecosystems* (eds Meema, K.M. and Hutchinson, T.C.), NATO ASI series, vol. G16, Springer-Verlag, Berlin.
54. Pearsall, W.H. (1968) *Mountains and Moorlands* (2nd edn), New Naturalist series, Collins, London and Glasgow.
55. Anderson, P. and Shimwell, D. (1981), *Wild Flowers and Other Plants of the Peak District*, Moorland Publishing, Ashbourne, UK.
56. Tallis, J.H. (1987) Fire and flood at Holme Moss: Erosion processes in an upland blanket mire, *Journal of Ecology*, **75**, 1099–1129.
57. Conway, V.M. (1954) Stratigraphy and pollen analysis of South Pennine blanket peats, *Journal of Ecology*, **42**, 117–147.

58. Tallis, J.H. (1964) Studies on southern Pennine peats. I. The general pollen record, *Journal of Ecology*, **52**, 323–331.
59. Tallis, J.H. (1985) Mass movement and erosion of a southern Pennine blanket peat, *Journal of Ecology*, **73**, 283–315.
60. Farey, J. (1813) *A General View of the Agriculture and Minerals of Derbyshire; with observations on the means of their improvement*, Board of Agriculture, London.
61. Moss, C.E. (1913) *Vegetation of the Peak District*, The University Press, Cambridge.
62. Ferguson, P. and Lee, J.A. (1983) Past and present sulphur pollution in the southern Pennines, *Atmospheric Environment*, **17**, 1131–1137.
63. Lee, J.A., Press, M.C., Woodin, S.J. and Ferguson, P. (1987) Reponses to acidic deposition in ombrotrophic mires in the UK. In: *Effects of Atmospheric Pollutants on Forests, Wetlands and Agricultural Ecosystems* (eds Hutchinson, T.C. and Meena, K.M.), NATO ASI series, vol. G16, Springer-Verlag, Berlin.
64. Ferguson, P., Lee, J.A., and Bell, J.N.B. (1978) Effects of sulphur pollutants on the growth of *Sphagnum* species, *Environmental Pollution*, **16**, 151–162.
65. Ferguson, P. and Lee, J.A. (1980) Some effects of bisulphite and sulphate on the growth of *Sphagnum* species in the field, *Environmental Pollution*, **21**, 59–71.
66. Woodin, S.J. (1986) *Ecophysiological effects of atmospheric nitrogen deposition on ombrotrophic* Sphagnum *species*. Ph.D. Thesis, University of Manchester.
67 Ferguson, P. and Lee, J.A. (1983) The growth of *Sphagnum* species in the southern Pennines, *Journal of Biology*, **12**, 579–586.
68 Baxter, R., Emes, M.J. and Lee, J.A. (1991) Short term effects of bisulphite on pollution-tolerant and pollution-sensitive populations of *Sphagnum cuspidatum* Ehrh. (ex. Hoffm.) *New Phytologist*, **118**, 425–431.
69 Baxter, R., Emes, M.J. and Lee, J.A. (1991) Transition metals and the ability of *Sphagnum* to withstand the phytotoxic effects of the bisulphite ion. *New Phytologist*, **118**, 433–439.
70 Montgomery, T. and Shimwell, D. (1985) *Changes in the Environment and Vegetation of the Kinder–Bleaklow SSSI, 1750–1840: Historical Perspectives and Future Conservation Policies*. A report to the Steering Group of the Moorland Restoration Project, Peak Park Joint Planning Board, Bakewell, UK.
71 Conlan, D.E. (1991) *An investigation of the historical and present day* Sphagnum *on Cowms Moor, southern Pennines*, Ph.D. Thesis, University of Manchester.
72 Hettelingh, J-P., Downing, R.J. and de Smet, P.A.M. (1991) *Mapping Critical Loads for Europe*, UNECE Coordinating Centre for Effects Technical Report 1, National Institute for Public Health and Environmental Protection, Bilthoven.
73 Hornung, M. and Skeffington, R.A. (eds) (1993) *Critical Loads: Concept and Applications*, Institute of Terrestrial Ecology Symposium No. 28, HMSO, London.
74 Hall, J.R. (1993) Critical loads mapping at the U.K. Critical Loads Mapping Centre – data requirements and presentations. In: *Critical Loads: Concept and Applications* (eds Hornung, M. and Skeffington, R.A.), Institute of Terrestrial Ecology Symposium No. 28, HMSO, London.
75 Longhurst, J.W.S. (1988) Some observations on the environmental implications of the U.K. flue gas desulphurisation programme, *Environmentalist*, **8** (2), 115–121.
76 Gibbs, D.C. and Longhurst J.W.S. (1989) Acid deposition abatement: assessing the economic and environmental impact of the UK FGD programme. In: *Acid Deposition: Sources, Effects and Controls* (ed. Longhurst, J.W.S.), British Library, London and Technical Communications, Letchworth, pp 309–327
77 Vernon, J. (1989) Technologies for control of sulphur dioxide emissions. In: *Acid Deposition: Sources, Effects and Controls* (ed. Longhurst, J.W.S.), British Library, London and Technical Communications, Letchworth, pp 287–299
78 Ogawa, M. (1971) Absorption cross section of O_2 and CO_2 continua in the Schumann and far UV regions, *Journal of Chemical Physics*, **54**, 2550.
79. Campbell, I.M. (1986) *Energy in the Atmosphere*, John Wiley & Sons, Chichester.
80. UNEP, United Nations Environmental Program (1991) *Synthesis of the Reports of the Ozone Scientific Assessment Panel, Environmental Effects Assessment Panel and Technology and Economic Assessment Panel*, UNEP, Geneva.

81. National Aeronautics and Space Administration (1992) Press release: *Ozone Depletion a Possibility over Northern Populated Areas*, NASA, Washington, DC, 3 Feb.
82. Van der Leun, J.C. (1988) Ozone depletion and skin cancer, *Journal of Photochemistry and Photobiology*, **1**, 493.
83. Smith, R.C. (1992) Ozone depletion – ultraviolet radiation and phytoplankton biology in Antarctic waters, *Science*, **255**, 952–959.
84. Houghton, J.T., Callander, B.A. and Varney, S.K. (1992) *Climate Change 1992, The Supplementary Report to the IPCC Scientific Assessment*, IPCC, Intergovernmental Panel on Climate Change, Cambridge University Press, Cambridge.
85. *United Nations Climate Change Bulletin* (1993) **1** (1), IUCC and UNEP, Geneva.

3 Water pollution

D. HAMMERTON

3.1 Introduction: water – the essential resource

> Without water, plants and animals cannot survive. All living organisms contain water and use it externally. Water is our most precious mineral. From the beginning of history it has been the key to civilisation and development. It is the largest single controlling factor in the growth of population. The available water supply is a boundary line beyond which no society or nation, agricultural or industrial, can go. Perhaps no greater conservation problem faces mankind today than that of keeping the waters clean and maintaining adequate and qualitatively useful supplies of this natural resource.[1]

This statement, by one of the world's great ecologists, sums up with admirable brevity the case for the conservation and quality control of the earth's most essential resource. Paradoxically, from the earliest times, human beings have probably tended to pollute the very sources from which their drinking water was taken. Hynes (1960)[2] has pointed out that ever since people ceased to be roving hunters and perhaps before they took up life in settled communities they altered the landscape by clearing trees, burning and cultivation, and by piling up middens – all activities which led to alterations in streams. Thus erosion caused by these activities resulted in heavier silt loads in streams, while the middens of early humans must have increased, at least by a small amount, the load of organic matter in the watercourses or groundwaters.

However, under natural conditions flowing waters have powerful self-purifying mechanisms including sedimentation and biochemical oxidation. Naturally occurring micro-organisms decompose organic wastes and are, in turn, consumed by larger organisms and fish while aquatic plants, in the presence of sunlight, help to restore the oxygen balance which is also aided by surface aeration. The main risk to human health under the simple conditions of small rural or semi-nomadic populations of early times was that pathogenic micro-organisms could get into the sources of drinking water and cause one of the many intestinal diseases which were to become a major scourge of humanity in the nineteenth and twentieth centuries.

The use of watercourses for the conveyance and disposal of waste liquids became of major importance only as recently as the first half of the nineteenth century in western Europe and the USA. According to the UK Rivers Pollution Commission set up in 1868 the principal factors were

(1) the rapid growth of manufacturing cities following the Industrial Revolution, (2) the building, alongside rivers, of factories which required water for power or processing or both and (3) the parallel development (from about 1810 onwards) of the water carriage system of waste disposal.

By the middle of the nineteenth century many of Britain's rivers had degenerated into an appalling condition for the simple reason that all effluents were discharged to the nearest watercourse without any form of treatment. Many European rivers were likewise becoming badly polluted but undoubtedly the worst affected was the river Thames in London. It was, no doubt, because of the proximity of this river to the Houses of Parliament that Britain produced the first legislation to control water pollution – the *Rivers Pollution Act* of 1876. This Act provided the basis of controlling river pollution for the next 75 years but it had been so seriously weakened in its passage through Parliament, due to powerful objections from industrial manufacturers, that it was, from the beginning, a highly defective measure.[3] Despite notable efforts by a few local authorities and some early river boards to enforce the Act it largely failed for three main reasons:

1. It made no provision for administrative organisations to enforce it.
2. It did not provide for the use of enforceable standards.
3. It severely restricted the circumstances in which controls could be placed on industry.

There is not space here to go into the long history of attempts to control water pollution in Britain, Western Europe and the USA but there is abundant evidence that all governments have been slow to take action on sewage pollution and particularly reluctant to impose adequate controls on industry. For example, successive British governments delayed the full implementation of the *Control of Pollution Act 1974* for more than 10 years on the grounds that the country could not afford the cost of new curbs on pollution during a recession while, in December 1986, President Reagan vetoed a major new Act to control environmental pollution in the USA for similar reasons. Indeed, at the time of writing, the British Government is trying to delay the impact of European Directives which it approved only a few years previously.

There can be little doubt, taking into account the complex environmental pathways of chemical pollutants and the evidence that some of these pollutants may have profound and seriously damaging effects in the early developmental stages of animal life, that we are at a position when yet more stringent standards will need to be imposed on discharges to the aquatic environment. Here, as in the other environmental sectors, there is now a need for a fundamentally new approach whereby pollutants are eliminated or minimised at each and every stage in the manufacturing process.

3.2 The present state of the aquatic environment

3.2.1 River pollution in the United Kingdom

Although a great deal of information is available on the pollution status of inland and coastal waters around the world there is enormous variation, from country to country, in the amount and quality of the data and in their accessibility. The repeated measurement or monitoring of physical, chemical and biological characteristics of rivers, lakes and coastal waters is an essential prerequisite of effective pollution control. Comprehensive monitoring systems were usually set up only as a result of legislation which established appropriate regulatory organisations.

In the United Kingdom routine monitoring of river water quality began around 1956–58 as a result of the *Rivers (Prevention of Pollution) Acts* of 1951 which established 33 River Boards in England and Wales and nine River Purification Boards in Scotland. Sampling stations were set up at key points along the major rivers and samples taken at mid-point and mid-depth, usually at monthly intervals. These samples were analysed for a range of physical and chemical parameters such as pH, suspended solids, dissolved oxygen, biochemical oxygen demand, ammonia, nitrate and phosphate. In the course of time as laboratory resources increased more parameters were included such as dissolved natural minerals and a growing range of pollutants including heavy metals, detergents, pesticide residues, oils and organic micropollutants. In classifying the results four categories of water quality were defined and used by the river boards to draw up remedial programmes for their areas. They were also used by the Department of Environment (DoE) and the Scottish Development Department (SDD) [now Scottish Office Environment Department (SOEnD)] to provide nationwide surveys of river quality at regular intervals.

The classification system used in England and Wales was modified for use in the 1980 and subsequent water quality surveys to take account of the suitability of water for abstraction according to the degree of treatment required. River water quality is divided into five categories:

Class 1a: Water of high quality suitable for potable supply abstractions; game or high class fisheries; high amenity value.

Class 1b: Water of less high quality than class 1a but usable for substantially the same purposes.

Class 2: Water suitable for potable supply after advanced treatment; supporting coarse fisheries; moderate amenity value.

Class 3: Waters which are polluted to an extent that fish are only sporadically present; may be used for low grade industrial abstraction purposes; considerable potential for further use if cleaned up.

Class 4: Waters which are grossly polluted and are likely to cause nuisance.

Scottish river waters, which are not greatly used for potable supply, are classed in four broadly similar categories.

The results for England and Wales (shown in Tables 3.1 and 3.2) demonstrate that considerable progress was made in reducing pollution between 1968 and 1980 but that a real net deterioration in river quality occurred in the subsequent decade to 1990. In Scotland, by contrast, where similar progress was made up till 1980 the surveys in 1980, 1985 and 1990 showed a continuing record of improvement as illustrated in Table 3.3.

The better performance in Scotland was attributed to the complete separation of the regulatory function from the operation of sewage works, whereas in England and Wales there was an extensive period (1974–89) when both these functions were carried out by the Regional Water Authorities. The setting up of the National Rivers Authority (NRA) in 1990 for England and Wales, modelled to a certain extent on the Scottish system, restored this separation which had existed before 1974 through the system of river authorities. Very detailed information on water quality in rivers, estuaries and coastal waters is available in the annual reports of the NRA and of the seven Scottish river purification boards while national

Table 3.1 River water quality in England and Wales, 1958–80

	1958		1970		1975		1980	
Class	km	%	km	%	km	%	km	%
Unpolluted	24 950	72	28 500	74	28 810	75	28 810	75
Doubtful	5220	15	6270	17	6730	17	7110	18
Poor	2270	7	1940	5	1770	5	2000	5
Grossly polluted	2250	6	1700	4	1270	3	810	2
Total	34 690		38 400		38 590		38 740	

Table 3.2 River water quality in England and Wales, 1980–90

	1980		1985		1990	
Class	km	%	km	%	km	%
Good 1a	13 380	34	13 470	33	12 408	29
Good 1b	14 220	35	13 990	34	14 536	34
Fair 2	8670	21	9730	24	10 750	25
Poor 3	3260	8	3560	9	4022	9
Bad 4	640	2	650	2	662	2
X	–	–	–	–	39	–
Unclassified	–	–	–	–	17	–
Total	40 630		41 390		42 434	

Table 3.3 River water quality in Scotland, 1980–90

	1980		1985		1990*	
Class	km	%	km	%	km	%
1	45 184	95.0	45 510	95.6	46 111	96.9
2	1981	4.2	1688	3.5	1177	2.5
3	256	0.5	266	0.6	233	0.5
4	162	0.3	131	0.3	70	0.1
Total	47 583		47 595		47 591	

* Excludes rivers on islands which were not included in previous surveys (3142 km Class 1; 4 km Class 2).

Source: Scottish Office Environment Department. Water Quality Survey of Scotland 1990.

surveys continue to be published by the Scottish Office Environment Department and the NRA every five years.

In the early years of river boards the pollution status of rivers was measured solely in physico-chemical terms but it was soon recognised that the ultimate test of a clean river is whether or not it supported the diversity of species and numbers of organisms characteristic of an unpolluted environment. The first biologists were appointed in the late 1950s and started developing biological water classification systems. However, it was not until the mid-1970s that the importance of biological quality became generally recognised and the results of biological surveys were published alongside the chemical classification system. During the last decade the design and interpretation of biological surveys has become much more refined and considerable research effort has been directed into correlating physical and chemical quality with biological quality by means of the River Invertebrate Prediction and Classification System (RIVPACS).

RIVPACS has been developed by the Institute of Freshwater Ecology to overcome a weakness of existing biological indices of pollution in that the latter do not take account of natural physical and chemical phenomena which may have a profound effect on aquatic communities unrelated to water pollution. The strength of RIVPACS is that it is based on a massive database derived from studies at 438 unpolluted sites on 80 rivers. On the basis of chemical indices which, as far as possible, are independent of organic pollution the RIVPACS software predicts the invertebrate fauna (families of macro-invertebrates comprising insects, molluscs, crustaceans, roundworms and flatworms) at each site which are then compared with the actual fauna to produce an Ecological Quality Index (EQI). At present the RIVPACS approach generates several indices of environmental quality which are then combined to produce a four-class (A–D) system of classification (Royal Commission on Environmental Pollution, RCEP, 1992).

The NRA has proposed a combined river water quality classification scheme whereby the chemical classification would incorporate a 'biological over-ride'.[4] However, the RCEP recommends, in its 16th Report, that a general classification scheme based on biological assessment should be developed for use throughout the UK in the 1995 and subsequent river quality surveys. The RCEP also pointed out that classification of river quality by length gives a misleading impression by exaggerating the amount of clean water and recommended that results should be reported volumetrically as well as by length.

In England, Wales and Scotland estuaries are classified in terms of aesthetic, biological and chemical quality, the only difference being that in England and Wales the classification is by length whereas surface area is used in Scotland. The results of the 1990 survey are shown in Table 3.4. In Scotland the inshore coastal waters were classified for the 1990 survey (Table 3.5) on the basis of aesthetic, biological, microbiological and chemical quality.

To the general reader it may seem that we scarcely have a pollution problem if, as indicated in the tables, 90–96% of estuarine waters and 86–99% of river lengths are of good or fair quality. These figures, as pointed out by the RCEP, are misleading because the badly polluted waters are, almost invariably, in the upper estuaries and the stretches just above the tidal limit, thus representing very large volumes of water at critical points in our river basins where barriers to migratory fish have existed, in some cases for over 100 years. There can be no doubt that the lower reaches of many of our rivers, especially near to centres of population, are still of poor quality and that substantial capital investment in effluent treatment plant will be required to restore our estuaries, coastal and bathing waters to a fully acceptable condition. Indeed, the government is committed to the expenditure of around £8 billion by the year 2005 to meet the minimum requirements of the Urban Waste Water Treatment and other EU Directives.

3.2.2 The global state of river pollution

Information on the quality of inland and coastal waters worldwide is available from two principal sources, i.e. the United Nations Environment Programme (UNEP) which was set up by the United Nations (UN) following the UN Conference on the Human Environment convened in Stockholm in 1972, and the Organisation for Economic Co-operation and Development (OECD) which was set up in 1961.

One of the main functions given to UNEP by the General Assembly of the UN was 'to keep under review the world environmental situation in order to ensure that emerging environmental problems of wide international significance receive appropriate and adequate consideration by

Table 3.4 Estuary water quality in Scotland (upper table) and England (lower table)

RPB	Area (sq km) in class shown (% of total in each class) A	B	C	D	Total
Clyde	56.8 (*63.6*)	30.0 (*33.6*)	2.5 (*2.8*)	0.0 (*0.0*)	89.3
Forth	28.6 (*32.9*)	38.5 (*44.3*)	19.6 (*22.5*)	0.3 (*0.3*)	87.0
Highland	220.6 (*84.2*)	32.4 (*12.4*)	9.0 (*3.4*)	0.0 (*0.0*)	262.0
North East	1.6 (*31.4*)	3.7 (*72.6*)	0.0 (*0.0*)	0.0 (*0.0*)	5.3
Solway	262.7 (*100.0*)	0.0 (*0.0*)	0.0 (*0.0*)	0.0 (*0.0*)	262.7
Tay	70.8 (*59.5*)	48.2 (*40.5*)	0.0 (*0.0*)	0.0 (*0.0*)	119.0
Tweed	0.0 (*0.0*)	0.0 (*0.0*)	0.0 (*0.0*)	0.0 (*0.0*)	0.0
Total	641.1 (*77.7*)	152.8 (*18.5*)	31.1 (*3.8*)	0.3 (*0.0*)	825.3

	Percentage of estuary length in each class						
	Good	Fair	Good and fair	Poor	Bad	Poor and bad	
Region	A	B	A and B	C	D	C and D	Total (km)
Anglian	79	14	93	7	0.4	7	514
Northumbria	34	38	73	17	10	27	135
North West	49	23	72	13	15	28	452
Severn-Trent	14	61	75	25	0	25	56
Southern	75	21	96	4	0	4	388
South West	92	8	100	0	0	0	350
Thames	45	55	100	0	0	0	112
Welsh	78	20	98	2	0	2	420
Wessex	45	51	96	4	0	4	120
Yorkshire	12	43	55	45	0	45	40
Shared estuaries:							
Humber	43	57	100	0	0	0	65
Severn	61	39	100	0	0	0	71

Source: (a) Scottish Office Environment Depart: Water Quality of Scotland 1990.
(b) National Rivers Authority: The Quality of Rivers, Canals and Estuaries in England and Wales, 1990.

governments'. In accordance with this requirement UNEP has published each year since 1974 a series of annual reports which initially concerned a wide spectrum of issues such as climate change, condition of the biosphere, stress, social tension and pollution, while from 1977 onwards the annual state of the environment reports focused on a number of selected specific environmental issues. In 1987, 15 years after the Stockholm Conference, UNEP published a comprehensive summary and update of the annual reports in a single volume, *The State of the Environment*.[5]

Table 3.5 Coastal water quality in Scotland, 1990

Inshore waters	Length (km) in class shown (% of total in each class)				
RPB	A	B	C	D	Total
Clyde	2037.8 (*78.7*)	228.5 (*8.6*)	325.8 (*12.3*)	11.7 (*0.4*)	2639.8
Forth	29.3 (*15.6*)	62.1 (*33.0*)	79.9 (*42.4*)	17.0 (*9.0*)	188.3
Highland	3378.9 (*96.5*)	47.7 (*1.4*)	68.4 (*2.0*)	5.0 (*0.1*)	3500.0
North East	47.1 (*18.7*)	118.4 (*47.0*)	68.3 (*27.1*)	18.3 (*7.2*)	252.1
Solway	249.6 (*89.5*)	25.1 (*9.0*)	4.2 (*1.5*)	0.1 (*0.0*)	279.0
Tay	48.9 (*60.0*)	21.0 (*25.8*)	11.6 (*14.2*)	0.0 (*0.0*)	81.5
Tweed	43.1 (*98.0*)	0.7 (*1.6*)	0.2 (*0.4*)	0.0 (*0.0*)	44.0
Total	5870.7 (*84.1*)	503.5 (*7.2*)	558.4 (*8.0*)	52.1 (*0.7*)	6984.7

Source: Scottish Office Environment Department Water Quality Survey of Scotland, 1990.

In 1989 the Paris G-7 Economic Summit asked the OECD to develop a set of selected environmental indicators, a request which was further reinforced by the Houston G-7 Economic Summit held in 1990. Work progressed rapidly and two reports were published in 1991 – a preliminary report on environmental indicators and a comprehensive report on the state of the environment in the member countries.[6,7]

According to UNEP[5] 97% of the earth's water is in the oceans and only 3% on land. Of the latter (freshwater) component 77% is stored in the polar ice caps and in glaciers and a further 22% as groundwater. Thus the available surface water in lakes and rivers comprises barely 1% of the total freshwater and much of this is polluted.

UNEP estimates that on the global scale 73% of water use is for irrigation, 21% for industrial purposes and only 6% for public use. There are, of course, wide variations in usage according to climate and industrial development. In the developed countries as much as 40% of usage is industrial while in developing countries the overwhelming bulk of consumption is for irrigation. In 1976 UNEP inaugurated the Global Environmental Monitoring System (GEMS) for rivers, lakes and groundwaters, but although this system has grown to 450 stations in 59 countries the data so far obtained are inadequate for the establishment of clear trends.

According to UNEP the situation in many developing countries is that water pollution is a growing problem and, not surprisingly, most serious in densely populated countries. Thus in China out of 78 monitored rivers 54 are seriously polluted with untreated sewage and industrial waste. Reports

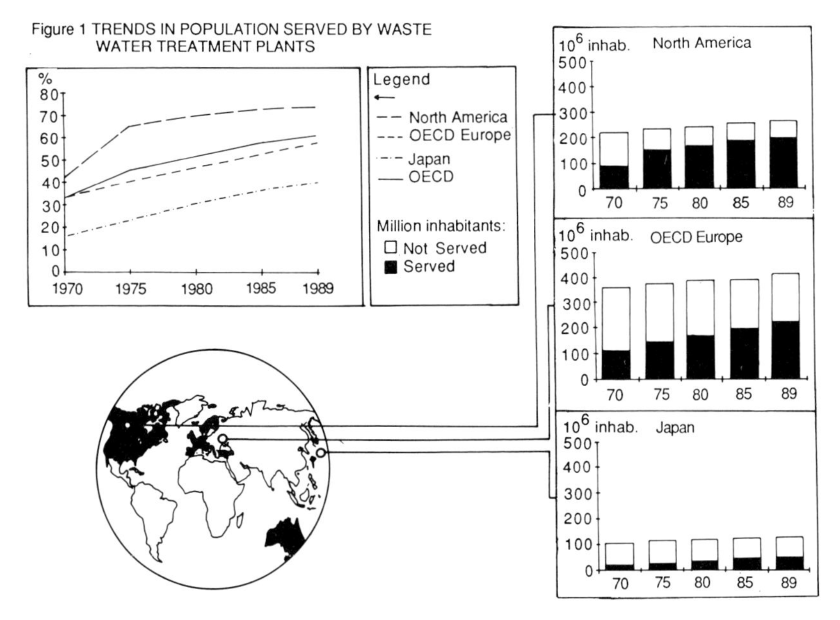

Figure 3.1 Trends in population served by waste water treatment plants. (Reproduced with permission from *The State of the Environment*, OECD, Paris, 1991.[7])

indicate that over 40 rivers in Malaysia are so badly affected that they are nearly devoid of fish and aquatic mammals. Here the main sources of pollution are stated to be oil-palm and rubber processing residues, sewage and other industrial wastes. In India 70% of total surface waters are polluted. Here, as witnessed by the writer, many rivers during the inter-monsoon period are totally dry and their channels are filled with raw sewage in sewered urban areas.[8]

India's biggest river, the Ganges, is dangerously polluted for 600 of its 2525 km length by human and animal wastes together with rising amounts of hazardous industrial and agricultural wastes. Of 132 industrial effluents discharged to the river only 10 have effective treatment plants while an even smaller proportion of cities along the river treat their sewage effluent.

The results of monitoring of water quality from the 24 member states of the OECD indicate that, after many decades of serious pollution, the major rivers have shown considerable overall improvements in the last 25 years. This is due in a large measure to a substantial investment in new wastewater treatment plants where none existed or to improvements to existing plants to produce better quality effluents. Figure 3.1 illustrates the increase in wastewater treatment plants while Figure 3.2 illustrates changes in water quality in major rivers in the OECD countries.

The overall picture from these and other data is that pollution by organic wastes which consume oxygen has decreased significantly as a result of

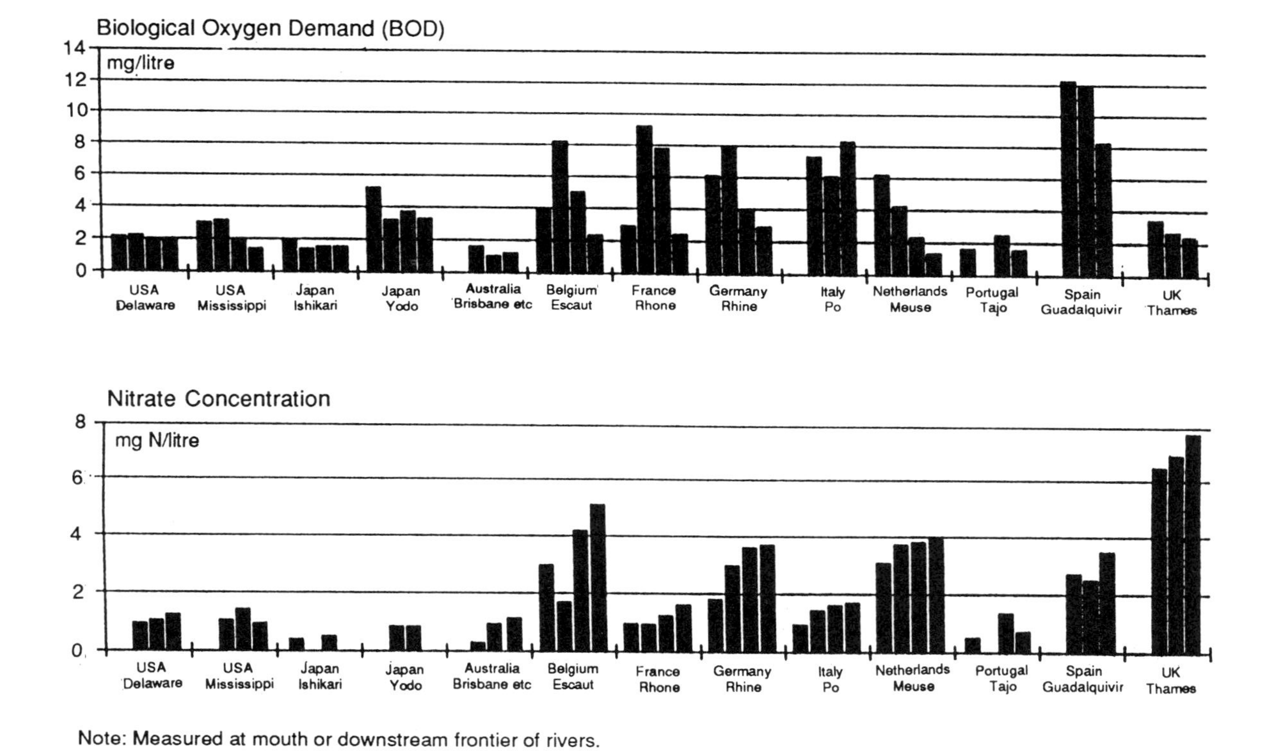

Figure 3.2 Water quality of selected rivers, 1970, 1975, 1980, late 1980s. (Reproduced with permission from *The State of the Environment*, OECD, Paris, 1991.[7])

increased treatment of urban wastewaters. This has resulted in higher oxygen levels in many rivers followed by improvements in fish populations and in the numbers of organisms on which fish feed.

The OECD report, however, mentions several areas of concern. For example, in Figure 3.2 it is obvious that there is a marked trend of rising nitrate levels which, in combination with continuing high levels of phosphates, has led to a major problem of algal growths, specific problems being mentioned in the Netherlands, Belgium, France, Italy and the UK.[7] The same report mentions two cases where phosphorus levels have been reduced, firstly in the Great Lakes where 20 years' joint effort by the USA and Canada has been very successful and secondly Sweden where removal of phosphorus in sewage treatment has reduced the concentration of this nutrient in many rivers in the period from 1971 to 1985.

Nutrients from fertilisers and animal wastes were found to be the main contaminants in nearly 60% of polluted lakes in the USA. In Europe, Lake Geneva and lakes in the lake districts of Friesland and Zuid-Holland in the Netherlands suffer from high amounts of agricultural chemicals which result in degraded water quality and fish deaths.[9]

Despite efforts in many countries to reduce the levels of heavy metals and other toxic substances the OECD concludes that little progress has been made except in the case of lead where measurements at the mouths of rivers have shown decreasing concentrations over the period 1970–85 for Canada, Japan, Belgium, West Germany and the UK. However, in the light of recent legislation referred to later, such as the EU Directive on Dangerous Substances, and the North Sea Ministerial Decision on 'Red List' substances it is likely that the next 5–10 years will see a significant overall reduction in the levels of metals and organic micropollutants in the aquatic environment.

The OECD report refers briefly to problems due to acid rain. In Sweden, for example, 16 000 lakes have been affected by acidification to the point where sensitive organisms have been eliminated and expensive liming programmes are in progress. Acidification of lake systems and streams is widespread, particularly in regions where the underlying rocks (e.g. granite) are deficient in calcium. Many water bodies are affected in North America (especially Canada), Scotland and Scandinavia. There is now evidence, from both North America and Europe, that recovery from acidification is taking place as a result of reduced emissions of sulphur dioxide but a considerable further tightening of controls will be necessary to restore vulnerable catchments to a satisfactory condition.

3.2.3 Pollution of groundwaters

Finally, the OECD report expresses serious concern about the pollution of groundwaters. The use of fertilisers and pesticides in agriculture, the inappropriate disposal of both domestic and hazardous wastes and the failure of storage devices such as underground petroleum tanks as well as badly designed or located landfills can all lead to contamination of groundwaters. Once this has happened the aquifers are liable to remain polluted and possibly unusable for very many years while remedial action is almost impossible in such situations.

UNEP[5] has also expressed concern at the worsening state of groundwaters and focused its attention on agriculture as a major source of pollution from fertilisers, pesticides and animal wastes. In Denmark the nitrate level in groundwater has trebled in the last 20–30 years resulting from the increased use of fertilisers and manure. In the USA groundwater contamination by pesticides has been confirmed in California, New York, Wisconsin and Florida. UNEP pointed out that the emergence of the problem of pesticide residues in groundwater 'adds a new dimension to the whole array of public health, environmental protection, pesticide innovation and marketing, and agricultural management'.

UNEP also drew attention to problems from active and abandoned landfills, deep well disposal of liquid wastes and the leakage of radioactive constituents arising from the burial of containers of low and intermediate level wastes from nuclear facilities. Its report drew attention to the need for groundwater management to be seen as an integral part of the overall water development plan of a country including groundwater monitoring. It is noteworthy here that the Royal Commission on Environmental Pollution, in its sixteenth report on 'Water Quality', stressed the need in the UK for a groundwater monitoring network (RCEP, 1992).[10]

3.2.4 Marine pollution in the United Kingdom

During most of the past century UK coastal waters have been exempt from the pollution control legislation which has governed discharges to inland rivers and lakes. This was largely due to the view that the marine environment had an almost inexhaustible power to absorb pollution without damage – a view which was prevalent throughout most of Europe and the rest of the world. As a result of this it was normal practice until the last two decades for sewage from coastal communities to be discharged untreated through short outfalls to just below low water mark, the location being chosen with the aid of a few float tests to ascertain that tidal currents would disperse the worst of the pollution out to sea. Industrial effluents, usually into the larger estuaries, were dealt with in like manner through fairly short outfalls. The result was hardly surprising: most of the popular bathing beaches were grossly polluted and industrial estuaries such as the Thames, Clyde, Mersey, Tyne, Tees and Humber were anoxic and impassable to migratory fish for all or part of each year.

The first signs of improvement were seen in Scotland as a result of Tidal

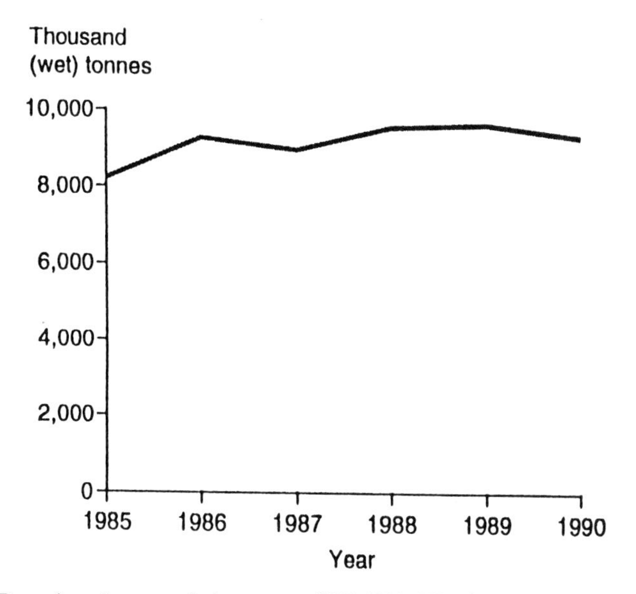

Figure 3.3 Dumping of sewage sludge at sea: 1985–1990, UK. (Reproduced with permission from *The UK Environment*, DoE, 1992.[11])

Waters Orders which brought about 40 estuary and coastal waters under the control of the river purification authorities between 1961 and 1970. The rest of the coastline including England and Wales was brought under control through the Control of Pollution Act 1974 (though implementation was delayed for 10 years). Progress in the past 10 years has been slow because of the economic recession and because of the long lead times for construction of long outfalls and associated treatment schemes. However, there has been a large increase in capital investment and many new schemes are under way. This is attributable in a large measure to European legislation, notably the Urban Waste Water Treatment (UWWT) Directive and the Bathing Water Directive. Under the UWWT Directive all discharges to tidal waters must be treated by the year 2005 in a phased programme while all dumping of sewage and industrial sludges at sea must cease by 1998. The cost of implementing this programme is estimated currently at about £8000 million for the UK. However, as a result there should be a major reduction in the pollution of coastal waters and over 450 identified bathing waters should meet the EU standards by 1997.

The following information is derived from a recent report.[11] In 1990, 9.3 million tonnes of sewage sludge (wet weight) were dumped in UK coastal waters, a figure which has remained fairly constant since 1985 but, thanks to tighter controls, the heavy metals content has been reduced (see Figure 3.3 and Table 3.6). In the same period liquid industrial waste dumped at sea declined from 300 000 tonnes to just over 200 000 tonnes with a marked

Table 3.6 Metal content of dumped sewage sludge at sea 1986, 1988 and 1990 (UK)

	Tonnes		
	1986	1988	1990
Zinc	542	429	288
Lead	196	172	129
Copper	171	166	147
Chromium	124	140	104
Nickel	25.7	22.2	20.7
Cadmium	3.84	2.92	2.06
Mercury	1.37	1.17	1.06

Source: MAFF

Reproduced with permission from *The UK Environment*, DoE, 1992.[11]

decline in heavy metals. Dumping of solid industrial waste increased from 3.9 million tonnes to 4.9 million tonnes with an increase in the amounts of chromium, nickel and cadmium. In 1990, 90% of estuary length was classed as 'good' or 'fair' quality in England and Wales while in Scotland the proportion was 96%. In 1991, out of 453 identified bathing waters, 343 (76%) met the EU imperative values for coliform bacteria.

3.2.5 The global state of marine pollution

As part of the Action Plan adopted at the UN Conference on the Human Environment held in Stockholm in 1972, it was recommended that the UN Joint Group of Experts on the Scientific Aspects of Marine Pollution (GESAMP) 'should assemble data and provide advice on scientific aspects of marine pollution'. Later, in 1977, at the request of UNEP, it was decided that periodic reviews of the state of the marine environment as regards pollution should become one of the main terms of reference for GESAMP. The first such review was published by UNESCO and UNEP in 1982 while a more detailed second review was published by UNEP in 1990.[12] This review was based on 16 Technical Annexes written by individual experts and also amplified by 12 regional reviews which were prepared by 12 UNEP sponsored Task Teams. These reviews are still in the process of being published by UNEP.

This report[12] expresses considerable anxiety about the worsening state of the marine environment and states that, unless co-ordinated national and international action is taken now, there could be a real deterioration in the next decade. The major problems are described as 'coastal development and the attendant destruction of habitats, eutrophication, microbial contamination of seafood and beaches, fouling of the seas by plastic litter, progressive build-up of chlorinated hydrocarbons, especially in the tropics

Table 3.7 Main sources of pollution (GESAMP Report)

Source	All potential pollutants (%)
Offshore production	1
Marine transportation	12
Dumping	10
Run-off and land based discharges	44
Atmosphere	33

and sub-tropics, and accumulation of tar on beaches'. It is of interest to note that while public perception around the world is that contaminants such as oil, radioactive substances and trace elements are of greatest concern and were particularly highlighted in the earlier 1982 GESAMP review, they are now considered to be of lesser concern. The report also draws attention to problems which need further study to determine their possible impact, namely climate change with a possible rise in sea-level if global warming continues and loss of ozone in the stratosphere which could damage marine life through increased exposure to ultra-violet light. Amongst many recommendations the report says that controls of the use of tributyltin (TBT) as an anti-fouling agent should be further extended. GESAMP's view on this matter has subsequently been reinforced by the results of a report sponsored by DoE[13] which revealed that the southern North Sea was so heavily contaminated with TBT that dogwhelks on the Belgian, Dutch and German coasts were sterile. A report by the North Sea Task Force recommends further reductions in the use of TBT or even a total ban.

Estimates of the main sources of pollution are essential to deciding the priorities for action. The GESAMP report produced the rough figures shown in Table 3.7.[12] These figures indicate that the major inputs come from land-based sources and the atmosphere and that these are the areas where controls are most needed. In fact, as seen later in this chapter, pollution control in the marine environment has initially focused on shipping and dumping at sea.

3.3 The impact of pollution on aquatic systems

3.3.1 Defining pollution

There have been many definitions of pollution but the best is perhaps the one adopted by the Royal Commission on Environmental Pollution (RCEP). 'The introduction by man into the environment of substances or energy liable to cause hazards to human health, harm to living resources and ecological systems, damage to structure or amenity, or interference with legitimate uses of the environment.'

In this definition the word 'pollution' is used both to describe the act of polluting and its consequences. The RCEP also pointed out that, for many substances, whether a particular discharge or emission is considered to be pollution depends not only on the nature of the substance, but also on the circumstances in which it occurs, and often on the attitude of the people affected and on value judgements. It should be noted that, in this definition, pollution is the result of human activity and that natural sources of the same substances or energy are excluded. Pollution thus results in a modification of the water quality which is determined by natural physico-chemical and biological processes.

There are many ways of classifying pollutants, for example by their nature, properties and sources, their usage, targets and effects. None are wholly satisfactory but here it is proposed to group pollutants by categories which are defined in recent legislation. The two most important categories are organic wastes which are broadly covered by the EU Urban Waste-water Treatment (UWWT) Directive, and toxic substances which are controlled by the Dangerous Substances Directive and its 'daughter' directives. Each of these directives has its own classification system. Nutrients are partly controlled by the UWWT Directive and partly by the Nitrate Directive.

3.3.2 Organic wastes

Organic wastes comprise domestic sewage, farm wastes including those arising from the intensive farming of pigs, cattle and chickens, fish farming and production of silage. Also included are food processing wastes, e.g. from abattoirs, meat processing, canneries, distilleries, breweries and vegetable processing factories.

Although organic wastes vary enormously in composition they all share one particular characteristic in that they are 'biodegradable' – that is to say that they contain organic compounds which can be decomposed by micro-organisms into simpler compounds. Some wastes such as raw sewage already contain large numbers of bacteria while others rapidly acquire and build up the necessary specialist aerobic bacteria to break down fats, starches, proteins, amino-acids, alcohols, cellulose and other substances of varying degrees of complexity into ammonia, nitrates, phosphates, sulphates and carbon dioxide. In these processes large amounts of oxygen are consumed and therefore, if large amounts of untreated or poorly treated organic wastes are discharged, eventually total de-oxygenation of the receiving water may take place.

Once de-oxygenation is complete heterotrophic aerobic bacteria are replaced by anaerobic bacteria which can produce methane, sulphuretted hydrogen, ammonia and black iron sulphide. These processes are particularly evident where discharges contain high levels of organic solids which settle out on the bed of the river, lake or coastal water producing conditions which are highly offensive to the eye and the nose.

Such effects as these are rarely seen now in western countries due to better pollution control policies but they illustrate, in cases of severe pollution, the first stages in natural self-purification. Lack of oxygen eliminates fish and a wide range of invertebrate fauna which can only survive in well oxygenated water. A highly diverse fauna is replaced by enormous numbers of a few species which live at low oxygen levels such as tubificid worms and particular species of chironomid larvae. Large numbers of protozoa feed on the high microbial population. As recovery proceeds downstream (or away from the source of pollution in static waters), bacteria and 'sewage fungus' begin to be replaced by algae, tubificids and chironomids diminish in number and are replaced by the water-louse *Asellus*. As oxygen levels further increase there is a gradual return of the typical 'clean-water' fauna, i.e. caddisfly larvae, mayfly larvae and finally stonefly larvae which require the highest levels of oxygen. These effects are best demonstrated diagrammatically in Figure 3.4.

The UWWT Directive sets out minimum standards of treatment for urban wastewaters (which largely comprise domestic sewage) according to the size of discharge, based on population, and the sensitivity of the receiving waters. More stringent standards can be imposed where this is deemed necessary to protect the receiving waters, for example where inadequate dilution is available. In the case of discharges to sensitive waters tertiary treatment to reduce nitrate and phosphate is required. The directive lists 11 types of industry, e.g. breweries, distilleries and food processing factories, whose effluents will be expected to achieve similar standards.

3.3.3 Dangerous substances

3.3.3.1 Around 100 000 chemical substances are in use in Europe today. Of these, a small proportion are harmful to living organisms and liable to enter the aquatic environment. The EC Framework Directive on Dangerous Substances provides two lists of substances for priority action: List I (the 'black' list) which are considered particularly dangerous and must be eventually eliminated and List II (the 'grey' list) where preventive action is required to reduce the potential to pollute.

3.3.3.2 List I comprises the following groups of substances:

1. Organohalogen compounds.
2. Organophosphorous compounds.

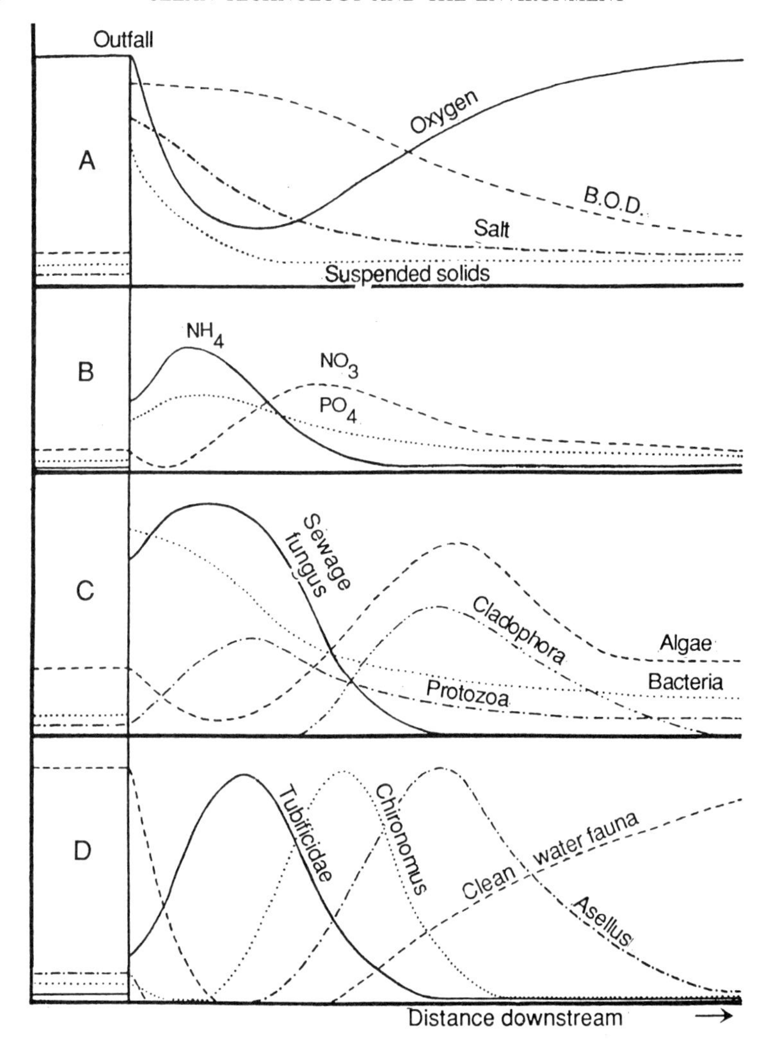

Figure 3.4 Diagrammatic representation of the effects of an organic effluent on a river and the changes as one passes downstream from the outfall. A and B, physical and chemical changes; C, changes in micro-organisms; D, changes in larger animals. (Reproduced with permission from Hynes.[20])

3. Organotin compounds.
4. Carcinogenic, mutagenic and teratogenic substances.
5. Mercury and its compounds.
6. Cadmium and its compounds.
7. Persistent mineral oils and hydrocarbons.
8. Persistent synthetic substances which may float, remain in suspension, and which may interfere with any use of the waters.

3.3.3.3 From the above list the EC Commission identified 129 'candidate' List I substances on the basis of their persistence in the environment, their toxicity and their capacity for bioaccumulation. In a series of daughter directives, regulations for the control of mercury and cadmium and their compounds and fifteen organic compounds have been approved. The regulations include emission standards and environmental quality standards for surface and coastal waters but not groundwaters which are covered by the separate Groundwater Directive.

3.3.3.4 List II comprises the following substances:

1. List I substances for which no standards have been set.
2. Metals, metalloids and their compounds: antimony, arsenic, barium, beryllium, boron, chromium, cobalt, copper, lead, molybdenum, nickel, selenium, silver, tellurium, thallium, tin, titanium, uranium, vanadium and zinc.
3. Biocides and their derivatives not in List I.
4. Substances which have a deleterious effect on the taste and/or odour of groundwater; compounds which are liable to cause the formation of such substances and render water unfit for human consumption.
5. Toxic or persistent organic compounds of silicon.
6. Inorganic compounds of phosphorus and elemental phosphorus.
7. Non-persistent mineral oils and hydrocarbons of petroleum origin.
8. Cyanides, fluorides.
9. Substances which have an adverse effect on the oxygen balance, particularly ammonia and nitrites.

3.3.3.5 Member states are required to set their own Environmental Quality Standards (EQSs) or emission standards for List II substances. In the UK the Department of Environment has already approved or proposed provisional EQSs for a considerable number of these substances.

3.3.3.6 By definition, toxic substances or poisons exert harmful effects on living organisms at very low doses and usually can be detected only by sophisticated methods of chemical analysis. Thus, whereas organic pollution by, say, a sewage effluent, is highly obvious to the senses by discoloration, by foam and by smell, toxic pollution may be completely invisible and only revealed by the sudden appearance of dead fish or by a biological survey which may show that species of insects living in the stream bed have been killed. Aquatic organisms are particularly vulnerable because the solvent capacity of water ensures that the toxic substance is dissolved and dispersed throughout the medium in which they live.

The most common toxic inorganic substances affecting inland waters are the salts of heavy metals such as copper, lead, nickel, zinc and chromium, soluble sulphides and the dissolved gases ammonia and chlorine. Very small amounts of these are lethal to fish and bottom-living invertebrates while lower concentrations may seriously reduce the rate at which natural recovery from pollution takes place. For example, the trace residual levels of chlorine in drinking water (around 0.1 to 0.3 parts per million) will kill most fish and there have been many examples of service reservoirs and water mains being drained to the nearest watercourse for cleaning or repair with dire consequences for the stream. Nowadays this is well recognised and such waters are always dechlorinated before discharge. Ammonia presents special problems because it exists in two forms in solution in an equilibrium between the un-ionised ammonia molecule (NH_3) and the ammonium cation (NH_4^+). The un-ionised form is extremely toxic to fish and the proportion of this form increases rapidly with both rising pH and rising temperature. It follows, therefore, that the concentration of ammonia in an effluent which may be safe to discharge to a river during most of the year may be lethal to fish on a few hot days in summer; on these occasions temperature and pH will be at a maximum and effluent dilution at a minimum.

In setting EQSs for discharges it is important to take into account not only fish but also invertebrates such as shrimps and water-fleas. Acute toxicity tests need to be carried out on the most vulnerable stages in the life cycle, e.g. eggs and larval stages, but also sub-acute, non-lethal effects such as bioaccumulation and biodegradability. Where possible, information on mutagenicity, carcinogenicity and teratogenicity are also taken into account. A further potentially serious toxic effect that has only recently come to light is that of oestrogen induced effects.

The chemicals of most serious concern in the aquatic environment are undoubtedly those used in agriculture, forestry and fish farming to control a vast range of pests and weeds. These include insecticides, molluscicides, rodenticides, nematocides, algicides, herbicides, fungicides and also a range of antibiotics used to control diseases of farm animals and fish. There is not space here to detail the impact of these poisons on aquatic organisms and ecosystems for which a vast literature exists. However, it is worth drawing attention to the recent findings of the International Joint Commission (IJC) on the Great Lakes. This body was set up by the Canadian and United States Governments for monitoring and assessing the implementation of the Great Lakes Water Quality Agreement which was signed in 1978. Included in this agreement was an undertaking to eliminate, as far as possible, persistent toxic chemicals. Despite the ever-increasing controls on pollutant releases to the environment in the last two decades, the Commission has concluded in its 1994 report[14] that the issue of toxic, persistent chemicals is 'the most significant problem facing the Great Lakes Basin'. The Commission draws attention to behavioural, metabolic, neurological and reproductive abnormalities in wildlife and also in humans following long-term exposure to persistent toxic substances. Its report also draws attention to the impact of a variety of chemicals on hormonal systems.[15] It states that PCBs, dioxins, atrazine, hexachlorobenzene and other organochlorines and polycyclic aromatic hydrocarbons are 'strongly implicated in the disruption of endocrine systems, including oestrogenic effects, in laboratory animals and wildlife'. Apparently '42 chemicals or classes of chemicals have been reported to affect the reproductive or hormone systems and 23 of these compounds contain chlorine as an essential ingredient'. From this evidence, the IJC concludes that although many effects of the persistent substances are still unclear, 'there is sufficient evidence now to infer a real risk of serious impacts in humans. Increasingly, human data supports this conclusion'.

3.3.4 Acids and alkalis

Aquatic plants and animals are sensitive to moderate differences in the acidity or alkalinity of their environment. Thus the flora and fauna of a stream or lake at around pH 6.0 will be very different from the typical biota of alkaline waters around pH 8.0.

Many organisms are adapted to living within a fairly narrow pH range and waters of differing natural pH values usually develop communities which are characteristic of 'acid' or 'alkaline' conditions. Excessive acidity may be derived from mine water drainage where the rocks are rich in sulphur; such discharges are often also rich in iron. The combined toxic effects of strong acid (pH 3–4) and ferric hydroxide are lethal to aquatic life other than iron bacteria. Acid rain falling on catchments which are deficient in calcium can also give rise to extremely acid surface waters. If the pH falls much below 5.0 aluminium may be leached out of the surface soils and cause the death of fish eggs and newly hatched fish fry. One or two episodes of acid rain falling in vulnerable catchments may be sufficient to cause the loss of a whole new generation of young fish.[16]

High alkalinity, which may arise from alkaline industrial discharges or by high photosynthetic rates in eutrophic waters, can be highly damaging or lethal to fish life. Many instances have been recorded in hot summer weather when a combination of high temperatures and high pH (above 9.5) has caused severe stress or death to fish.

3.3.5 Biological detection of pollution

The impact of pollutants on aquatic ecosystems often remains evident for weeks or months after the pollution has ceased. Thus single, short episodes of pollution may escape detection if only chemical monitoring of discharges at, say, weekly or monthly intervals is employed. However, biological surveys carried out long after the pollutant has disappeared will often provide valuable information as to the nature of the pollutant and the location of the discharge. Finally, it should be pointed out that only biological surveys can determine whether rivers, lakes and tidal waters have been restored to full health and whether the consent conditions on particular discharges are sufficiently stringent to prevent any adverse impact. The forthcoming EC Directive on Ecological Quality of waters will, hopefully, provide a uniform basis for biological assessment of river and coastal water quality across Europe.

3.4 Legislation and control systems

3.4.1 European legislation

The experience of the past century in attempts to control water pollution has shown that effective protection of the aquatic environment requires four key ingredients:

1. Appropriate, enforceable legislation.
2. Independent enforcement agencies whose areas are based on whole river basins.
3. Scientifically based standards of water quality.
4. A uniform approach to control both nationally and internationally.

Since 1972 the European Economic Community (EEC) has become the dominant force in legislative action to protect both the natural environment and the workplace environment. In 1973 the EEC adopted the first Community Action Programme with the following specific aims: (i) to reduce pollution and nuisances; (ii) to husband natural resources and the balance of ecological systems; and (iii) to improve the quality of life and working conditions. The first action programme emphasised the need to establish uniform quality objectives across the member states. The Council defined these as a set of requirements which must be fulfilled at a given time, now or in the future, by a given environment or particular part thereof. Such quality objectives which, once adopted, 'are binding, as to the result to be achieved, upon each member state to which it is addressed, but shall leave to the national authorities the choice of form and methods'. Under the first action programme quality objectives were to be set for the following uses and functions of water: drinking, swimming, farming, fisheries, industry, recreation and aquatic ecosystems.

Since 1973 the EEC has promoted further Environment Action Programmes each lasting for 4 to 5 years and focused on a particular range of policy targets. The 5th Action Programme was adopted unanimously by the 12 member states in 1992 and sets a new strategy and new policy targets for the period 1993–2000. Its purpose is to strengthen and review existing controls (where thought necessary) and to increase cooperation between the public sector and industry by the introduction of voluntary measures to reduce environmental impact. This is a major change of direction which, it is hoped, will prove more acceptable and successful than the regulatory approach. Industry has tended to see regulation as a threat to profitability whereas in future it will, hopefully, find it commercially advantageous to be seen to be promoting sound environmental policies and to be setting steadily improving performance targets for the release of toxic substances to the environment and for conservation in general. The 5th Action Programme will also promote policies which aim to restore natural ground and surface waters to an ecologically sound condition and ensure that water demand and supply are brought into equilibrium on the basis of a more rational use and management of water resources.

Since 1973 some 300 items of EC environmental legislation have been adopted covering the following eight sectors: water pollution, air pollution, chemicals, wildlife, noise, environmental assessment, information and finances, and international conventions. In the field of water pollution the principal measures include the following Council Directives:

1976 Bathing Water Directive.
1976 Dangerous Substances Directive.
1977 Exchange of Information Decision.
1978 Freshwater Fisheries Directive.
1979 Shellfish Growing Water Directive.
1991 Nitrate Directive.
1991 Urban Waste Water Treatment Directive.

(Note: The Dangerous Substances Directive is a 'framework' directive which has so far generated 10 'daughter' directives covering 13 substances and their compounds or isomers.)

It can be seen that in the past 20 years the EEC action plans have resulted in a comprehensive system of uniform environmental controls which, however, have to be implemented through national legislation and national enforcement systems which differ in the various member states.

3.4.2 UK legislation and control systems

In the UK there is separate legislation for England and Wales, for Scotland and for Northern Ireland. In England and Wales the *Water Industry Act 1991* sets out the functions of the water and sewerage undertakers, the Director General of Water Services and local authority responsibilities for water supplies. The *Water Resources Act 1991* sets out the functions and powers of the National Rivers Authority which is responsible to the Secretary of State for the Environment for all matters concerning pollution control and water quality in surface waters, groundwaters and coastal waters. The National Rivers Authority is thus responsible for the enforcement of all relevant EC legislation and monitoring to demonstrate that the requirements are met.

In Scotland EC legislation is enforced by seven river purification boards and three islands councils with the Secretary of State for Scotland holding overall responsibility for water pollution control in the freshwater and marine environment. Here the *Control of Pollution Act 1974*, as amended by the *Water Act 1989*, is the main legislation which enables EU directives to be enforced. In Northern Ireland the control of water pollution is directly administered by the Department of Environment for Northern Ireland.

Under the *Environmental Protection Act 1990* a new system of Integrated Pollution Control (IPC) was introduced to control emissions to air, water and land from certain industries with complex, potentially harmful processes. IPC is administered by HM Inspectorate of Pollution (HMIP) in England and Wales while in Scotland it is jointly administered by HM Industrial Pollution Inspectorate (HMIPI) and the river purification authorities. The advantage to industry is that only a single application is required for each process even though that process may give rise to multiple emissions to two or three media. In granting authorisation the Best Practicable Environmental Option (BPEO) is used to specify the medium (air, water or land) to which the discharge must be made (as well as the limits on each component of the discharge) in order to minimise the environmental impact as a whole.

Integrated Pollution Control is seen by the government as an intermediate step towards the creation of a completely unified system of control. Legislation is now imminent which will establish, by 1996, an Environmental Protection Agency for England and Wales and a similar Scottish Environment Protection Agency in Scotland.

An important development in recent years in Britain and Europe has been public involvement in environmental control. For example, all significant applications for discharges of effluents to water must be advertised to the public so that objections can be considered before deciding whether to grant consent. Likewise all applications for authorisation under IPC are advertised and members of the public can make representations. Details of all applications, consents and authorisations are placed in accessible public registers. The results of all compliance monitoring and environmental monitoring must be placed in the registers and also details of enforcement action such as warnings and results of prosecutions.

3.4.3 Control in other EC member states

It would be quite impossible within the space of this chapter to describe the different legislative and institutional arrangements for water pollution control within the 12 member states of the EC (shortly to be enlarged to 15 members). Although all members are required to implement the EC Directives there are great variations in the nature of legislation and in the administrative structures which are utilised for enforcement of Directives and other EC laws. Several countries have fairly devolved systems of government which creates potential problems when a uniform international system of control is introduced.

Germany and Denmark are examples of states with highly devolved administrations. In Germany water resources management (including pollution control) is governed by the Federal Water Management Act which provides a legislative framework implemented through State Water Acts of the 11 states or Länder. Since the unification of East and West Germany there are now in addition six Länder from former East Germany. The former environmental laws of the German Democratic Republic have been superseded by those of the Federal Republic.[17]

In Denmark most environmental functions are carried out by the 14 County Councils, two Metropolitan Councils and smaller local councils. The Environment Protection Act provides the basis of pollution control for inland and coastal waters while regulations under this Act are made by the Ministry of Environment on the advice of its own Environment Agency.

In France control of the water environment is largely in the hands of six large river basin agencies set up in 1973 to control discharges into water and also the use of water. Water usage taxes imposed by the agencies are used in a variety of methods to finance effluent treatment plants. A new law in 1992 has provided heavier penalties for polluters and for the introduction of 'integrated planning and management schemes'. An overall Environment Code now nearing completion will bring all existing environmental legislation up to date in one single Act.[18]

3.4.4 Water pollution control in the USA

Prior to 1948 water pollution control was the preserve of individual states and few had effective, enforceable legislation. Regulation at the national level commenced with the *Water Pollution Control Act 1948*, which sought to establish responsibility for regulation and the funding of municipal sewage works in the hands of the Federal Government. However, further legislation was required over the next 24 years in the form of amendments to the 1948 Act to finally establish the Federal Government's responsibility to set discharge limits and to provide 75% funding assistance for municipal sewage works. The passing by Congress of the *Water Pollution Control Act Amendment 1972* (more conveniently referred to as PL 92–500) proved a landmark in US environmental legislation. It extended to all municipal and industrial discharges the existing permit system for control with uniform discharge limits. At first the permits were to be issued by the newly created

(1972) Environmental Protection Agency (EPA) with individual states taking over this function in future years. The 1972 Act required each state to produce water quality standards for rivers and lakes and to obtain EPA approval for these. A further requirement of the Act was for the EPA to produce discharge standards for 129 priority pollutants. The progress of control under this Act has had a chequered history and further legislation in the form of the *Clean Water Act 1977* and the *Clean Water Act Amendments 1982* has proved necessary.[16]

3.4.5 Pollution control in developing countries

Many developing countries have pollution control legislation in place and a considerable degree of awareness of environmental problems at government level. However, they have a serious lack of financial resources and of trained personnel. Regulatory bodies, if they exist, are rarely able to enforce effluent standards or even to monitor adequately the waters they are expected to safeguard.

In Egypt, because of the almost total dependence on the river Nile, the conservation, distribution and use of water is of fundamental importance. There are separate laws governing protection of the Nile river and its waterways from pollution, wastewater drainage, collection and disposal of solid wastes. The administration of the legislation is in the hands of various government departments and agencies, mostly based in Cairo, which means that enforcement tends to be lax and very uneven. In 1993 a comprehensive Environment Law was under preparation and likely to be approved by the National Assembly at an early date. This law covers the pollution of fresh water and sea water, the atmosphere and the disposal of solid waste.[19]

In India, since 1974, there has been in operation a soundly based system of state boards for the prevention and control of water pollution backed up by a powerful Central Board based in the capital, Delhi. The boards were set up under the *Water Act 1974*, which is closely modelled on UK legislation. The boards have enthusiastic staffs comprising engineers and scientists but are seriously under-resourced for the huge areas under their control. In general they have been reasonably effective in controlling discharges from industries but 90–95% of pollution is due to domestic sewage from urban communities.[8]

During the 1980s the Syrian Government established a series of water pollution control centres in three river basins and in the coastal zone. The first of these was located in the capital, Damascus, and primarily responsible for the Barada river. However, the Director is in overall control of the other centres based at Homs (Orontes river), Al-Raqqa (Euphrates river) and at Latakia (marine and freshwater pollution in the coastal zone). Resources are slender both financially and in terms of pollution control staff and remedial action is unable to deal with the backlog of severe pollution in the Barada and Orontes rivers let alone with the growing numbers of new industries.

3.5 The way ahead

3.5.1 Recent progress

In the highly industrialised western countries we are approaching the point where the gross, visible, pollution of rivers and estuaries has either already been eliminated or is within sight of being properly controlled. It has been found that the restoration of even moderately clean conditions in terms of the physical and chemical quality of water is often followed surprisingly quickly by the return of benthic flora and fauna and the return of fish life. The degree of recovery depends not only on water quality but also to a considerable extent on the condition of the stream bed. If the sand and gravel has been seriously impacted by solid particles, benthic invertebrates will have difficulty in re-colonising the bed and several years of spates may be necessary to restore the bottom sediments to a healthy condition.

However, as rivers recover from past pollution, they are increasingly vulnerable to sporadic pollution incidents which previously would have passed unnoticed. Small spillages of chemicals or trace amounts of poisonous substances in effluents, although visually not noticeable, may damage bottom-living plants or animals, an effect which will be detectable by biological monitoring for many weeks afterwards. However, if fish are killed the incident will be usually noticed and investigated immediately. Furthermore, with recovery the beneficial changes such as reduction in organic load, in suspended solids and in ammonia are counterbalanced not only by increased levels of dissolved oxygen, which is highly desirable, but also by rising levels of nitrates and phosphates which are much less desirable. Small increases in these nutrients may increase productivity of stream organisms and fish without much harm to the stream communities but larger amounts, especially in the sunnier months, can promote excessive growths of aquatic vegetation such as *Cladophora* (blanket weed). If lakes are affected large numbers of blue-green bacteria may appear in dense surface blooms which can be highly toxic.

3.5.2 The quest for higher standards

During the last few decades, partly for the foregoing reasons, there has been a slow but steady tightening of standards imposed on effluents and there can be little doubt that the trend will continue in the future. There are several reasons for this. First, our knowledge of the impact of pollutants on the aquatic environment has increased enormously in the last three to four decades, as has our ability to identify and measure low concentrations of toxic substances. Second, toxicological research now enables us to set environmental quality standards for a growing number of harmful substances such as pesticides, other organic micropollutants, heavy metals and inorganic nutrients. Third, our greater understanding of aquatic ecosystems has led to a reappraisal of water quality objectives by a number of national and international environmental bodies.

The Royal Commission on Environmental Pollution in the UK, in its 16th report (RCEP, 1992)[10] on freshwater, has, for example, substantially revised its earlier views on water quality objectives. Taking as its theme the sustainable use of water resources, the Commission states *inter alia* that 'processes which give rise to a cumulative long term deterioration of freshwater must be brought under firmer control' and that 'a precautionary approach to pollution control should be maintained'. This approach, first developed in Germany in the 1970s, was adopted by the UK Government and expressed in a White Paper (HM Government, 1990) in the following terms:

> Where there are significant risks of damage to the environment the Government will be prepared to take precautionary action to limit the use of potentially dangerous materials or the spread of potentially dangerous pollutants, even where scientific knowledge is not conclusive, if the balance of likely costs and benefits justifies it.

Following this approach the Commission takes the view 'that progressively less reliance should be placed on the environment as a mechanism for processing anthropogenic waste'. Specifically two changes in policy are recommended: first, the tightening of environmental quality objectives 'to reflect increasingly ambitious targets for water quality' and second by the use of technology-based emission limits in accordance with Best Available Techniques Not Entailing Excessive Costs (BATNEEC). In relation to eutrophication, the Commission considers that 'where the trophic status prior to significant enrichment resulting from human activities can be identified . . . it should form the long term target for flowing and standing waters'. Among other recommendations the Commission also states that, with a view to encouraging industry to reduce all emissions beyond what is required by law 'a programme based on the US Toxics Release Inventory could usefully be developed by the European Community'.

In 1987, at the Second International Conference on Protection of the North Sea, concern was expressed at the growing range of toxic substances which were being found in these waters. A Ministerial Decision was taken requiring all the member countries around the North Sea to take steps to reduce the inputs of those pollutants considered to be the most dangerous to water in terms of their toxicity, long term persistence in the environment, or their capacity for bio-accumulation. Each country was required to draw up its own list and to prepare an Action Plan to reduce the loads reaching tidal waters by 50% by 1995 as compared with 1985. At the Third Conference held in 1990, the more stringent target of a 70% reduction was adopted for selected pollutants.

In 1992 the Oslo and Paris Commissions, which were set up to protect the marine environment of northern Europe, announced an accord which was also designed to further reduce the input of toxic, persistent and bioaccumulative substances but with priority given to organohalogens. Initially, through its Working Group on Industrial Sectors (INDSEC), the Paris Commission is focusing research on the environmental aspects of polyvinyl chloride (PVC) manufacture including the emissions of ethylene dichloride and vinyl chloride monomer, the disposal of sludges which contain chlorinated hydrocarbons such as dioxins and the effects of phthalates.[14] More recently, in February 1994, the Environmental Protection Agency in the United States announced its intention to study the impact of chlorine compounds with the promise that 'the Administration will develop a national strategy for substituting, reducing or prohibiting the use of chlorine or chlorinated compounds.[14] In the same month this move has been reinforced by evidence gathered by the International Joint Commission (IJC) on the Great Lakes. In its seventh biennial report this body, which reports to both the US and Canadian governments, considers that the problem of persistent, toxic substances is the most important issue facing the Great Lakes Basin.[14] Both the EPA and the IJC are seriously concerned at the growing evidence that chlorinated compounds in particular are implicated in damage to the function of the reproductive, endocrinal, hormonal, immune and nervous systems.

The IJC report shows strong support for the precautionary principle, criticising the view that proof of harm must be demonstrated before action is taken. It now holds the view that 'persistent toxic substances are too dangerous to the biosphere and to humans to permit their release in any quantity . . . All such substances are dangerous to the environment, deleterious to the human condition and can no longer be tolerated in the ecosystem, whether or not unassailable scientific proof of acute or chronic damage is accepted'. The IJC calls for the onus of proof to be reversed, so that it would be for the proponents of a new chemical's production and use to prove that it is not harmful to the environment or health.[14]

These moves towards even tighter restrictions on emissions to water and other environmental media are, of course, only part of a wider concern for the global environment which led to the UN Conference on Environment and Development (the 'Earth Summit') which was held in Rio de Janeiro in 1992. The most important products of this conference were:

- Agenda 21: a world-wide programme of action towards the achievement of sustainable development in the next century.
- Convention on biodiversity: an international agreement to protect the diversity of living species and habitats.
- Convention on climate change: an international framework for action to reduce the risks of global warming.
- Sustainable forestry: a statement of principles for management, conservation and sustainable development of all the world's forests.

Member states are expected to prepare their own national strategies and action plans to implement these agreements. The UK government, for example, published its strategies and action plans under these four headings in January 1994 and also played a leading role within the European Union in preparing the Fifth Environmental Action Plan referred to earlier in this chapter. In the Strategy for Sustainable Development the government proposes to further improve the quality of surface and groundwaters and, of particular importance here, to introduce measures to encourage environmentally friendly farming and to minimise the use of pesticides and the impact of agricultural wastes on the environment (*Sustainable Development – the UK Strategy*, 1994).

3.5.3 The green image versus the industrial lobby

Today the industrial community finds itself in a serious dilemma. Whereas for more than a century it has held sway with governments in Europe and North America and managed to delay or seriously weaken pollution control legislation, it now has to recognise the enormous pressure of public opinion in environmental matters. The turning point in this sphere was the publication by Rachel Carson of her book *Silent Spring* in 1962. Based on her many years experience as a marine biologist in the US Fish and Wildlife Service, Carson wrote a meticulously documented account of the devastating effects of modern organochlorine pesticides on wildlife. Overnight the book became a bestseller in the English-speaking world and, within a year, was translated and published in no fewer than 14 industrial countries. This happened despite the fact that the powerful pesticides industry tried to stop publication of the work and withdrew advertising and financial support from scientific journals and media programmes which publicised the book with favourable reviews.[20] Although primarily concerned with the pesticides industry, Carson's work led to the foundation in 1970 of the Environmental Protection Agency and to the banning of DDT in 1972, followed later by the banning of other organochlorine-based pesticides in most western countries. It is moreover recognised that Carson's influence led, soon after her death in 1964, to the founding of highly influential environmental reform bodies such as the Environmental Defence Fund,

Friends of the Earth, Greenpeace and others too numerous to mention. The initial concern with pesticides led on to every aspect of environment concern, especially conservation, biodiversity and the concept of sustainability.

Throughout the western and in much of the developing world, public concern for the environment now vies with unemployment as the most important problem of the immediate future. It is this concern which has led to an unprecedented amount of new pollution control legislation and international agreements to protect the environment. It is this same public concern which has impelled almost every industry to demonstrate its concern for the environment. This has been done in a variety of ways. Thus food manufacturers and supermarkets chains have proclaimed that their products are free of preservatives, colourings, flavourings and other additives, while a growing range of fresh fruit and vegetables are stated to be organically grown without the use of pesticides or chemical fertilisers. In a similar way toiletries are claimed to be ozone friendly and household tissues are now unbleached or bleached by a peroxide process which does not produce dioxins. Paper manufacturers consider it essential to advertise the fact that their supplies of timber for pulp are drawn from sustainable forestry while furniture makers claim that they are not damaging tropical rain forests or exploiting tropical hardwoods.

The large chemical manufacturers feel that their image has particularly suffered at the hands of the environmental lobby and most have taken determined steps to improve their record. Environmental performance is now a top priority at boardroom level and several of the large manufacturers require each of their factories not only to meet all the standards imposed by the regulatory authorities but to show a significant improvement year on year. However, the chemical industry's concern to promote a green image has been strained to the limit by recent pronouncements on the dangers of chlorinated compounds in the environment by bodies such as the EPA (USA), the IJC and the Paris Commission as quoted earlier. The Chlorine Chemistry Council of the US Chemical Manufacturers Association and its European counterpart, Eurochlor, have both strongly attacked the views of these bodies, arguing that they were unscientific, that they did not take into account the chemical industry's own views on science, health, risks, etc., and that 'a realistic and thorough assessment of the socio-economic role of chlorine chemistry is the only way forward'.[14]

3.5.4 The need for clean technology

There is now a sufficient weight of evidence from reputable bodies to ensure that not only will more stringent environmental quality standards be required in the future but certain chemicals and classes of chemicals will need to be phased out if our aquatic resources are to be protected on a sustainable basis and if species diversity is to be maintained. It follows that the principles of clean technology will have to be applied increasingly in manufacturing industry to replace the 'end of pipe' treatment which is generally used at present. However, it is self-evident that there would be little point in applying expensive techniques in the manufacture of a chemical to prevent its escape if in its subsequent storage and use significant amounts still manage to enter the environment. Equally it is not sufficient to ban the manufacture of organochlorine pesticides in the northern hemisphere when they are allowed to be manufactured on an increasing scale in south-east Asia and widely used in the tropics from whence they are evaporated into the atmosphere and condensed out again in temperate and polar latitudes.

In conclusion, to return to the opening theme of this chapter – that water is our most important mineral – it needs to be constantly borne in mind that to protect water we need to protect all environmental media. If we pollute the soil by dumping material in landfills we risk polluting water; if we apply chemicals to the soil they will leach out into groundwater and streams. Pollutants which escape into the atmosphere are washed by rain or by dry deposition on to vegetation and soil and hence into water. Indeed, as we have tightened our controls on direct, piped discharges to surface waters it is already the case that non-point sources from air or soil are often the largest sources of pollution of water as exemplified by pesticides, nitrates and acid rain. It follows, therefore, that if we are to achieve the twin goals of sustainability and biodiversity clean technology is only one component of the required strategy. What will also be required is the application of life-cycle studies of the most important pollutants in a 'cradle to grave' approach to ensure that environmental releases are prevented or adequately controlled at every stage through to final disposal.

References

1. Curry-Lindahl, K. (1972) *Conservation for Survival*, Gollancz, London.
2. Hynes, H.B.N. (1960) *The Biology of Polluted Waters*, Liverpool University Press.
3. Hammerton, D. (1987) The impact of environmental legislation, *JIWPC*, **86** (2), 333–344.
4. National Rivers Authority. Proposals for Statutory Water Quality Objectives, Water Quality Series No. 5.
5. El-Hinnawi and Hashmi (1987) *The State of the Environment*, United Nations Environment Programme, Butterworths, London.
6. Organisation for Economic Co-operation and Development (1991a) *Environmental Indicators – A preliminary set*, OECD, Paris.
7. Organisation for Economic Co-operation and Development (1991b) *The State of the Environment*, OECD, Paris.
8. Hammerton, D. (1979) Report to WHO on water pollution in India.
9. Lean, G. and Henrichsen, D. (1990) *Atlas of the Environment*, Helicon, London.
10. RCEP 1992.
11. Department of Environment (1992) *The UK Environment*, HMSO, London.

12. Group of Experts on the Scientific Aspects of Marine Pollution (1990) Reports and Studies No. 39: UNEP, The State of the Marine Environment.
13. Environmental Data Services (1993) TBT pollution from ships threatens North Sea Species, *ENDS Report*, No. 227.
14. Environmental Data Services (1994) New twists in the battle over chlorine, *ENDS Report*, No. 229.
15. *ENDS Report*, No. 226.
16. Ellis, K.V. (1989). *Surface Water Pollution and its Control*, Macmillan, London.
17. Scherer, J. (1993) Environmental regulation in the Federal Republic of Germany. In: *Regulating the European Environment*, Baker and McKenzie, London.
18. Dowding, A. (1993) Environmental regulation in France. In: *Regulating the European Environment*, Baker and McKenzie, London.
19. Hamza, S.M. and Campbell, R. Environmental regulation in Egypt. In: *Regulating the European Environment*, Baker and McKenzie, London.
20. Hynes, H.P. (1989) *The Recurring Silent Spring*, Pergamon Press.

4 Bioremediation: A practical solution to land pollution

Z.M. LEES and E. SENIOR

4.1 Introduction

The production and subsequent storage or transportation of hazardous materials are integral parts of our economy. Consequently, there is an enormous variety of synthetic organic chemicals (xenobiotics) which enter our environment by way of leakages or spillages from pipes and tanks; deposition of airborne emissions; storage and disposal of raw materials or unwanted wastes and residues (for example, sludge lagoons, mixed landfills, slag areas, etc.); use of contaminated fill material; application of sewage or industrial sludge to land; and spraying of pesticides. In most instances, these problems involve the contamination of soils and/or groundwater and may also involve the contamination of sediments either on-site or in nearby water bodies (i.e. drainways, rivers, lakes). Furthermore, point and non-point discharges of pollutants result in the contamination of sediments in natural surface water bodies. Although the majority of polluted sites have been identified in the United States, the European continent and the United Kingdom, few (if any) such records exist for most Third World countries where numerous uncontrolled waste disposal sites represent a critical problem. In addition, there are untold numbers of operating industries, as well as commercial, residential and agricultural lands, with on-site contamination problems. Amelioration is thus a daunting and often expensive challenge. Although many clean-up strategies may be considered, *in situ* bioremediation represents, arguably, the most cost-effective and ecologically friendly solution.

Bioremediation is the use of natural, enhanced or genetically engineered organisms to improve environmental quality by exploiting their ability to treat hazardous (including toxic) or merely offensive compounds at contaminated sites. This natural process has been used for decades to treat wastes such as municipal sewage and effluents from industrial processes such as oil refining and chemical manufacture. It is also emerging as an extremely attractive alternative technology for the economic treatment of a wide range of environmental contaminants. Natural biological processes involved in bioremediation can be selectively enhanced and focused on the rapid removal of noxious or hazardous chemicals from soil, water and, even, the atmosphere. Bioremediation can mineralise waste products and hazardous chemicals into water, carbon dioxide, biomass or other innocuous products and thus negate the need to move the contaminants from one site or medium to another. In many cases, this may also be accomplished with considerably less environmental and worker exposure to hazardous substances.

The range of compounds treatable by bioremediation is immense. In addition to municipal waste and process waters, micro-organisms can degrade pesticides, industrial chemicals, jet fuel, gasoline and many components of crude oil. Even compounds that, until recently, were not regarded as biodegradable, such as chlorinated solvents, polychlorinated biphenyls (PCBs), chlorofluorocarbons and other stable synthetic organics, are being biodegraded in the laboratory by selected strains or associations of microorganisms. It now appears that, given the right conditions, most organic compounds, both natural and man-made, can be catabolised by micro-organisms, either through direct utilisation or by co-oxidation. Bioremediation may also be used either directly or indirectly to treat inorganic wastes through, for example, the transformation of nitrogen or sulphur-containing compounds, or the mobilisation/immobilisation of metals.

A significant advantage of bioremediation is that it can be very cost-effective. Adding fertiliser, other nutrients and even selected micro-organisms to contaminated soil/water, or otherwise manipulating the microbial environment, can often be much less expensive than alternative processes such as incineration, the use of adsorbents and catalytic destruction. Bioremediation also offers inexpensive treatment options for materials such as PCBs and other highly toxic contaminants which present a daunting liability to responsible industries. In addition, we now have the tools of genetic engineering to manipulate the genetic material that encodes degradative enzymes and pathways and so create vastly superior organisms which may have future use in bioremediation.

There are a number of types of bioremediation treatments. These include modifications of techniques used historically in conventional waste treatment as well as innovative new methods. Approaches include enhancing indigenous microbial activity by adding specially formulated fertilisers to soils or sediments contaminated with, for example, oil or other carbon-rich wastes; landfarming and composting techniques for degradation of refinery wastes and military explosive materials; the degrading of chlorinated compounds in soils by the intervention of microbial enzymes; treating PCBs in soils and sediments with micro-organisms; and degradation of recalcitrant compounds from lagoon sludges and contaminated soils in slurry reactors or other biological processing systems. Groundwater and surface water contamination can be treated by encouraging the growth of indigenous micro-organisms to degrade the waste *in situ*, by the addition of oxygen or by an alternative ‘pump-and-treat’ technology. Treatment of toxic or noxious substances (as vapours) is showing promise with bioreactors containing solid supports for micro-organisms which degrade airborne contaminants as they pass through.

All of these technologies have their applications, either alone or in combination. In choosing an appropriate biotechnology for remediation of a specific hazardous waste problem, it is useful to make a selection based on reliable information. It should be the goal of the waste technologist to base (and record) every decision on a sound evaluation of the alternatives, by an unbiased assessment of their applicability from both a technical and non-technical perspective. Successful bioremediation techniques must address both the heterogeneous nature of the many contaminated waste sites and the complexity of using living organisms. Not all sites can be treated effectively. Bioremediation is not a panacea. At the same time, strong evidence is emerging that the right application of biotreatment, combined with a thorough knowledge of the limitations of this technology, can be highly effective in remediating contaminated sites.

Treatment technologies are selected by considering three factors: (1) cost; (2) time to completion; and (3) familiarity.[1] When choosing an appropriate strategy selection should be made for the approach that:

(a) most effectively degrades the target compound(s);
(b) best addresses remediation of the matrix or phase in which the contamination is present (dissolved, absorbed or both);
(c) has proven to be the most effective in transporting nutrients and oxygen (or other electron acceptors) to the micro-organisms at the point of contamination;
(d) will make use of as many existing site conditions, utilities and/or labour as possible;
(e) has generated positive case histories for this type of matrix and contaminant;
(f) is least disruptive to the facility;
(g) can be most readily permitted by the authorities involved, or is deemed acceptable by those regulators;
(h) can be most rapidly deployed and initiated;
(i) best suits the site layout or is most mobile for use at other site locations;
(j) will close the site (i.e. reach minimum concentrations);
(k) will allow salvage of the capital equipment once the project is completed;
(l) will offer the most flexibility in system operations;
(m) will most readily complement other technologies to be utilised at the site;
(n) will not require excessive treatment of off-gas sludges, or other residuals; and
(o) will best reduce or eliminate client liability[1].

4.2 Microbial nutrition and environmental requirements

It has been said that every biochemically-synthesised organic compound is biodegradable and that somewhere there is a class of micro-organism which possesses the metabolic and enzymatic potential to degrade it[2]. This availability of a 'work force' for bioremediation makes most sites suitable for treatment. All sites contain micro-organisms but the challenge is to harness their metabolic potentials. The key factors are the environmental parameters at a given site and the nutritional factors which will facilitate the biodegradation of target contaminants and so produce the desired remedial end point.

The mineralisation, or complete degradation, of an organic molecule in soil/water is mostly a consequence of microbial activity. Few abiotic mechanisms in nature totally convert organic compounds of any kind to inorganic products and mineralisation sequences characterise the microbial metabolism of several classes of synthetic compounds. As they convert the organic substrate to inorganic products, they make use of much of the carbon in the substrate and convert it to cell constituents. At the same time, in most cases, energy is released and the populations increase in number and biomass as they assimilate some of the carbon and acquire energy for biosynthesis. Detoxification is a common outcome of mineralisation except when one of the products itself is of environmental concern[3].

Together with the above requirement for carbon, there are at least 11 essential nutrients that must be present in the soil in correct proportions and forms to sustain microbial growth[4]. These include nitrogen, phosphorus, potassium, sulphur, iron, calcium, manganese, zinc, copper, cobalt and molybdenum. It is not usually necessary to supplement soil/water with the latter nine elements as they are normally present in sufficient concentrations[5] and are rarely limiting factors. In contrast, nitrogen and phosphorus are of concern since they may become the limiting factors and, thus, supplementation is usually necessary. As a general rule, nitrogen must be present in the form of nitrate[6], although some species can utilise ammonia or nitrogen gas, while orthophosphate must be available as the source of phosphorus. Micro-organisms also require moisture for their growth and metabolism.

Elemental uptake is by absorption and transport across cell membranes of solubilised molecules. Thus, the amount of water present in the treatment matrix is critical in making the target molecules available to the micro-organisms. Optimal water content for aerobic bioremediation of soils is usually 10–20% by mass.[7] Overwatering will, however, saturate the soil and promote anoxic conditions which, in most cases, are undesirable.

During aerobic biodegradation, carbon atom oxidation releases electrons which are passed directly to molecular oxygen. When free oxygen is not available, many micro-organisms can oxidise carbon through anaerobic pathways, utilising nitrate or sulphate as a terminal electron acceptor. In general, aerobic degradation proceeds at a faster rate than anaerobic degradation and, as a result, most remediation technologies have focused on an aerobic solution.

Together with nutritional requirements, microbial activities are influenced by environmental conditions. Although it may be said that micro-organisms are found in most extreme situations, the majority prefer pHs of between 6 and 8. Similarly, they metabolise at temperatures between 0° and 80°C.[8] Bioremediation rarely takes place at temperatures beyond these limits although it must be realised that temperature profoundly affects the rates of degradation and substrate availability[9] and must, therefore, be seriously considered.

A familiarity with the basic responses of micro-organisms to their physical and chemical environments, and their nutritional needs, is fundamental to the application of bioremediation. An exhaustive discussion of the many degradative pathways known to exist for organic compounds can be found in other texts and will receive no attention here.

4.3 Bioremediation options

Numerous technologies to remediate sites exist and many can be used in process control in industries. Some technologies have more applications than others while some are more specialised or contaminant-specific. The following discussion is not an exhaustive review of all the available clean-up methods and design approaches. Instead, the salient aspects of biological remediation are given. No one technology is suitable for all sites, since every site and contaminant situation is different.

4.3.1 Bioreactors

Bioreactor technology is broad in application, competitive and, still, innovative. Researchers are constantly striving to achieve the design of the absolute bioreactor: the reactor capable of fitting every possible clean-up situation. The bioreactor sector of the bioremediation industry is one in which engineers and scientists can, through manipulation of the elements of nature, create a near-perfect environment for biodegradation. This environment of 'total control' is achievable because the matrix (e.g. soil or sediment) does not, as in other technologies, govern the success of the process and, thus, the treatment tends to produce quick results under both aerobic and anaerobic conditions. The close ties of this approach to

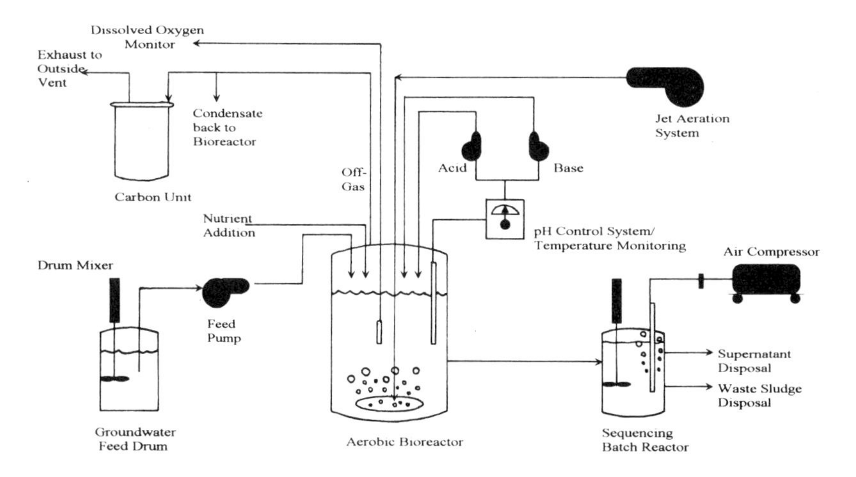

Figure 4.1 Typical aerobic bioreactor design (after Usinowicz and Rozich[13]).

conventional wastewater treatment also aid its favourable perception by both environmentalists and engineers.

Simplistically, a bioreactor is a reaction vessel which has a system for delivering oxygen and nutrients and devices for thorough mixing and the adjustment/maintenance of pH. Also, the reactor is normally fitted with influent and effluent pumps and can be run in a batch or continuous mode (Figure 4.1). All bioremediation bioreactors use water to provide an aqueous matrix, whatever their treatment purposes, and they have the flexibility of providing primary, secondary or tertiary treatment. An added benefit of their use is that they are mobile and can, therefore, be used on site. Two of the major disadvantages of bioreactors are that excavation of soil or the pumping of groundwater is necessary and, during treatment, they all produce a certain amount of sludge or biomass, and/or a volume of off gas (carbon dioxide, methane, hydrogen sulphide, etc.) which must be correctly disposed of. These, unfortunately, increase the treatment costs considerably.

The types of bioreactors used in commercial soil and groundwater bioremediation are numerous and include submerged fixed-film, plug flow, fluidised bed and sequencing batch reactors.[1] There are also slurry reactors and reactors which are designed to treat vapour phase contaminants. These all bear a resemblance to other bioremediation options where micro-organisms, contaminants, nutrients and (usually) oxygen are all brought into contact in a favourable environment for biodegradation to occur. The preferred above-ground bioreactor designs used suspended microbial growth on a fixed solid support. The principal advantages are reduced sludge production and a more rapid degradation rate, probably due to the greater surface area of the biofilm. However, caution must be exercised when using these techniques as non-biodegraded compounds may be sorbed onto the biofilm and the support matrix (especially charcoal), making the generated sludge a potentially hazardous material.[10]

Although it is always preferable to use indigenous microorganisms for the treatment of soil and groundwater, they are not always successful. Bioreactor systems do, however, lend themselves well to genetically-engineered micro-organism technology because of the enclosed design that facilitates greater control of the organisms.

4.3.2 *Aerobic bioreactor technology*

4.3.2.1 Slurry bioreactors. One reactor which has particular significance in the bioremediation arena is the soil slurry reactor. Few publications have emerged using this type of reactor primarily because they are very expensive[11] and the soil needs to be pretreated before entry into the vessel. However, they show great promise.

Silts and clays, or sludges, are most suitable for this type of treatment as the soil must be smaller than 60 mesh[1] so that an adequate slurry can be maintained against gravity. As treatment is completed during several days' retention time, the soils must be dewatered by centrifugation or a belt filter press. They are then, depending on the regulations and clean-up levels attained, reused on site as fill material, landfilled or thin-spread in road construction. The resulting water may be recycled or discharged.

Since bioreactors operate at near-ideal conditions for biodegradation, contaminant removal proceeds at a rapid rate and, therefore, offers significant advantages compared with conventional techniques such as land farming.[12] However, as with other methods, the biokinetics depend upon the type of contaminant, its concentration and the treatment standard which must be attained. Typical treatment times in a slurry bioreactor range from less than a month to more than 6 months.

An interesting pretreatment technique that may be considered for hydrophobic contamination involves the use of a soil washing process with biodegradable non-ionic or anionic surfactants. This removes surface contamination and aggressively scrubs soils so that strongly adsorbed contaminants are released. In doing so, soil volume reduction is achieved and there is a separation step in which washed coarse particles are removed and fines exit the system to a slurry reactor. There are good reasons to believe that the combination of soil washing and slurry bioreactors will be a likely process for future development.

4.3.2.2 Thermophilic bioreactor treatment. Aerobic thermophilic biological treatment suggests an operating temperature between 50 and 80°C. These temperatures are not usually considered in the application of aerobic biological treatment although recent reports have suggested that high temperatures offer potential benefits for organic destruction. First, under thermophilic conditions, the aerobic metabolic rate is increased and so the reactor size required for treating any organic waste concomitantly decreases. Second, the generation time for thermophiles is less than a tenth of that for mesophiles, resulting in a significantly reduced biomass sludge for disposal[13].

Aerobic thermophilic biological treatment can be used as an alternative to anaerobic biological treatment for sludges or sediments that are highly contaminated with organic compounds, hot streams containing biodegradable organics, or highly contaminated (organic) groundwater. As bacterial metabolic activities are, typically, exothermic, research has shown that aerobic reactors operate in a thermophilic range as long as the chemical oxygen demand (COD) is at least 30 000 mg l^{-1}, the reactor is insulated and covered to capture the heat generated, and a relatively efficient aeration system is provided[13].

4.3.2.3 Vapour phase bioreactors (biofilters). Another area of bioreactor technology receiving a significant amount of development time and funding is the vapour phase filter[14]. In Europe, biofiltration has become an accepted technology for treating volatile organic compounds (e.g. air emissions during *in situ* bioremediation, composting and bioreactor off-gas) and odour-containing industrial exhaust[15,16]. It is now a proven technology that is very economical for high volume emissions with low concentrations of pollutants. It is especially attractive because of its low energy consumption and low maintenance requirements.

Biofiltration harnesses the processes of decomposition with immobilised micro-organisms. The bioreactors are packed with solid material and the micro-organisms are attached to the surfaces as a biofilm. Gaseous wastes are passed through, by an induced or forced draught, and the micro-organisms catabolise the organic component of the vapour[17]. Not only do they provide a truly destructive process but they have none of the landfilling or regenerative problems associated with competing processes (incineration and carbon adsorption). Also, they do not generate sludge. They may also be used at the end of a treatment train to deal with organic emissions that could arise from the application of another technology, such as bioventing.

There are two types of biofilters: the soil filter and the treatment bed/disk.[18] The former is the simpler in design. Contaminated air is passed through a nutrient-supplemented compost pile which facilitates catabolism by indigenous mesophilic bacteria. In the treatment bed, the waste air stream and filter are humidified as gas is passed through one, two, or more beds made up of compost, wood chips, refuse, sand or diatomaceous earth.

In the disk approach, a series of humidified disks, containing activated charcoal, nutrients, peat, micro-organisms and compost, is placed inside a reactor vessel. Biodegradation of the contaminants occurs as the gas is vented through the system. Spent filters can be used as fertiliser as they present no hazard. Biofilters are in use throughout Europe, receiving and detoxifying emissions containing aldehydes, esters, amines, ethers, acetone, mercaptans, hydrogen sulphide, organic acids, a host of solvents and aromatic compounds.[14]

A carefully engineered biofilter must be enclosed, insulated and have a pre-treatment humidification system together with monitors for temperature and pressure of the waste stream. The filter medium must contain the essential nutrients, carbon, nitrogen and fibre to be properly operational. Although the initial capital costs are high, biofiltration is still an attractive option because of the low operating costs. Biofilters have been built to receive up to 90 000 ft^3 min^{-1} of air flow using filters up to 20 000 ft^2 in wetted area[1] with retention times of between 6 and 80 seconds.

4.3.2.4 Submerged fixed-film, plug-flow reactors. This reactor type (Figure 4.2) has the reputation of being highly effective and adaptable. As a result, it may be successfully used to treat low concentrations of organics found in groundwater and in treatment of wastewater where the organic loading can exceed 1000 mg l^{-1}.[19] Its secret lies in its design flexibility. It can withstand extreme fluctuations in organic loading while maintaining an active biomass for long periods of time.[20] The critical factor is the C : N : P ratio which must be 100 : 5 : 1 to effect biodegradation.

To apply bioreactor technology to groundwater remediation, a water quality study and, perhaps, a pilot study should be made. As with most bioremediation processes, the system requires a multidisciplinary approach[21] to identify potential problems such as metal precipitation or toxicity. The degradation capabilities, pH, and nutrient preferences of the introduced or indigenous micro-organisms should also be investigated.

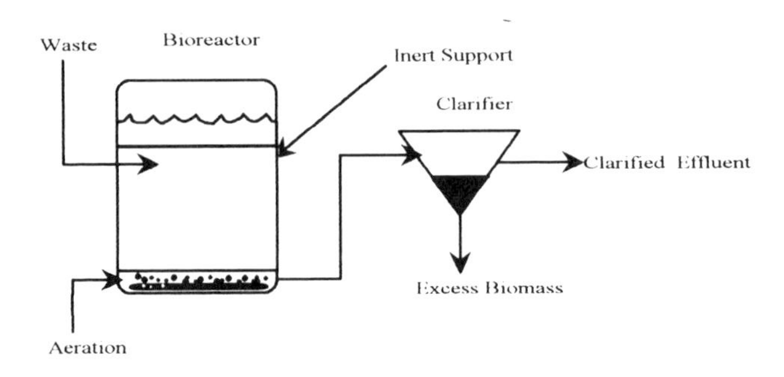

Figure 4.2 Diagrammatic representation of a submerged fixed-film reactor.

4.3.3 Anaerobic bioreactor treatment

This has many advantages. Anaerobic systems not only retain their biomass longer than aerobic treatment systems but the integrity of the self-sustaining biomass is so strong that contamination is obviated. This means that the biomass produced is extremely efficient in its capacity to degrade high concentrations of waste without high consumption of nutrients. Also, the sludge produced, typically, has a high commercial value as fertiliser.

This type of bioreactor operates under the same principles as the aerobic bioreactors and is extremely useful for the treatment of water or soils contaminated with halogenated pesticide residues and chlorinated solvents.[22–24]

As alternatives to bioreactor technologies, solid-state fermentation (composting), land treatment (on-site treatment) and *in situ* bioremediation may be considered.

4.4 Composting

Composting is a biological process which depends on the optimal growth, activity and interaction of a mixed population of mesophilic and thermophilic micro-organisms. Thus, the composting environment is characterised by elevated temperatures (>50°C), plentiful nutrients, high moisture content (>50%), non-limiting oxygen and a neutral pH. In a sense, this method combines many of the good points of incineration and land farming and minimises their disadvantages.

The technology involved in composting hazardous wastes falls primarily into three classes: turned windrow systems, static windrow systems and in-vessel systems. The process is aerobic, therefore the extent of aeration determines the rate of destruction of the waste and the level to which the temperature will rise in the composting mass. Insufficient aeration leads to anaerobiosis, and an accompanying generation of objectionable odours, although certain pesticides or haloaromatic compounds may be broken down under these conditions[25]. The most important consideration in the selection of a system for composting a hazardous waste is the control of emissions from the composting operation. The three broad types of composting systems are equally amenable to the effective control of solid and liquid discharges although this equality may not prevail with respect to gaseous emissions.

Williams and Keehan[26] pointed out that in composting, the capability of micro-organisms to biodegrade specific contaminants may not differ significantly from their capability in the ambient soil environment. The principal difference is the susceptibility of composting to increased control so that the rate and extent of activity can be significantly better than those in land farming.

The transformation potential differs for a variety of reasons. First, the elevated temperatures increase the process enzyme kinetics. Second, the opportunity for co-oxidation may be enhanced due to the range of alternative substrates present. Thirdly, modifications in the physical/chemical micro-environments within the composting mass can increase the diversity of the microflora to which the contaminant is exposed. Finally, the high temperatures will, typically, increase the solubility and mass transfer rates of the contaminants, making them more available to metabolism[25].

There is growing interest in the feasibility of composting as a process for detoxifying, degrading or inactivating hazardous wastes. Unfortunately, little research has been done. Composting is thought to provide a controlled system for containment of toxic constituents which may be subject to volatilisation or leaching during degradation. Rose and Mercer[27] found that the insecticides Diazinon, Parathion and Dieldrin degraded rapidly when composted with cannery wastes. The results of composting petroleum refinery sludges were reported by Deever and White,[28] who found that there was a significant reduction in the extractable grease and oil content after composting. Plant potting mixes were prepared from the compost and satisfactory growth was obtained. Studies by Snell[29] and Epstein and Alpert[30] showed that several polynuclear aromatic hydrocarbons were degraded during composting while soil contaminated with chlorinated aromatic compounds was successfully decontaminated during a study at the University of Helsinki[31]. Research by others[32,33] showed that by interposing specific conditions (e.g. a period of anaerobiosis) two insecticides, initially found to be recalcitrant, could be decomposed.

The methodology involved in the composting of hazardous wastes does not differ greatly from the composting of non-hazardous wastes. Hence, the parameters (particularly aeration, moisture content, temperature and pH) to be optimised are the same. Again, it is essential that nutrients are available in non-limiting concentrations. In the case of hazardous waste, however, emission control is often necessary and, as a consequence, an in-vessel system may be required.

4.4.1 Co-composting

Waste chemicals from industrial processes often contain compounds which are highly toxic, even in small quantities. Direct application to soil is, therefore, undesirable. Moreover, some wastes may have adverse physical, chemical or microbiological properties which would prevent or limit the extent to which they could be composted alone. It may then be feasible to mix them with, for example, sewage sludge, refuse or even manure or crop residues to provide a more compostable mixture.[25]

The co-composting of toxic chemicals with sewage sludge offers several advantages. Sludge contains an active indigenous population of micro-organisms, with broad degradative potentials, and provides a unique buffering and/or dilution system for at least some neutralisation of acid or alkaline wastes. It also provides labile carbon, energy and other nutrients to sustain a high level of biodegradative activity.

This approach could also be applied in land farming sites where composted sewage sludge could be applied to the soil before the addition of chemical wastes. The net result would be minimisation of both possible biocidal effects on soil microflora and soil structure damage.

If retention time is not a consideration then all but the most recalcitrant of organic wastes would be suitable for composting.[34] The problem is that for some wastes the process is so slow that it entails extended retention times. Despite this, effective composting technology, co-composting of wastes and, perhaps, pre-conditioning of land treatment sites with compost prior to direct waste application could markedly reduce the amount of land required for land treatment systems and greatly minimise the potential environmental impacts of the wastes.

4.5 Land treatment

Land treatment is a technology developed from the historic refinery practice of land farming, a method by which oily waste was spread over the soil surface to facilitate subsequent natural degradation. Although it is now outlawed in many countries, it was the first demonstration of the use of aerobic microbial processes to surface remediate hydrocarbons. Land treatment has been most successful with diesel and crude oils[35].

Modern soil bioremediation systems, unlike conventional land farming techniques, do not rely on large surface areas for spreading contaminated oil and sludge. Instead, solids (i.e. soils or sludge) are placed in windrows or lined treatment cells (Figure 4.3) and atmospheric oxygen is supplied by tilling, forced aeration or negative-pressure systems. Inorganic nutrients (fertilisers) are applied simultaneously, either manually or by automated systems installed in the treatment cells. The advantages of biotreatment

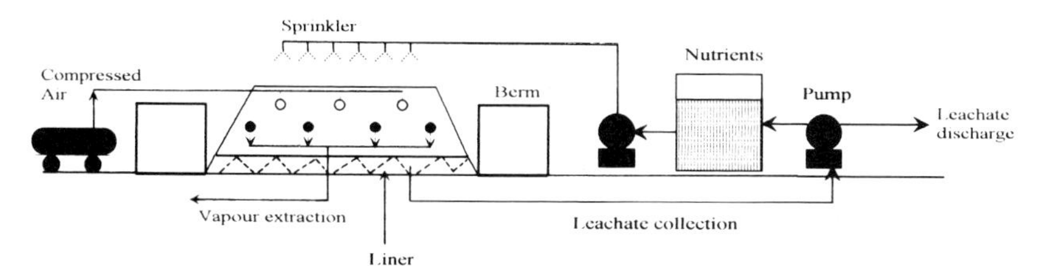

Figure 4.3 Diagrammatic representation of a typical land treatment design (after Hildebrandt and Wilson[36]).

systems like this are numerous and include a significant reduction in the surface area required for treatment, reduced remediation time because of improved design and increased system control, and the ease of applying treatment to gas emissions if necessary[36].

The successful operation of a land treatment system is based on a programme that incorporates three phases: a feasibility study, the design and construction of facilities and the operation and maintenance of the treatment cells.

4.5.1 The feasibility study

This is usually undertaken to decide whether the chemical, microbiological and hydrogeological conditions are favourable for remediation of the contaminated soil/sediment with a biological treatment system. Given the simplicity of the technique, it is not surprising that testing to determine whether land treatment will work is relatively straightforward. Usually, a field sampling programme, laboratory analysis and a bench-scale optimisation study are included in the test. Samples are assayed for nitrogen and orthophosphate content, bacterial enumerations of total heterotrophs and specific degraders, soil pH and buffer capacity, moisture content and contaminant concentrations.

The most important part of the feasibility study is confirmation of the presence of indigenous catabolic micro-organisms. This is done by bacterial enumeration analyses of aqueous extracts of the solids, following cultivation under carbon-limited conditions. The appropriate nutrient requirements are determined by soil microcosm optimisation testing where various types and ratios of nitrogen and phosphorus compounds are added to the contaminated soil and their stimulation effects evaluated. The soil pH is, typically, adjusted to between 6.5 and 8.0. If necessary, water is added to raise the moisture to, approximately, 50% of the field capacity.

Once an active microbial population has been confirmed, tests should be made to identify potential problems regarding air flow and nutrient percolation through the soil matrix. If inadequate mass transfer exists to stimulate the required biodegradation effects, flow rates may be improved through the addition of ameliorants such as wood chips, pine bark (composted) or rice hulls. An accurate estimate of the volume of soil to be treated is also essential.

4.5.2 Design and construction of above-ground biodegradation facilities

The simplest design is the windrow, which is a proven design used extensively in the composting industry. Soils for treatment are laid on a synthetic or clay lining in long row-like 'biopiles' about 2 ft thick. Slotted piping is connected to the air system and is installed so that a 'cylinder-of-influence' encompasses each windrow. Oxygen is either drawn outwards from the centre of the pile (forced-air) or is drawn inwards from the outside atmosphere (towards the piping in a negative-pressure system). Pressure is applied by a regenerative blower to provide air flow. Control of air emissions may sometimes be required if volatiles are displaced by the introduction of air. Nutrients are either mixed with the soil during the pile construction or are supplied by periodic spraying.

Another design which has been used extensively is based on the old land farming scheme where soil is spread in a number of 'cells' to a depth of up to 18 in. over an impermeable surface/liner. The cells slope towards a sump, for the collection of leachates or runoff water, and a pump redistributes the water back to the treatment cells. Like windrows, all cells are surrounded by retaining walls. A knowledge of the annual rainfall for the treatment area should indicate whether extra allowance must be made for drainage.

4.5.3 Operation and maintenance of biotreatment cells

The soil/sludge is usually tilled, to provide exposure to air, after every significant rain (> 25 mm day^{-1}) or after 2 weeks without significant rain. The soil water may be maintained at 50% of field capacity by pumping water from the sump, if necessary. It is important to recognise that inadequate moisture disturbs the osmotic balance between the bacterial association and the medium, while too much moisture reduces air transport by waterlogging.

The condition of liners, pile covers and berms must be checked periodically, and field analyses and laboratory assays must also be made regularly. Soil nutrient, pH and buffering capacity (lime requirement) parameters are monitored and adjusted to ensure optimal bioremediation. Treatment progress is most easily monitored by determining chemical concentrations from samples collected at different intervals and comparing them with the initial concentrations. When the appropriate chemical concentration reduction has been reached, the final sampling and closure steps should be initiated. These steps are different for each country but, generally, data proving that target levels have been reached are required to be submitted to the appropriate authorities for consideration.

The costs associated with soil pile bioremediation systems are related primarily to construction. As with bioreactors, most of the costs are for earthmoving and construction management. Generally, operation and maintenance represent 25–30% of the total remediation cost, depending on the volume of soil treated.[36]

4.6 *In situ* bioremediation

In situ bioremediation processes all have the common objective, i.e. to use micro-organisms (usually aerobic) to degrade contaminants in soil and/or water with least disturbance, and are, therefore, recommended in cases where excavation or disposal of soil or groundwater is not possible or economical. This is typically achieved through manipulation of the environmental conditions on site, to stimulate indigenous catabolic microbial associations.[37]

The effectiveness of the technique is primarily dictated by the site characteristics. More specifically, the site hydrogeology, climatic factors, soil types and properties, microbiological presence, and the concentration and physical/chemical characteristics of the waste to be treated are key variables. Important hydrogeological characteristics include the direction and rate of groundwater flow, the depth to the water table and/or the contaminated zone, and the heterogeneity of the soil.

4.6.1 Preliminary site investigation

The initial approach is to explore the site history. Information relating to whether site spills were documented, which chemicals or waste materials were stored, used or handled at the site, and whether there were/are leaking tanks present must be gained. Visual observations, such as stained soil, odours and signs of phytotoxicity must all be recorded and can, subsequently, provide the basis for a sampling programme. The initial investigation will usually lead to the installation of groundwater monitoring wells to determine the groundwater characteristics as well as the extent (vertical and horizontal) of the contamination. Cores and water samples collected during well installation are also useful in the feasibility tests. Well installation must be made under the strict guidance of a qualified geologist/hydrogeologist, who should present a reliable log of the soil stratigraphy and character using a standardised method such as the Unified Soil Classification System (USCS). Logs provide an indication of the sedimentary layers encountered with depth and an estimate of the soil's ability to transport air, water, nutrients, chemical contaminants and by-products through the contaminated zone. Finally, soil samples must also be tested for the presence of heavy metals and other elements which could preclude bioremediation.

4.6.1.1 Bioassessment testing. The investigations so far will give an idea of what is contaminating the site, where it is located and the extent of the contamination. The next step is to determine if bioremediation is feasible by making a bioassessment study. This research may consist of a simple nutrient study or a detailed degradation and stimulation evaluation. The choice depends on the funds and facilities available, and the time constraints.

The main objective of a bioassessment study is to identify factors, other than those realised during the site investigation, which might militate against bioremediation. Initially, soil and/or water nitrogen and orthophosphate concentrations must be determined with microbial characterisation tests. Details of the total heterotrophic population and specific degraders are necessary to decide whether there is a suitable indigenous population present *in situ* and if essential nutrients are in plentiful supply. If nutrient supplements, or liming, are necessary then these must be determined by the methods described above. A test to determine nutrient permeation/passage through soil would also be appropriate at this stage. Finally, if a chemical oxygen source is required, stability tests of such molecules in the presence of different nutrient concentrations are needed to identify possible transport and decomposition difficulties.

4.6.1.2 Pilot testing

(a) Microbiology. If the initial bioassessment is favourable, a detailed series of site-specific tests is then required. This treatability study includes a comprehensive investigation to determine the catabolic rates of the target compounds under site (soil and water) conditions. Catabolic optimisation is normally achieved through intermittent nutrient and oxygen supplementation in conjunction with target molecule assay and/or by-product appearance in comparison with abiotic controls to monitor non biological removal rates. Since comprehensive analyses are very expensive, only selected samples should undergo detailed characterisation while the rest should be subjected to routine, inexpensive tests. Other test parameters such as pH, nutrient content, oxygen consumption/carbon dioxide evolution may also be useful in monitoring the biological response.

Laboratory degradation rates are likely to be faster than *in situ* rates since microcosm studies (jars, pans, columns or lysimeters) are maintained at near ideal conditions. Realistic extrapolation to field conditions, therefore, necessitates the use of a 'fudge factor'.[1] This is dependent upon research experience, test conditions and the degree of confidence in site homogeneity. The factor usually varies between 2 and 6. Alternatively, a more detailed study close to site conditions can be made.

(b) Hydrogeology. Should the treatment employ groundwater, or if the groundwater is contaminated, then several other tests will be necessary to investigate groundwater behaviour and aquifer characteristics. Simple percolation tests or slug tests provide an estimate of the vertical and horizontal permeability of the aquifer and its ability to convey nutrient-laden water to the region, while a static ring test also provides information

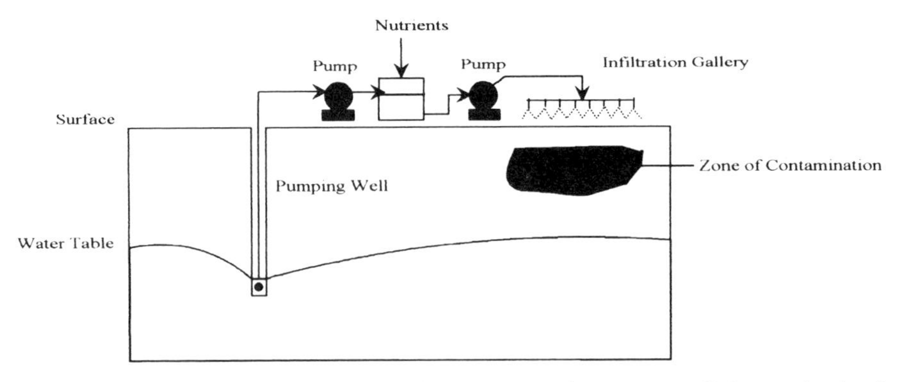

Figure 4.4 Classical *in situ* bioremediation where nutrients are supplied to recirculated groundwater which percolates into the subsurface (after Morgan and Watkinson[58]).

about the infiltration rate of water into the soil. Some form of pump test will also be required on at least one well installed in the contaminated zone. Usually, the well is pumped for an extended period, ranging from 6 h to several months, depending on the site hydrogeology and the response of the aquifer to a sustained pumping rate. These tests yield information on the transmissivity of the system, which is a measure of the volume of water which can be passed through a given cross-section of the subsurface.[1] Data obtained from the tests are then used to calculate an approximate flow velocity across the site which is critical for the design of an aqueous injection and recovery system across the contamination plume.

Geochemical testing is recommended at this point to determine the response of the water system to the supplements that will be added during remediation. For example, the intended injection concentration of orthophosphate could cause precipitation of calcium phosphate and the presence of iron, copper and manganese could catalyse the decomposition of hydrogen peroxide.

The treatment system must be designed to raise contaminated water to the surface, remove the contaminants, for example, by ion exchange, carbon adsorption or bioreactor treatment, add the right proportions of necessary nutrients and oxygen to stimulate microbial activity in the contaminated area, and reinject some of the extracted water by well injection or infiltration galleries (Figure 4.4). It is, therefore, necessary to be able to calculate the pore volume to be treated (i.e. the water volume existing between the injection and recovery systems). Dividing the length of the plume by the groundwater flow velocity gives an approximation of the time required to process one pore volume of water, which is usually targetted to be between one and three months.[1] Multiplying the expected number of pore volumes by the initial time estimate produces a preliminary estimate of the time required for remediation. It must be clearly recognised, however, that this is only a *prediction*, as few sites are completely homogeneous geologically.

The injection wells are, typically, placed at the perimeter of the contamination plume, at the highest point of the groundwater gradient, which raises the water table in the immediate area ('mounding'). The water recovery, by contrast, typically occurs at the perimeter of the contaminated area at the lowest point of the groundwater gradient. These recovery wells can lower the water table in the immediate area, forming a 'cone of depression', which may pull clean water from areas adjacent to the recovery zone, a situation that should be avoided if possible[38].

If time and cost factors allow, an *in situ* pilot test of the remediation system is appropriate, especially if the client is inexperienced with bioremediation technology. This allows an opportunity to identify potential problems and modify design/analytical parameters to obtain the best operating conditions. At the same time, confidence is gained to execute the full-scale programme.

4.6.2 Full-scale implementation and monitoring

At this stage, the treatment/process design should have been optimised by the expertise of engineers and hydrogeologists. The purchasing and installation/assembly of the necessary equipment is the next phase.

Details of the monitoring phase are different for every situation although there are some common principles. Briefly, field monitoring consists of recording data of soil and water pH, dissolved oxygen and nutrient concentrations, microbial numbers of heterotrophs and specific degraders (as colony forming units), and contaminant concentrations collected at frequent intervals, particularly during the first few weeks of operation. These will progressively give the project manager a clear indication of the progress of the remediation. Any nutrient or pH adjustments will then be made on the basis of these data. As the remediation proceeds, the sampling will become less frequent until the target contaminant concentrations are reached. At this point, detailed analyses of several samples must be made to confirm that the project objectives have been met.

In situ treatment of contaminated zones is not yet a proven remedy. Although the advantages of attractive economics, minimal site disruption and eliminated liability make it the most promising of all remediation options, it is also the most difficult treatment strategy to control and is chosen when time is not a constraint. Extra investigative work is required to predict its success because of the heterogeneity of most contaminated sites. Neglecting the need for treatability and pilot studies is, therefore, risky. It must be stressed that bioremediation does not work at every site, is not necessarily the least expensive method and may not reach the required clean-up standards in the given time.

4.7 The oxygen question

In situ bioremediation was one of the first technologies with the potential to address both dissolved- and adsorbed-phase organic contamination. Early research demonstrated the feasibility of bioremediation using simple infiltration of nutrient-supplemented water and in-well aeration, and a key finding of this work was the importance of the rate of oxygen delivery.[39,40] Where bioremediation was not effective, the failure was often found to be a lack of sufficient oxygen.

4.7.1 *Use of hydrogen peroxide in bioremediation*

Oxygen supply was thus identified as the central issue to be resolved. This challenge led to the first major innovation in bioremediation, the use of H_2O_2 as an oxygen 'carrier'. Hydrogen peroxide was considered promising because it is miscible with water, and each mg l^{-1} can supply ~0.5 mg l^{-1} of oxygen. This meant that the available oxygen could, potentially, be increased beyond the 8–10 mg l^{-1} limit of in-well aeration.

Laboratory studies which investigated the use of H_2O_2 in *in situ* bioremediation[41,42] showed that the growth of aerobic bacteria in general, and hydrocarbon-degraders in particular, was significantly enhanced by H_2O_2. Complementary field tests showed that the injection of H_2O_2 solutions into groundwater could increase the dissolved oxygen content at distances of 6–15 m from the injection point. Loss of injection as a result of biofouling could be rectified by adding periodic 'spike' concentrations in excess of 2000 mg l^{-1} to the wells. In 1987, the American Petroleum Institute demonstrated the successful use of H_2O_2 in a full-scale bioremediation system, where 4400 kg of hydrocarbons were removed in 160 days of treatment.

Despite the significant improvement in oxygen supply compared to in-well aeration (air sparging), H_2O_2 has many limitations. Because a basic issue has always been its cost, uncontrolled decomposition or the loss of oxygen equivalents is a serious concern. Therefore, the key to the successful use of H_2O_2 is controlling its decomposition. The conversion to oxygen is most often catalysed by metals such as Fe and Mn and by the enzyme catalase (which is secreted by many micro-organisms). Too rapid a decomposition can result in supersaturated water and the subsequent loss of oxygen. Such conditions may result in degassing which causes gas blockage and reduced permeability around injection points.

The potential toxicity of H_2O_2 to micro-organisms could be problematic in bioremediation. This issue of toxicity is complex and is influenced by a number of factors which make it impossible to predict the specific threshold toxicity concentration. Most treatments to date have used concentrations <2000 mg l^{-1}.

A final concern is the precipation of iron in contaminated aquifers after the introduction of H_2O_2. Such precipates can plug the aquifer in the vicinity of the injection well, thus necessitating frequent maintenance. The precipitation potential can be minimised by adding tripolyphosphate as the phosphorus source or by flushing the aquifer with water which has a low metal content. This may be obtained from a deeper aquifer[43].

Due to the above concerns with H_2O_2 in the unsaturated zone, it was quickly superseded by the development and use of soil vapour extraction technology.[44,45] The original focus of this extraction was to remove volatile organic compounds. Early developmental work,[46] however, showed that increased biodegradation rates were effected in response to the oxygen supplied.[47] This technology is now commonly referred to as bioventing.

4.7.2 *Bioventing*

Bioventing is an *in situ* remediation technique that combines the physical processes of conventional soil venting with enhanced biodegradation, thus providing speed and flexibility. Soil venting involves the application of a vacuum on vapour extraction wells installed in the unsaturated zone within, or adjacent to, the zone of contamination[48] (Figure 4.5). The negative pressure which develops in the soil pores accelerates volatilisation of sorbed compounds with high vapour pressures. By lowering the groundwater table simultaneously, by placing dewatering points at and just beneath the water table, additional vapour phase removal can result beneath the normal water table. In addition, the dewatering points

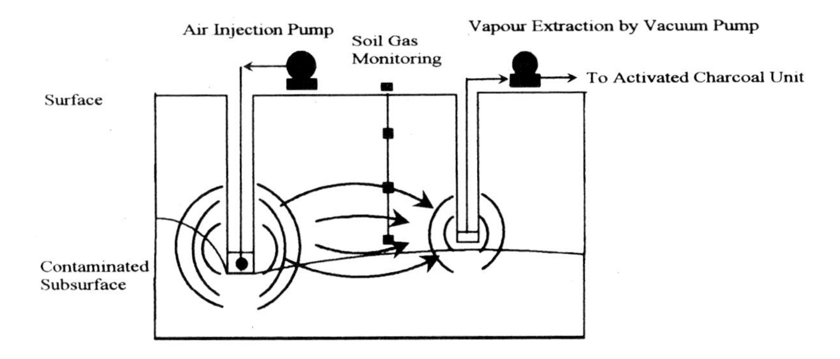

Figure 4.5 Enhanced bioreclamation through soil venting where air is withdrawn from clean soil (after Hoeppel *et al*[49]).

facilitate simultaneous removal of groundwater, free product and the vapour phase[49].

Soil venting increases the rate of air diffusion into the subsurface which, in turn, helps to satisfy the vacuum. Consequently, the air has the potential to diffuse through the soil, displacing the fraction with higher vapour pressures and stimulating aerobic degradation of low volatility compounds.[47] This is particularly significant in soils with low water permeability because of the greater diffusivities of gases compared with liquids. Since air contains >200 000 ppm oxygen, soil venting can overcome the oxygen deficits that often occur in heavily contaminated soils[50].

The limitations of air sparging currently centre on control factors. A primary operational concern is overpressurising the sparge system. This can displace both vapours and water and cause the undesired dissemination of contaminated vapours to 'clean' or low pressure areas such as building basements[45]. Thus, extreme care must be exercised in the design, implementation and control of an air sparging system. As with classical *in situ* bioremediation, the collection of site data and field pilot testing are critical in identifying any irregularities that might restrict airflow or cause accelerated vapour migration.

The number of field demonstrations and pilot applications of bioventing reported in the literature is limited. It has, however, been successful at many sites contaminated with fuels[7,44,51] and shows promise for sites contaminated with halogenated solvents, although the effects of environmental variables on bioventing treatment rates are poorly understood. It is generally known that compounds that are very soluble in water (e.g. alcohols), and which tend to partition in groundwater, are not good bioventing candidates[52]. Also, the tendency of a compound to adsorb to the matrix into which it has been introduced has a bearing on its suitability for soil venting, while physical factors such as the thickness of the unsaturated zone, the soil permeability, soil moisture content and the macronutrient availability each exert an influence on the success/applicability of this technique.

4.7.3 Alternative oxygen sources or electron acceptors

In designing a bioremediation system, it is important to be aware of the benefits and limitations of all of the available electron acceptors so that the most cost-effective system can be implemented.

The injection of water sparged with liquid or pressurised oxygen, instead of with air, is probably the most obvious alternative to the above techniques and effects a five-fold increase in the concentration of dissolved oxygen. Another novel approach was investigated by Michaelson and Lofti[53] who used oxygen microbubbles ('colloidal gas ephrons') created by mixing oxygen under pressure with water supplemented with a surfactant.

Other electron acceptors include nitrate, sulphate, carbon dioxide and iron, of which nitrate has been the most extensively researched. This is because it is inexpensive, highly soluble, is not adsorbed to soil matrices and does not decompose as rapidly as hydrogen peroxide. Unfortunately, besides the limitations imposed by water quality standards, nitrate is not as effective as oxygen for the biodegradation of most classes of compounds[54].

4.7.4 Anaerobic in situ *bioremediation*

There have been few attempts in commercial bioremediation to use anaerobic conditions for site reclamation. This is an area that has been largely unexplored, and even rejected, because aerobic biodegradation is more energy efficient than anaerobic degradation. Yet, anaerobic systems are prevalent in soil and aquifer environments, and a wide array of reaction mechanisms, including reduction, hydrolysis, dealkylation and dehalogenation, are possible in the absence of molecular oxygen. Research indicates that anaerobes are much more nutritionally diverse than previously imagined and capable of catalysing novel biotransformations which are of environmental and commercial interest[55]. In fact, many compounds (for example, PCBs, chlorinated solvents and pesticides) have only been shown to be degraded under anaerobic conditions. Thus, anoxic conditions may be of potential use in bioremediation programmes where the redox potential could be selectively adjusted to favour the degradation of a particular contaminant, and/or pH adjustment may be made to promote sulphate reduction and denitrification[38,56].

When organic matter enters an oxic environment, the indigenous microflora use the contaminants as electron donors to support heterotrophic microbial respiration. Since oxygen is used as an electron acceptor, it is rapidly depleted. This consumption of oxygen provides the selection pressure for other catabolic species. Under anoxic conditions, nitrate, sulphate and carbonate are used as alternative electron acceptors. As a consequence of self-generating redox gradients, one can often observe a spatial separation of dominant metabolic processes, depending on the availability of electron acceptors, the presence of suitable micro-organisms and the energetic benefit of each process to the microbial communities. Typically, nitrate reduction occurs first, followed by sulphate reduction and then methanogenesis[49]. Not surprisingly, different types of biodegradative activities can be observed within depth-related redox zones and research has shown that it may prove possible to stimulate desirable metabolic sequences through the intentional introduction of electron donor or acceptor combinations[56,57].

Although *in situ* anaerobic treatment would be slow and may be questionable in terms of its overall efficiency in meeting remediation targets, the anaerobic biodegradation of pollutants offers potential advantages over aerobic bioremediation approaches. Despite the fact that aerobic *in situ* biorestorations provide a cost-effective method of cleaning up contamination, much of the expense associated with these proceedings is accounted for by the costs of air, ozone, hydrogen peroxide or pure oxygen. Furthermore, biofouling is often a consequence of these treatments. Anaerobic processes, by contrast, are low energy producers and, therefore, generate less biomass which limits biofouling of the system. Also, anaerobic biotransformations sometimes result in metabolic products which are less toxic and more amenable to subsequent aerobic metabolism.

4.8 Conclusions

It is important to understand that there is unlikely to be a single technology that will work in all situations or that will by itself totally remediate a complex waste. Most likely, waste site clean-ups will require 'treatment trains', a sequence of applications of different technologies. The classic use of serialised anaerobic and aerobic degradations in a sequencing batch reactor is a good example. Also, a treatment train for the remediation of creosote- or oil-contaminated soil and groundwater may involve product removal using a pumping system, flushing with water and surfactants using pump-and-treat technology and, finally, *in situ* biodegradation of the residual contamination. It should be expected that the uses of complementary technologies will be a common application as a part of a sequenced approach to a site remediation programme[58].

Soils can be treated *in situ* or excavated for bioreactor or land treatment. If time is not a constraint, excavation is impractical, or the contaminants are degradable/moderately volatile, the most cost-effective method is *in situ* bioremediation with soil vapour extraction as an additional option. However, if the quantity of soil to be remediated is relatively small, the soil may be excavated for more timely bioreactor or land treatment. When time is short, conventional landfilling may be considered although with the threat of continuing liability. It is rarely necessary to incinerate soils. From a cost standpoint it is not feasible and there is great public opposition to the potentially toxic gas emissions.

Sludges containing organic or hydrocarbon compounds are amenable to land treatment, bioreactor processing or composting. Two difficulties associated with sludges are their high moisture content and amorphous structure although, in composting, these difficulties can be ameliorated through the use of a bulking agent such as wood chips. Landfilling or incineration is rarely recommended, although, if it appears imminent, a biofeasibility study should be pursued in the interest of economics.

Surface waters and wastewaters containing soluble compounds (organic or inorganic) are particularly suited to biotreatment *in situ* or by means of a treatment train employing bioreactors (aerobic or anaerobic) and/or pump-and-treat technology. Groundwater, by contrast, should always be assessed for *in situ* treatment before consideration of any other alternative as surface biotreatment, solely, of extracted groundwater could prove to be very expensive due to the volumes of water involved. Simultaneous surface and *in situ* biotreatments are becoming more popular, where product recovery and bioreactor treatment can be coupled with nutrient injection, and subsequent recycling of the produced water.

A very real impediment to applying environmental biotechnology can be the developmental time and costs. Deadlines must be met and frequently the public and the clients want the quickest and cheapest treatment available. However, this attitude can be changed through communication and demonstration that microbial treatment, provided that mineralisation is possible, is usually the most rapid and complete treatment system. The true limitations of applying biotechnology to environmental problems are our imaginations and the willingness of microbiologists and engineers to work together.

Microbiological clean-up is a developing technology founded upon basic principles of microbial ecology and physiology. The ability of microorganisms to reduce the potential toxicity of substances to higher organisms through processes of biodegradation, biotransformation and bioaccumulation can easily be demonstrated in the laboratory. In theory, there is no reason why biotechnologies based on such capabilities cannot be successfully developed and applied. In the field, however, these processes may be limited by environmental conditions. Hence, the bioengineering challenge is to realise *in situ* the potential observed in the flask so that biotreatment will become the *first* technology to be considered, not the last.

Acknowledgements

We are grateful for the financial assistance of Shell S.A. (Pty) Ltd and the South African Foundation for Research, Development for research in bioremediation of oil contaminated soils.

References

1. King, R.B., Long, G.M. and Sheldon, J.K. (1992) *Practical Environmental Bioremediation*, Lewis Publishers, CRC Press Inc., Boca Raton, FL.
2. Chakrabarty, A.M. (1982) *Biodegradation and Detoxification of Environmental Pollutants*, CRC Press, Boca Raton, FL, 127–140.
3. Alexander, M. (1981) Biodegradation of chemicals of environmental concern, *Science*, **211**, 132–138.
4. Alexander, M. (1977) *Introduction to Soil Microbiology*, John Wiley and Sons, New York.

5. Clarke, E.A. (1979) Soil microbiology. In: *The Encyclopedia of Soil Science*, Part 1 (eds E.A. Fairbridge and C.W. Finkl, Jnr), Dowden, Hutchinson and Ross, Stroudsburg, PA.
6. Swindell, C.M., Aelion, C.M. and Pfaender, F.K. (1988) Influence of mineral and organic nutrients on biodegradation and the adaptation response of subsurface microbial communities, *Applied and Environmental Microbiology*, **54**, 212–217.
7. Hinchee, R.E., Downey, D.C., DuPont, R.R., Aggerwal, P. and Miller, R.N. (1991) Enhancing biodegradation of petroleum hydrocarbons through soil venting, *Journal of Hazardous Materials*, **27**, 315–325.
8. Higgins, J. and Burns, R.G. (1978) *The Chemistry and Microbiology of Pollution*, Academic Press Jovanovich, London, pp. 1111–1162.
9. Dragun, J. (1988) *The Soil Chemistry of Hazardous Materials*, Hazardous Materials Control Research Institute, Silver Spring, MD.
10. Tzesos, M. and Bell, J.P. (1988) Significance of biosorption for the hazardous organics removal efficiency of a biological reactor, *Water Research*, **22**, 391–394.
11. Ross, D. (1991) Slurry-phase bioremediation: case studies and cost comparison, *Remediation*, **1**, 61–74.
12. Catallo, W.J. and Portier, R.J. (1992) Use of indigenous and adapted microbial assemblages in the removal of organic chemicals from soils and sediments, *Water Science and Technology*, **25**, 229–237.
13. Usinowicz, P.J. and Rozich, A.F. (1993) Thermophilic process cuts biomass wastes, *Environmental Protection*, **4**, 26–34.
14. Standefer, S. and Van Lith, C. (1993) Biofilters minimise emissions, *Environmental Protection*, **4**, 48–58.
15. Duncan, M., Bohn, H.L. and Burr, M. (1982) Pollutant removal from wood and coal flue gases by soil treatment, *Journal of the Air Pollution Control Association*, **32**, 1175–1179.
16. Prokop, W.H. and Bohn, H.L. (1985) Soil bed system for control of rendering plant odours, *JAPCA*, **35**, 1332.
17. Kambell, D., Willson, J.T., Read, H.W. and Stockdale, T.T. (1987) Removal of volatile aliphatic hydrocarbons in a soil bioreactor, *Journal of the Air Pollution Control Association*, **37**, 1236–1240.
17a. Kamnikar, B. (1992) Bioremediation of contaminated soil, *Pollution Engineering*, **24**, 50–52.
18. Holusha, J. (1991) Using bacteria to control pollution, *The New York Times*, C6, March 13.
19. Hamoda, M.F. and Al-Haddad, A.A. (1989) Treatment of petroleum refinery effluents in a fixed-film reactor, *Water Science and Technology*, **20**, 131–140.
20. Lewandowski, G.A. (1990) Batch biodegradation of industrial organic compounds using mixed liquor from different POTWs, *Water Pollution Control Federation*, **62**, 803–809.
21. Bourquin, A.W. (1989) Bioremediation of hazardous wastes, *Hazardous Material Control*, Sept-Oct.
22. Genig, R.K., Million, D.L., Hancer, C.W. and Pitt, W.W. (1979) Pilot-plant demonstration of an anaerobic fixed-film reactor for waste-water treatment, *Biotechnology and Bioengineering*, **8**, 329–344.
23. Bouwer, E.J. and McCarty, P.L. (1985) Utilisation rates of trace halogenated organic compounds in acetate-supported biofilms, *Biotechnology and Bioengineering*, **27**, 1564–1571.
24. Calmbacher, C.W. (1991) Biological treatment gaining acceptance, *Environmental Protection*, **2**, 38–140.
25. Crawford, S.L., Johnson, G.E. and Goetz, F.E. (1993) The potential for bioremediation of soils containing PAHs by composting. In: *Proceedings of the Second International On Site and In Situ Bioreclamation Symposium*, San Diego, CA (in press).
26. Williams, R.T. and Keehan, K.R. (1993) Hazardous and industrial waste composting. In: *Science and Engineering of Composting: Design, Environmental, Microbiological and Utilisation Aspects* (eds Hoitink and Keener), Renaissance Publications, Worthington, OH, pp. 363–381.
27. Rose, W.W. and Mercer, W.A. (1968) *Fate of Pesticides in Composted Agricultural Wastes*, National Canners Association, Washington, DC.

28. Deever, W.R. and White, R.C. (1978) *Composting Petroleum Refinery Sludges*, Texaco Inc., Port Arthur, TX.
29. Snell, J. (1982) *Rate of Biodegradation of Toxic Organic Compounds while in Contact with Organics which are Actively Composting*, Snell Environmental Group, NTIS, CA.
30. Epstein, E. and Alpert, J.E. (1980) Composting hazardous wastes. In: *Toxic and Hazardous Waste Disposal*, Vol 4, Ann Arbor Sciences Publishers, The Butterworth Group, Ann Arbor, Mic, pp. 243–252.
31. Valo, R. and Salkinoja-Salonen, M. (1986) Bioreclamation of chlorophenol-contaminated soil by composting, *Applied Microbiology and Biotechnology*, **25**, 68–75.
32. Guenezi, W.D. and Beard, W.E. (1968) Anaerobic conversion of DDT to DDD and aerobic stability of DDT in soil, *Soil Science Society of America Proceedings*, **32**, 322–324.
33. Wilson, G.B., Sikora, L.J. and Parr, J.F. (1983) Composting of hazardous industrial wastes. In: *Land Treatment of Hazardous Wastes* (eds J.F. Parr, P.B. Marsh and J.M. Kla), Noyes Data Corporation, New Jersey, pp. 268–270.
34. Savage, G.M., Diaz, L.F. and Golueke, C.G. (1985) Disposing of hazardous wastes by composting, *BioCycle*, **26**, 31–34.
35. Wilson, G.B., Parr, J.F., Taylor, J.M. and Sikora, L.J. (1982) Land treatment of industrial wastes: principles and practices, *BioCycle*, 37–42.
36. Hildebrandt, W.W. and Wilson, S.B. (1991) On-site bioremediation systems reduce crude oil contamination. In: *Proceedings of 1990 SPE California Regional Meeting*, Ventura.
37. Kaufmann, A.K. (1986) *In situ* biodegradation: the viable alternative. *HAZ-NEWS*, June–July, Hazardous Waste Association of California.
38. Lee, M.D., Thomas, J.M., Borden, R.C., Bedient, P.B., Ward, C.H. and Wilson, J.T. (1988) Biorestoration of aquifers contaminated with organic compounds, *CRC Critical Review of Environmental Control*, **18**, 29–89.
39. Floodgate, G.D. (1973) The microbial degradation of oil pollutants, in *Publ. No. LSU-SG-73–01* (eds D.G. Ahearn and Meyers, S.P.), Center for Wetland Resources, Louisiana State University, Baton Rouge, LA.
40. Zobell, C.E. (1973) The microbial degradation of oil pollutants, in *Publ. No. LSU-SG-73–01* (eds D.G. Ahearn and Meyers, S.P.), Centre for Wetland Resources, Louisiana State University, Baton Rouge, LA.
41. API (1987) Field study of enhanced subsurface biodegradation of hydrocarbons using hydrogen peroxide as an oxygen source, *American Petroleum Institute Publication 4448.*
42. Brown, R.A. and Crosbie, J. (1989) *Oxygen Sources for In Situ Bioremediation.* Hazardous Materials Control Research Institute, Baltimore, MD.
43. Brown, R.A., Dey, J.C. and McFarland, W.E. (1991) Integrated site remediation combining groundwater treatment, soil vapour extraction and bioremediation. In: *In Situ Bioreclamation: Application and Investigation for Hydrocarbons and Contaminated Site Remediation* (eds R.E. Hinchee and R.F. Olfenbuttel), Butterworth-Heineman, Stoneham, MA.
44. Brown, R.A. and Jasiulewicz, F. (1992) Air sparging used to cut remediation costs. *Pollution Engineering*, July, 52–57.
45. Brown, R.A., Hicks, R.J. and Hicks, P.M. (1993) Use of air sparging for *in-situ* bioremediation. In: *Proceedings of the Second On Site and In Situ Bioreclamation Symposium*, San Diego, CA (in press).
46. Thornton, J.C. and Wooten, W.L. (1982) Venting for the removal of hydrocarbon vapours from gasoline contaminated soil, *Journal of Environmental Science and Health*, **A17**, 31–44.
47. Hinchee, R.E. and Arthur, M. (1991) Bench-scale studies of the soil aeration process for bioremediation of petroleum hydrocarbons, *Journal of Applied Biochemistry and Biotechnology*, **28/29**, 901–906.
48. English, C.W. and Loehr, R.C. (1991) Degradation of organic vapours in unsaturated soils, *Journal of Hazardous Materials*, **28**, 55–63.
49. Hoeppel, R.E., Hinchee, R.E. and Arthur, M.F. (1991) Bioventing soils contaminated with petroleum hydrocarbons, *Journal of Industrial Microbiology*, **8**, 141–146.
50. Connor, J.R. (1988) Case study of soil venting, *Pollution Engineering*, **20**, 74–78.
51. Downey, D.C., Hinchee, R.E., Westray, M.S. and Slaughter, J.K. (1988) Combined

biological and physical treatment of a jet-fuel contaminated aquifer. In: *Proceedings NWWA-API Conference for Petroleum Hydrocarbon and Organic Chemicals in Groundwater – Prevention, Detection and Restoration*, National Water Well Association, Worthington, Ohio, 627–644.

52. Reisinger, H.J., Johnston, E.F. and Hubbard, P., Jnr (1993) Cost effectiveness and feasibility comparison of bioventing vs conventional soil venting. In: *Proceedings of the Second On Site and In Situ Bioreclamation Symposium*, San Diego, CA (in press).
53. Michaelson, D.L. and Lofti, M. (1990) Oxygen microbubbles for *in situ* bioremediation: possible field scenarios. In: *Innovative Hazardous Waste Systems*, Technotric Publishing Co., New York, NY.
54. Brown, R.A. and Norris, R.D. (1993) The evolution of a technology: hydrogen peroxide in *in situ* bioremediation. In: *Proceedings of the Second On Site and In Situ Bioreclamation Symposium*, San Diego, CA (in press).
55. Suflita, J.M., Gibson, S.A. and Beeman, R.E. (1988) Anaerobic biotransformations of pollutant chemicals in aquifers, *Journal of Industrial Microbiology*, **3**, 179–194.
56. Ramanand, K., Balba, M.T.M. and Duffy, J. (1993) Anaerobic metabolism of chlorinated benzenes in soil under different redox potentials. In: *Proceedings of the Second On Site and In Situ Bioreclamation Symposium*, San Diego, CA (in press).
57. Kaake, R.H., Roberts, D.J., Stevens, T.O., Crawford, R.L. and Crawford, D.L. (1992) Bioremediation of soils contaminated with the herbicide 2-*sec*-butyl-4,6-dinitrophenol (Dinoseb). *Applied and Environmental Microbiology*, **58**, 1683–1689.
58. Kearney, P.C., Karns, J.S., Muldoon, M.T. and Ruth, J.M. (1986) Couthaphos disposal by combined microbial and UV-ozonation reactions. *Journal of Agricultural Food Chemistry*, **34**, 702–706.
58. Morgan, P. and Watkinson, R.J. (1989) Microbiological methods for the cleanup of soil and groundwater contaminated with halogenated organic compounds, *FEMS Microbiology Reviews*, **63**, 277–300.

5 The economics of pollution

D.W. PEARCE

5.1 Introduction: environmental economics

Environmental economics brings the discipline of economic analysis to environmental issues such as pollution, the rate of use of renewable and non-renewable natural resources, conservation of living species and resources, and the choice of policy to achieve environmental ends. It is now a huge subject (for a survey see Pearce[1] and for the standard textbooks see Pearce and Turner[2] and Tietenberg[3]).

This chapter focuses on just one aspect of environmental economics: the economics of pollution. It sets out the elements of the theory of pollution control, shows how that theory can be put into practice, and finally focuses on the issue of 'clean technology'.

5.2 The basic analytics

While scientists tend to identify 'pollution' with some metabolic change in living systems brought about by the emission of waste into receiving environments, economists tend to identify pollution in terms of any loss of 'human wellbeing' arising from physical environmental change. 'Human wellbeing' refers to a state of mind, a feeling of pleasure or pain. Its more traditional term in economics is 'utility' or 'welfare'. The provision of goods and services in the economy generates such wellbeing, as does a clean environment, the existence of fine views, wildlife and so on. Conversely, pollution generates losses in wellbeing.

Pollution is the result of waste disposal to the environment, whether the wastes in question are solid wastes (e.g. municipal waste), liquid effluent (sewerage) or gaseous wastes (air pollutants such as nitrogen oxides, sulphur oxides, etc.) In turn, waste is endemic to any economic system. It is the result of the first law of thermodynamics: matter and energy can be neither created nor destroyed. Hence whatever is taken out of natural environments and used in the economy must, eventually, become waste. This essential 'materials balance' idea is illustrated in Figure 5.1. Notice that the materials balance concept implies that phrases like 'zero waste society' and 'clean technology' are nonsense. There can only be cleaner technologies, those that use less input per unit of economic output, or

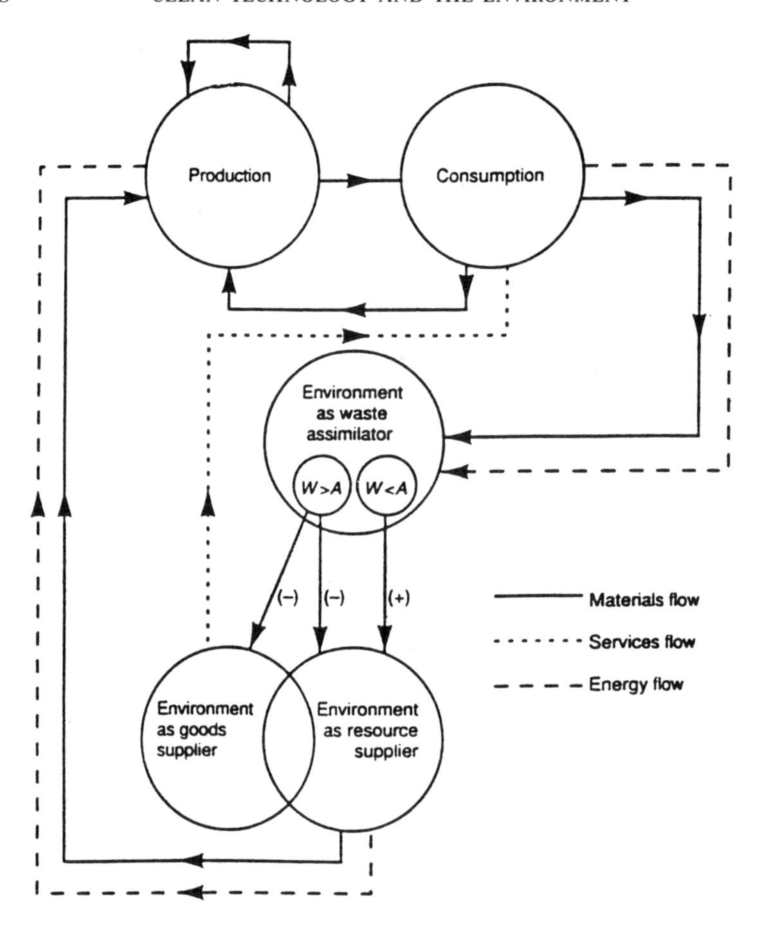

Figure 5.1 The materials balance perspective.

which use a different technology containing fewer components with zero or limited counterpart assimilative capacity.

Natural environments have some capacity to deal with waste flows, depending on their quantities and qualities. This 'assimilative capacity' is easily breached, however, either by emitting more waste than the environment can assimilate, or by emitting types of waste that have no counterpart assimilative capacity. Examples of the former problem are excess sewerage discharged to rivers. Examples of the latter are heavy metals such as mercury and cadmium.

Once assimilative capacity is breached, pollution problems arise in the technical sense. In so far as human wellbeing is affected, *economic* problems also arises. Notice that economics deals with gains and losses of wellbeing. It is not the case that economics is confined to the treatment of money, or even to the treatment of actual markets. This will become clear

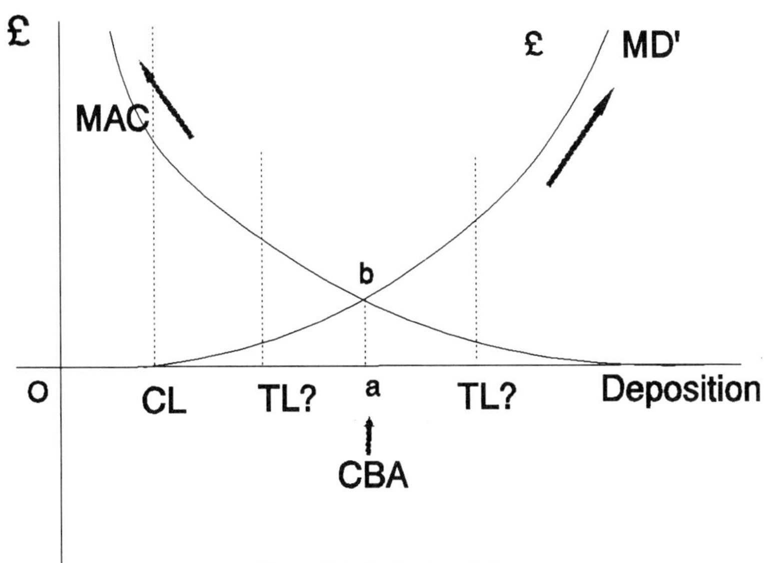

Figure 5.2 Optimal pollution.

shortly. One way of illustrating the pollution problem is in terms of Figure 5.2. Here the maximum amount of waste that the environment can handle without physical change occurring is referred to as the 'critical load'. Critical loads are in fact measured for a wide class of pollutants but especially for acidic depositions caused by 'acid rain', i.e. emissions of sulphur and nitrogen compounds. To make the analysis simple, assume that as soon as physical damage occurs, there is some loss of wellbeing because of, say, loss of biological diversity, odour, ill-health and so on. Then, the curve £MD′ measures the damage done by that pollution in terms of the loss of wellbeing.

In fact, £MD′ has a specific meaning. First, it is a *marginal* curve. That is, it is the *extra* damage done by an *extra* amount of deposition of pollutants. In more formal terms, marginal curves are simply the slopes of total curves, their first derivatives. Working in marginal terms is common in economics since it facilitates the identification of economic 'optima', as we shall see. The second important thing about £MD′ is that it is measured in *money* terms. Having said that economics is not just, or even predominantly, about money, this requires some explanation.

Human wellbeing is measured by referring to people's preferences. Basically, if I prefer situation A to situation B my wellbeing is higher in A than in B. Can the *amount* by which I am better off be measured? Economic theory shows us that this 'better offness' can be measured in terms of my 'willingness to pay' to be at A rather than B. Willingness to pay is familiar enough in market places – it is revealed through our demand for goods and services. Indeed, the 'demand curve' familiar in all elementary economics textbooks is a (marginal) willingness to pay schedule. So, as long as we accept that the criterion for choosing alternatives is the effect they have on human wellbeing, willingness to pay (WTP) is a good measure. It happens to be expressed in money because that is a common measuring rod. The use of money implies nothing about the 'worship' of money, capitalism or any of the charges often, but mistakenly, brought against economics.

MD′ is therefore measured by people's WTP to avoid pollution damages. It can also be measured by their 'willingness to accept compensation' (WTAC) to tolerate the pollution. These measures may differ and there is a debate in environmental economics as to which is to be preferred. MD′ is also given another name. Provided the polluter does not pay for the costs of the damages – they are known as 'externalities', or 'external costs'. This is because they are external to the polluter and are not accounted for by him unless some form of environmental regulation exists.

Pursuing Figure 5.2, we note that the damage function begins at the critical load, CL. It cannot begin earlier than this because the environment manages to assimilate the waste up to that point. Now we can superimpose a cost curve showing us the cost of reducing depositions, e.g. by 'end of pipe' technology such as flue gas desulphurisation or fuel switching or energy conservation. In general, this 'marginal abatement cost' curve will slope upwards from right to left, showing that the marginal cost of reducing pollution tends to be get more and more costly the cleaner the environment becomes. This is not always true, but is generally true.

Now look at the point marked 'CBA'. At this point, marginal damages just equal marginal abatement costs. The meaning of this intersection is important. To the left of CBA it will cost more to reduce a unit of pollution than the value of the damages reduced. Since avoided damage can always be thought of as a *benefit*, we can say that to the left of CBA marginal benefits are less than marginal (abatement) costs. Points to the left of CBA will therefore be characterised by 'too much' pollution control. To the right of CBA, on the other hand, $MAC < MD'$ so it will pay to go to the left. The intersection at CBA then turns out to be an economic optimum – the point where $MAC = MD'$ is in fact the point where the *net* benefits of control are maximised.

Note that CBA is *not* at the point of critical loads. This is problematic in the real world because critical loads, or 'zero damage' points, are widely entertained as the desirable end of environmental policy. This disparity between critical loads and the CBA (cost–benefit analysis) point does much to explain why scientists and economists often conflict. But too much should not be made of this difference. First, two different criteria are being used. The advocate of critical loads tends to be more concerned about the unknown effects of exceeding critical loads. He or she tends to adopt a 'precautionary approach'. The economist tends to worry about what we are getting for the extra cost of adopting the precautionary approach: nothing is for free and the money spent on reaching critical loads could, after all, have been spent on something else – roads, hospitals, other environmental improvement. Second, it may actually be that in the long term critical loads are the right objective after all. This is because point CBA in Figure 5.2 has positive levels of damage. This damage could give rise to a dynamic process whereby the critical loads themselves are adversely affected. That is, in the future, critical loads might begin to decline precisely because the 'optimal' amount of pollution was aimed at in earlier periods. This inconsistency of 'static' cost–benefit solutions with 'dynamic' cost–benefit solutions is not unusual. It does not make the economic approach irrelevant because one would still be interested in knowing at what rate to move to critical loads.

Finally for this very brief overview of the basic analytics, observe the CBA solution point in Figure 5.2 again. How would point CBA be obtained? There are various ways of achieving it. Environmental regulators could simply set an environmental standard so that no more than the optimal level of pollution depositions occurred. These 'ambient standards' would in turn imply limits to emissions ('emission standards'). Often, the standard is set not by direct emissions control or ambient concentration or deposition control, but by prescribing the kind of technology that can be used. This is the technology-based control option and the technology is often described as 'best available technology' (BAT). There are many variants of BAT, including 'BATNEEC' where the 'NEEC' means 'not entailing excessive cost'. These options comprise what is known as the 'command and control' approach (CAC) and it is the approach traditionally used in most countries to regulate pollution. Unfortunately, CAC is inefficient. The reason why is intuitively easy to see. Consider the BAT approach. All polluters have to adopt BAT regardless of any other options available to them. For example, BAT may be a flue gas desulphurisation (FGD) plant. But it may be cheaper to switch to low sulphur fuels, or engage in energy conservation, than to adopt an FGD plant. BAT tends to be inflexible and unnecessarily expensive. The same goes for most CAC measures – they leave the polluter with little choice about *how* to comply, and that is simply inefficient.

Notice that the same risk applies if 'clean technology' is prescribed. It is always important to leave polluters with options to meet given environmental standards. How they meet them should be up to them, not least because they tend to have far more information about their options and their costs than regulators have.

This unnecessary cost argument explains why economists tend to prefer 'market based instruments' (MBIs). MBIs take many forms – such as environmental taxes and tradeable emission permits. In terms of Figure 5.2, an environmental tax could be set equal to the distance ab. Consider the effects of such a tax on polluters. If they can abate pollution for less than the tax they can avoid the tax altogether and adopt an abatement measure. Note that *they choose* what measure to adopt: the regulator does not have to specify what is done. If, on the other hand, abatement is more expensive than the tax, the polluter will pay the tax. In this way, the optimal amount of abatement will take place. The tax secures the economic optimum. The same goes for tradeable permits. Here the idea is to issue permits for emissions, say one permit for one tonne of SO_x. The total number of permits can be set equal to the number necessary to secure the optimum in Figure 5.2. A pollution diffusion model will determine the links between target ambient quality and the emissions associated with that target. The number of permits is then set equal to that level of emissions, in, say, tonnes of sulphur or nitrogen oxides. In this way, the emission target is met because, in aggregate, emissions cannot exceed the number of permits. Now the permits need to be made 'tradeable' – i.e they should be bought and sold on the open market. The reason for this is that those polluters with high costs of abatement will be able to buy permits to pollute, avoiding the higher costs of control. But low cost polluters will have an incentive to sell their permits, which means they will have to abate emissions. In this way, abatement is concentrated in low cost sources, minimising the cost of complying with the regulation. MBIs generally have this property – they keep compliance costs down, but they do not sacrifice overall environmental quality.

Both CAC and MBI approaches 'internalise' the externality. What they do is to turn the external cost into a cost that the polluter bears, just like any other cost.

The basic analytical elements of the economics of pollution are now in place. We see that these comprise:

(a) estimating the economic value of pollution damage through willingness to pay studies;
(b) comparing this with the costs of controlling pollution;
(c) finding the optimum;
(d) adopting a market based approach for achieving the optimum.

This is the 'ideal'. In practice a great many problems arise.

5.3 Illustrating the economic approach – 'energy adders'

All energy production yields benefits in the form of deliverable heat, light and power. As a very rough first approximation, we can measure the benefits of energy supply by the price that is charged for it in the market place. This measure assumes that energy in sold in free markets, i.e. that they are not distorted by subsidies. Where such subsidies exist, and we know they are extensive in both the OECD and developing world, careful adjustment has to be made to estimate the true benefits.

All energy production also imposes costs on society. Those costs take two forms. The first is the cost of producing and delivering the energy. This is what most people usually understand by the 'costs of supply' or 'costs of production'. The second form of cost arises from the fact that energy production produces unwelcome side effects. This may be the visual intrusion or noise associated with wind power farms. It may be the acidic pollutants and particulate emissions associated with coal-fired electricity, the risks of health damage from routine radiation, the risk of a nuclear accident, the damage caused by a tanker oil spill, the impacts on biological diversity caused by tidal barrages, the risk of explosion from gas exploitation, and so on. This second category of 'externalities' has become known in the energy literature as 'adders' – because they are *additional* costs to the costs of production. The important thing about these additional costs is that, in the absence of regulation, they will be ignored by the producer of energy. The builder of the tidal barrage will not be concerned with the displacement of bird feeding grounds, for example. The coal-fired power station owner will not be concerned with the acid rain emissions he causes, emissions that become depositions both in the country of origin and, often, abroad as well (the phenomenon of 'transboundary' pollution). Similarly, the owner of a fossil fuel boiler will not be concerned with the carbon dioxide emissions that contribute to the possibility of global warming.

In an unregulated world, then, energy prices will be too low. For they will reflect the costs of production, including the desired profit of the energy producer. But they will not reflect the externalities. Yet the externalities are as 'real' a cost as the costs of production. Therefore an unregulated, free market system will 'fail' to achieve the most desirable allocation of energy resources. Not only will too much energy be consumed, the mix of energy sources will also, very probably, be wrong.

In reality, most of these externalities are regulated in one way or the other. There are national regulations, often based on the idea of adopting BAT to control emissions. There are regional regulations, such as the Large Combustion Plants Directive of the European Union. There are wider international regulations, such as the Sulphur and Nitrogen Protocols under the Convention on the Long Range Transport of Air Pollution. And now there is a global restriction on carbon dioxide under the Framework Convention on Climate Change, negotiated at Rio de Janeiro in 1992, and effective in March 1994. The existence of these regulations means that, in practice, market forces are not allowed a free rein. We regulate the externalities. The issue is, do we regulate them as much as we should? If not, then one way of providing additional regulation to reflect externalities is to impose some form of pollution tax as explained in Section 5.2.

There are also taxes on energy sources, taxes that invariably have nothing to do with the externalities, but which exist to raise revenue for the government. These taxes raise the price of energy, and reduce its consumption. However, these taxes are rarely related to the pollution impact of energy production and hence do not produce the right mix of energy sources. Moreover, although the arguments are quite complex, environmental taxes should be on top of, not instead of, general revenue raising taxes.

The correct mix of energy production should be determined by the production costs *plus* the external costs. This grand total is usually also known as the 'social' cost of production. The basic equation is therefore:

$$\text{Price} = \text{Marginal Production Cost} + \text{Marginal External Cost} = \text{Marginal Social Cost}$$

If the tax is to be related to the amount of damage done by each energy source, then there is an immediate problem. The tax is in money terms, i.e., money per kWh or per BTU, etc. The production costs are also in money terms. To relate the tax to damage therefore requires that we also estimate the pollution damage in money terms. This then takes us into the problem of placing money values on the pollution damage, or as it is often but misleadingly termed, 'valuing the environment'.

The rationale for finding monetary values of externalities is as follows. The costs of production of energy are expressed in money terms. The cash value of the resources used to produce energy reflects the scarcity of those resources. Generally, the scarcer the resource relative to the demand for it, the higher the price. So, production costs tell us something about society's perception of scarcity. The prices tell us something about the value of those resources if they were used in some alternative way, what economists call the 'opportunity cost'. In turn, that opportunity cost reflects society's valuation of the benefits of that alternative use, a valuation that reveals itself in the willingness to pay for the benefits yielded by those resources. So, the first observation is that production costs themselves reflect the willingness of society to pay for the benefits foregone by using scarce resources in energy production.

As shown in Section 5.2, the monetary value of the externality is also a reflection of society's willingness to pay, this time for avoiding pollution damage, or the amount of money that is required by way of compensation to tolerate the damage. In this way, all the components – demand, costs of production, and environmental impacts – are linked through the concept of willingness to pay. When economists 'value the environment', then, they are not engaged in some strangely immoral behaviour. They are measuring people's preferences for or against environmental change. In doing this they are seeking to allocate environmental resources in the same way that many other goods and services are allocated – according to market values. It is perfectly reasonable to argue that the environment should not be allocated in this way. Perhaps it is, in some sense, sacrosanct. If so, we might develop other rules for its allocation. But what those rules are is not obvious, and any rule faces as many problems as the willingness to pay criterion.

This discussion enables us to proceed to the issue of placing actual money values on environmental damage from energy production, i.e. to estimate 'externality adders'. As it happens, environmental impacts are not the only kind of 'adder' that is relevant. There are many non-environmental impacts that give rise to distortions in energy markets. For example, some sources of energy may need to be imported from areas of the world where there is risk of political disturbance. If resources are spent, on, for example, military installations in the area in an effort to protect those resources, then those expenditures can rightly be thought of as extra costs of that energy supply. Yet the military expenditures do not get reflected in the price charged for the resource, oil say. This is a further distortion: the oil price is effectively being subsidised by military expenditures.

Estimating adders involves some monetary estimate of the externality. In the case of energy security, for example, it might be the cost of maintaining armed forces in the Middle East or in readiness for use in that area. In the case of environmental externalities, monetary estimates involve the still controversial but now widely accepted procedure of estimating the willingness to pay for avoiding or cleaning up damage. This might be done by asking people (contingent valuation), by observing expenditures on travel to recreational sites (travel cost method), by looking at the way property prices respond to differences in environmental quality (hedonic property price method), and by using market or near-market values to value physical impacts such as crop losses, forest damage, buildings damage, etc. (the dose-response or production function approach).

Money estimates for adders to electricity systems have been estimated in a number of studies (e.g. Pearce *et al.*,[4] ETSU,[5] Resources for the Future[6]). We describe briefly below a procedure for estimating adders for the UK electricity system. These estimates develop those in Pearce *et al.*[4] and Pearce.[7] More detail for some of the pollutants can be found in CSERGE *et al.*[8]

The Pearce *et al.*[4] study is typical of what might be called 'top down' approaches in which some estimate of national damage is divided by total pollutant depositions to obtain a damage per unit of pollutant. These top down approaches may be contrasted with 'bottom up' approaches in which damage from a single emitting source – actual or hypothetical – is traced or simulated (ETSU[5], Resources for the Future[6]). Both procedures have shortcomings. The top down approaches clearly obscure impacts in one overall average. Bottom up approaches have little relevance for national damage estimates unless the single source can be regarded as being very typical of all such sources.

The procedure for estimating adders is first illustrated for conventional air pollutants. Results for carbon dioxide and nuclear power are reported only briefly.

5.3.1 Health: mortality

Pearce *et al.*[4] estimate health damage by electricity fuel cycle, but do not allocate the costs to pollutants. The following procedure yields an approximation of damage in p/kg of pollutant.

Damage costs for *mortality* are given as:

Dc = 0.32 p/kWh for coal-fired electricity
Do = 0.29 p/kWh for oil-fired electricity
Dg = 0.02 p/kWh for gas-fired electricity

We compute *total* mortality costs in the UK due to fossil-fuel electricity (Df) as:

Df = $\Sigma Di.Qi$ (Pearce[1])

where Qi is the output of the ith source of electricity and Di is the damage from the ith source. For 1991

Qc = 195 000 GWh, Qo = 23 000 GWh, Qg = 0

so that

Df = $\Sigma Di.Qi$ = £624m + £67m = £691m

Using shares of 13%, 50% and 37% (SO_2, NO_x, and TSP respectively) to allocate damage to pollutants (Norway Central Bureau of Statistics[9]), the resulting damages are:

SO_2 £90m
NO_x £345m
TSP £256m

Two adjustments are necessary.

First, we divide by (1990) total emissions from power stations to obtain:

SO_2 = £90m/2.72m tonnes = £33/tonne = 3.3p/kg
NO_x = £345m/0.78m tonnes = £442/tonne = 44.2p/kg
TSP = £256m/0.027m tonnes = £9481/tonne = 948.1p/kg

The second adjustment makes allowance for the fact that emissions and depositions differ due to the long range transportation of SO_2 and NO_x. The figures above are for damage per kilogramme deposited. A unit of emission of sulphur, for example, will result in a fraction of that emission being deposited in the UK. The remainder will be deposited in the sea or on other countries. Assuming damage to other countries is ignored, we multiply by a further factor of 0.36 for SO_2 and 0.13 for NO_x to allow for the ratio of depositions to emissions (CSERGE *et al.*[8]) to obtain:

SO_2 = £11.9/t = 1.19p/kg
NO_x = £57.5/t = 5.75p/kg
TSP = £9481/t = 948p/kg

The 'value of a statistical life' used in Pearce *et al.*[4] is £2 million. While this is warranted in light of a review of the literature, it is above that used in current UK government practice. A figure of £0.7 million would be consistent with UK Department of Transport practice, for example.

Assuming unit deposition damage gives rise to the same levels of health damage in Europe generally, the unit values of 3.3p/kg for SO_2 and 44.2 p/kg for NO_x need to be multiplied by 0.56 for SO_2 and 0.51 for NO_x to obtain European health damage arising from UK emissions. Note the rationale for including these wider geographic damages is that the UK is party to a number of European environmental agreements on transboundary air pollution. Europe is defined here as the UN Economic Commission for Europe region, i.e. it includes Scandinavia and the East European countries, as well as the western countries of part of the old Soviet Union. This gives the following results for UK adders:

SO_2 = 1.85p/kg
NO_x = 22.5 p/kg
TSP = 948 p/kg

5.3.2 Health: morbidity

In Pearce *et al.*[4] damage costs for morbidity are given as 0.12 p/kWh and are taken from PACE,[10] a study relating to the USA. These costs are directly attributable to particulates (TSP). This figure translates to £7.5 per kg of TSP. More recent work by Brajer *et al.*[11] and Hall *et al.*[12] suggests that morbidity damage from TSP is likely to be less than mortality damage. Evaluating the impacts of the control of TSP and ozone in the South Coast Air Basin centred on Los Angeles, they conclude that restricted activity days (RADs) and illness symptoms have an economic value of 50–70% of mortality gains. Since their work is the most thorough analysis to date we take the 50% value here, but the use of such figures raises the general issue of 'benefits transfer', i.e. the extent to which damage and benefit estimates in one context can be transferred to another context.

It is worth noting that the estimates here for health damage are very considerably below those quoted in ECOTEC[13] for health damages.

ECOTEC's estimates suggest something of the order of 185 p/kg (£1.85/kg) damages.

5.3.3 *Crop damage*

Pearce *et al.*[4] use a NAPAP estimate for the USA equivalent to 20 p/kg of NO_x for crop damages on the assumption that ozone damage can be ascribed to NO_x. The only adjustment required here is that which allows for the transboundary nature of NO_x. Multiplying by 0.13 produces a damage figure for the UK of 2.6 p/kg and by 0.51 gives 10.2p/kg for Europe. ECOTEC[13] suggest a UK figure of some 10 p/kg.

5.3.4 *Forests*

The damage figure of 60 p/kg SO_2 deposited comes from a study of forest damage in Europe by Nilsson.[14] This study is controversial in that it ascribes damage to SO_2 alone, whereas the role of acidic pollutants in forest damage is disputed – see especially NAPAP.[15] It also takes a fairly arbitrary multiplication factor of 2.7 of commercial timber values to obtain a total forest value inclusive of recreational values. Multiplying by 0.36 and 0.56 for the emissions/deposition ratio relevant to the UK and European damage figures gives 22 and 33.6p/kg, respectively. We reduce this further to 10 p/kg for the UK and 20 p/kg for Europe to lower the exaggerating effect of the recreational use 'multiplier'. This figure remains controversial. ECOTEC[13] suggest a figure of around 73p/kg.

5.3.5 *Buildings*

The original damage estimate in Pearce *et al.*[4] was based on provisional estimates of buildings damage in the UK by ECOTEC. Since the publication of Pearce *et al.*,[4] ECOTEC have revised their estimate of damage downwards in a significant way, from some £2.30/kg SO_2 deposited to some £0.37/kg (ECOTEC).[13] Taking this at face value, and multiplying by 0.36 and 0.56 to go from deposition to emissions gives a figure of 13.3 p/kg for the UK and 20.7 p/kg for Europe.

The results (rounded) are summarised in Tables 5.1 and 5.2.

To convert these to 'adders' we need to know the relationship between emissions and electricity produced. Table 5.3 provides information on emission coefficients. They relate to the whole fuel cycle and not just electricity generation. They are based on data from the UK Energy Technology Support Unit.[16] Table 5.4 shows the adders in p/kWh for selected main sources of electricity generation – coal, gas and nuclear. The effect is clearly to militate against conventional coal-fired plant, i.e. plant without gaseous emission control. The adder for a new coal fired

Table 5.1 Average unit damage costs, UK

	UK p/kg		
	SO_2	NO_x	TSP
Health:			
mortality	1.2	5.7	948.0
morbidity	–	–	474.1
Crops	–	2.6	–
Forests	22 adjusted to 10.0	–	–
Buildings	13.3	–	–
Water	na	na	na
Total	24.5	8.3	1422.1

Table 5.2 Average unit damage costs, UNECE

	UK p/kg		
	SO_2	NO_x	TSP
Health:			
mortality	1.8	22.5	948.0
morbidity	–	–	474.1
Crops	–	10.2	–
Forests	33.6 adjusted to 20.0	–	–
Buildings	20.7	–	–
Water	na	na	na
Total	42.5	32.7	1422.1

Table 5.3 Emission factors for UK fuel cycles

	Grams emissions/kWh			
Cycle	SO_2	NO_x	Part.	GHGs as Cequ
Old coal	14	5.3	0.16	352
New coal	1.2	2.7	0.16	296
Old oil	16.4	2.5	0.16	300
Gas	0.5	0.9	?	139
Nuclear	0	0.2	0	14
Landfill	0.5	0.9	?	139
Waste	4.1	2.7	6.4	?
CHP	0.6	1.3	0.08	150
Renews	0	0	0	0

Source: Adapted from Eyre[16]. Estimates for the nuclear fuel cycle are adapted from Table 8 of Eyre[16] on greenhouse gas emissions. CO_2 emissions from the nuclear fuel cycle are disputed. This controversy relates mainly to the pollution impacts from the *construction* of the generating station. No account is taken here of the relative emissions from plant construction other than through human health effects.

Table 5.4 Illustrative adders for 'conventional' air pollution in the UK and UNECE region

	p/kWh			
	SO_2	NO_x	TSP	Total
Old coal				
UK	0.34	0.04	0.23	0.61
ECE	0.59	0.16	0.23	0.98
New coal				
UK	0.03	0.02	0.23	0.28
ECE	0.05	0.08	0.23	0.36
Gas				
UK	0.01	0.01	neg	0.02
ECE	0.02	0.03	neg	0.05
Nuclear	zero	neg	zero	neg

Note: ECE includes the UK.

plant, i.e. with pulverised fuel and flue gas desulphurisation, is seen to be very much smaller.

In this way we can build up a picture of the adders that are relevant to each electricity system. Pearce *et al*[4] estimate adders for (a) carbon dioxide emissions, i.e. global warming damage, and (b) radiation hazards.

(a) Global warming. Since the publication of Pearce *et al*[4] further and more credible estimates of global warming damage have been provided by Fankhauser[17], reported in *The Social Costs of Greenhouse Emissions*. These have the effect of raising the CO_2 damage estimates from £5.8 to £17.3 per tonne carbon to a single best estimate of £14 per tonne carbon.* These estimates are based on assessments of likely economic damage from sea-level rise and other impacts of warming on crops, forests, etc. Recent scientific work also suggests that sulphur aerosols may actually have a *cooling* effect on global climate, offsetting to some extent the effects of the warming greenhouse gases. If so, then, strictly, sulphur emissions have a global warming *benefit*. No account of this is taken here but it clearly should enter into future extensions of the adders work – see Maddison.[18]

(b) Nuclear radiation. Radiation hazards are problematic since human health risks for routine radiation result in almost negligible adders for nuclear power. But the public's perception of radiation hazards from a major accident, especially since Chernobyl, mean that some account must be taken of the anxiety and concern shown. The 'objective' risk versus risk perception issue turns out to be important. Consider the objective risk of a nuclear power plant accident in the UK. Taking *design* risk factors for new

* Pearce *et al*[4] used an average figure of £11.55 tonne carbon in their summary tables. The adjustment reported here therefore raises the adder by a factor of 14/11.55 = 1.2.

PWRs such as that planned for Hinkley Point, and multiplying by the £2 million life value gives a 'nuclear accident adder' of 0.0000125 p/kWh, which is negligible. However, these risks are for accidents in which tens or hundreds of deaths occur because of the one accident. There is evidence to suggest that people do not value risks of 'group' accidents in the same way as they value individual deaths. Indeed, casual empiricism suggests this is so since of the many road deaths each year those that are reported by the media tend to involve only multiple deaths. This suggests that there should be some multiplication factor for 'group accidents' reflecting 'disaster aversion'. But just what the factor is, is open to debate and there appears to be no consensus in the literature.

The number of exceeded fatal cancers per reactor year for a degraded core accident are taken from NRPB. These give the second row in Table 5.5 (f for frequency, N for number). The last three rows of Table 5.5 show how various 'disaster aversion' functions might be used to reflect the fact that society tends to weight group losses more heavily than single deaths. The 'square rule' (row 3) is suggested by several authors but it is not based on empirical studies (nonetheless, one public utility is known to use it in its safety planning).

Analysis of the available evidence suggests that the risk of low probability, high consequence PWR accidents may be valued comparably with as much as about eight individual deaths per reactor year. At £2 million per statistical life, this would produce an adder of £16 m across 6000 m kWh per GWyr, i.e 0.27p/kWh.

In Table 5.5 (row 4) a disaster aversion function suggested by Rocard and Smets (R–S) is used. Self-evidently, the R–S rule gives higher values for 'damage' up to a value of N = 300. After that, the R–S rule gives *lower* damage values compared to the square function rule. If one reworks Table 5.5 with f.300N instead of the square rule, then the lives lost are very few, the highest value would be £0.21 m and the lowest £1000 or so. At £0.2 m the adder would be negligible at 0.0033p/kWh, and at £1 m it would be 0.016p/kWh. On the other hand, R–S is linear in N and this seems to offend the general intuition in the risk aversion literature. Looking at non-

Table 5.5 Disaster risk factors

f =	10^{-8}	10^{-9}	10^{-10}	10^{-11}
N =	11 000–35 000	46 000–150 000	110 000–350 000	180 000–580 000
fN^2	1–12	7–22	4–12	1–3
300 fN	0.03 to 0.11	0.01 to 0.05	0.003 to 0.01	0.0005 to 0.002
$f.N^{3/2}$	0.01 to 0.07	0.01 to 0.06	0.004 to 0.02	neg to 0.004

Figure 5.6 Provisional adders for the UK (p/kWh)

	Old coal	New coal	Gas	Nuclear
Conventional air pollution				
UK damage only	0.61	0.28	0.02	neg
All ECE damage	0.98	0.36	0.05	neg
Global warming	0.48	0.41	0.19	0.01
Routine radiation	0.02	–	–	–
Disasters	?	?	?	0.02–0.27
Sum				
UK only	1.11+	0.69+	0.21+	0.03–0.27
All ECE	1.48+	0.77+	0.24+	0.03–0.27*

Note: * assumes no contamination of countries other than the UK in the event of an accident.

linear functions, we illustrate a further possible function of N to the power 3/2. This produces similar small adders to the R–S rule.

Overall, then, a 'disaster aversion' adder remains very uncertain. Use of a square function could make the adder as high as 0.67p/kWh, but the square function has no apparent empirical basis. Use of a R–S function, which is linear in the number of people in the group accident, produces adders of 0.02p/kWh and perhaps as high as 0.05p/kWh. The issue is unresolved in the absence of fully fledged risk perception studies for the UK.

As far as nuclear accident costs are concerned, the health damage costs noted here do not take into account any property and output losses, nor any other economic costs of land sterilisation.

It is also important to note that 'major accidents' occur in other fuel cycles, notably with coal (mining accidents), oil and gas (offshore disasters, gas terminal explosions), and hydropower (dam bursts). These were not estimated in Pearce *et al.*[4]

Table 5.6 shows the results of combining the global warming and nuclear adders with those for conventional air pollution. The ranking is very much as one would expect: existing coal attracts the highest adder, new coal next, then nuclear and gas in equal positions if allowance is made for disaster aversion, and in the order gas then nuclear if aversion is ignored. But, of course, there are many caveats that can only be minimised with further research. We have omitted disaster costs from coal and gas; all the environmental damages are uncertain; non-environmental externalities are not estimated at all; the top-down approach probably exaggerates adders anyway; future private costs of generation are themselves uncertain, and so on. But the estimates do perhaps narrow the policy debate somewhat.

5.4 An application to recycling and waste disposal

The increasing 'scale' of economic activity has inevitably led to huge increases in waste in the form of low entropy material and energy flows. This large mass of redundant goods, by-products and a variety of other organic and inorganic residues must be re-used and/or disposed of somehow and at a cost. The environment has a large waste assimilation capacity, but it is not infinite. The inappropriate disposal of too much waste will generate a range of negative externalities. An additional, and often overlooked, problem is the cost of disposing of waste. This cost, running into billions of dollars for developed economies, is a drain on local government finance. In the developing world where the local tax base is often extremely small, waste disposal costs represent a serious financial burden.

Total waste arisings in the UK are estimated to be around 400 million tonnes of which mining and quarrying waste accounts for 27%. About a third of total waste arisings is defined as 'controlled waste' and includes household and commercial waste, some sewage sludge, demolition and construction waste and industrial waste. Special wastes in the UK (i.e. controlled wastes which are 'dangerous to life') are estimated to be around 2.5 million tonnes and growing. The remainder of the total waste flow is made up of material from agricultural premises and radioactive waste. In addition, some 40 to 50 million tonnes of hazardous waste are imported (for specialised treatment) into the UK each year. The modes of disposal in the UK are illustrated in Table 5.7.

Landfill is the main method of disposal for controlled waste accounting for 85% of the total, it also accounts for 70% of special waste; other disposal routes include incineration, sea dumping and physical or chemical treatment. About half of the sewage sludge produced is used as a soil conditioner and nutrient on farm land and horticultural areas.

In accordance with the Laws of Thermodynamics, production and consumption activities will always generate some amount of pollution and waste which will require proper (socially acceptable) disposal. Recycling systems are also not panaceas. Such systems impose their own costs and benefits on society and have theoretical and practical limitations in terms of the percentage of a given waste stream that they can utilise.

Nevertheless, society does have a choice over the total amount of waste that its economic system generates. Policy makers could decide to intervene in the economic process in order to change/modify production processes, products, or packaging and distribution methods. Policy makers should be searching for a socially acceptable balance between source reduction (often misleadingly referred to as 'waste minimisation'), re-cycling, reuse and final disposal of waste. Once this balance has been

Table 5.7 Disposal routes for waste in the UK

Waste category	Route	Percentage of total waste
Controlled waste	Landfill	85
	Incineration	4
	Sea dumping	4
	Other	7
Special waste	Landfill	70
	Incineration	5
	Sea dumping	10
	Other	15
Sewage sludge	Landfill	13
	Sea dumping	28
	Farmland	46
	Incineration	7
	Other	6
Imported hazardous waste	Incineration	29
	Solidification	7
	Physical/chemical treatment	58
	Other (including recycling and reprocessing)	6

Source: Department of the Environment (1992).

struck, policy makers would then have a set of consequential decisions to make about whether to recycle or dispose of the residual wastes that would continue to arise. This second stage of the waste management policy process essentially involves identification of the least-cost (social) configuration of recycling and disposal options.

The available evidence suggests that the industrial countries are finding it increasing difficult (and costly) to dispose of waste. In some countries – for example, the USA, Netherlands and Germany – there is a growing physical shortage, either regionally or nationally, of environmentally acceptable landfill disposal sites. For other countries, such as the UK, the 'shortage' of disposal sites (for landfill or incineration) has more to do with social constraints and the NIMBY (not-in-my-backyard) syndrome. National policy makers in the European Union (EU) face the added complexity that the European Commission and Council have laid down that 'the proximity principle' should be an accepted part of all member countries' waste management policy. Under the 'proximity principle' individual countries are encouraged to aim at self-sufficiency in waste disposal.

However, a simplistic and uniform application of this waste disposal principle across the EU would produce market distortions and inefficiency. This is because for some waste materials (secondary materials) there is an already functioning international market which facilitates recycling activities (collection, processing and utilisation as production inputs). From an economic perspective the 'proximity principle' should only apply to those materials for which there is no secondary materials market and consequent positive market price.

The full social costs of waste disposal have traditionally been disguised or underestimated. Thus in the UK, for example, the financial costs of landfill disposal continue to be relatively low (£5/tonne to £20/tonne for municipal solid waste) representing an under-pricing of the waste assimilating service of the environment. The full economic price for landfill should include all relevant costs; for example, pretreatment and proper pollution containment measures plus any remaining environmental impact costs. This economic price is much higher than the lower-bound financial cost.

Thus households do not typically pay the proper cost of disposal for their waste. They tend to be charged only for the financial operating costs of the collection and disposal system, rather than the full costs including site/facility environmental safeguarding costs. Further, many households do not pay the marginal cost (i.e. for each bag or bin of waste), only some notional average cost not tied to the amount of waste they generate. The result is that households face a marginal cost of refuse disposal equal to zero and demand too high a level of waste disposal service because they are not charged according to the amount of waste they produce.

Producers also fail to minimise and/or recycle as much waste as they would if they faced a fuller more efficient set of cost/price incentives. Industrial waste generators are controlled in terms of technology-based standards which lay down often quite rigid process/technique requirements and emission/discharge limits. In the European Union certain waste generators have to adopt the BATNEEC approach. The NEEC portion of BATNEEC means that the presumption in favour of BAT can be modified by two sorts of considerations, i.e. whether the costs of applying BAT would be excessive in relation to the environmental protection achieved, and whether they would be excessive in relation to the nature of the industry (e.g. its age and competitive position). The cost implications are usually interpreted (with no great consistency) as financial concepts, rather than as economic concepts, in which the wider social costs of pollution and waste disposal are included.[20]

The inefficient nature of such regulatory control is compounded by the information requirements that such a system imposes on the regulatory agency (paid for by the taxpayers); and by the fact that often the standards are imposed on a uniform basis, regardless of the actual pollution control cost situation that individual plants face. A good example of the uniform regulatory target-based approach to waste and pollution control is provided by the policies on packaging waste that a number of countries have, or will shortly, adopt. It has rapidly become clear that the cost implications of these measures are very significant.

The physical and thermal properties of the various types of waste such as calorific value, ash, moisture and bulk density provide a reasonable indication of the likely environmental response to processes such as collection, transport and incineration. Much more uncertainty surrounds the biodegradation processes in landfill sites and the transfer of pollution (leachate) plumes into surface water and groundwater around such sites.

All the available disposal options (landfill, incineration with and without energy recovery, sea dumping, composting and physical or chemical treatment) carry with them 'externalities'. The external cost effects include social costs such as disamenity – noise, smell, unsightliness – because of the presence of a disposal site/facility in or near to a neighbourhood; as well as air and water pollution, health impacts and congestion costs.

But the recovery of energy that occurs if methane is captured at a landfill site or from an incineration process is an example of an external benefit. The value of the energy is not an external benefit because it is already accounted for in the costs and revenues of the site owner, but the energy recovered will displace energy elsewhere in the economic system, e.g. by displacing power generated at an older and less efficient coal fired power station. The pollution avoided (i.e. that would have been emitted by the coal station) is then an external benefit of the methane capture or energy recovery at the landfill site or incinerator facility, respectively.

A number of factors contribute to the generation of external costs during the waste disposal process:

- the composition of the waste stream;
- the size of the disposal site, or facility;
- the physical characteristics of the disposal site;
- the age of the disposal site or facility;
- the spatial location of the site or facility;
- the degree of engineering (i.e. containment and abatement measures) adopted, or planned for new sites.

The net externalities from waste disposal (assuming energy recovery is practised) can be calculated in monetary terms as follows:

$$\text{Waste disposal externality} = \text{Site/fixed externality} + \text{Variable externality}$$

The site externality is 'fixed' in the sense that it is not closely related to the amount of waste passing through the site. There is a cost whatever the scale of the site. The variable externality is related to the amount of waste going through the site and therefore increases as the tonnage of waste increases.

The fixed element of the externality impact relates to the overall site or facility disamenity. The variable externality is made up of:

- global pollution costs (related to greenhouse gas emissions such as carbon dioxide (CO_2) and methane (CH_4)) +
- conventional air pollution costs (related to sulphur dioxide (SO_2), nitrogen oxides (NO_x) and particulates (TSP)) +

- 'air toxics' (such as dioxins) +
- water pollution costs +
- transport costs (air pollution, congestion and accidental costs) −
- displaced pollution damage (because of energy recovery systems).

Depending on the kind of waste to be minimised, economic incentives, in the form of say charges, could be levied on producers or consumers. Under certain conditions these two approaches would produce similar incentives. Consumers would internalise the increased cost of disposal by reconsidering their purchasing decisions, and by re-using or recycling some of their product purchases. Producers would not be able to pass on all their increased disposal costs to consumers and would therefore have to think seriously about recycling, process modifications and product re-design.

In principle, in the absence of government intervention in the form of an environmental quality protection/management policy, materials recycling or energy recovery will take place up to the level where the marginal cost of an additional unit of recycled material or recovered energy just equals its market value in a reclaimed condition. The condition for optimal recycling is:

$$MB_R = MC_R$$

or

$$[P_R] + [C_{DW}] = [C_{RE}] + [C_{CR}]$$

where

MB_R = the marginal (extra) benefits of recycling an extra tonne of waste
MC_R = the marginal cost of recycling
P_R = price of recycled material
C_{DW} = marginal social cost of disposal
C_{RE} = marginal external cost of recycling
C_{CR} = marginal cost of separate collection and processing for recycled material

Figure 5.3 illustrates the situation. Some amount of recycling (R) makes a financial profit (π) and the marginal profit function is $\delta\pi/\delta W_R$, where W_R is the quantity recycled. If there are only two disposal options, landfill (L) and incineration (I) and they are 'cost only' options, i.e. no revenues accrue, then the disposal cost functions are $\delta C/\delta W_L$ and $\delta C/\delta W_I$, where W_L is the quantity sent to landfill and W_I is the quantity incinerated. Landfill disposal is assumed to be less costly than incineration initially but becomes more expensive later due to perceived or actual scarcity of sites.

To find the optimal level of recycling (R), construct the aggregate disposal cost function $\delta C/\delta W_L = \delta C/\delta W_I$. The intersection, Y, of the aggregate disposal cost and the marginal profit function for recycling determines optimal recycling at OR^*. There is a financial loss, OX. The optimal amount of landfill is then L^*Q^- and the remainder R^*L^* is the optimal amount of incineration.

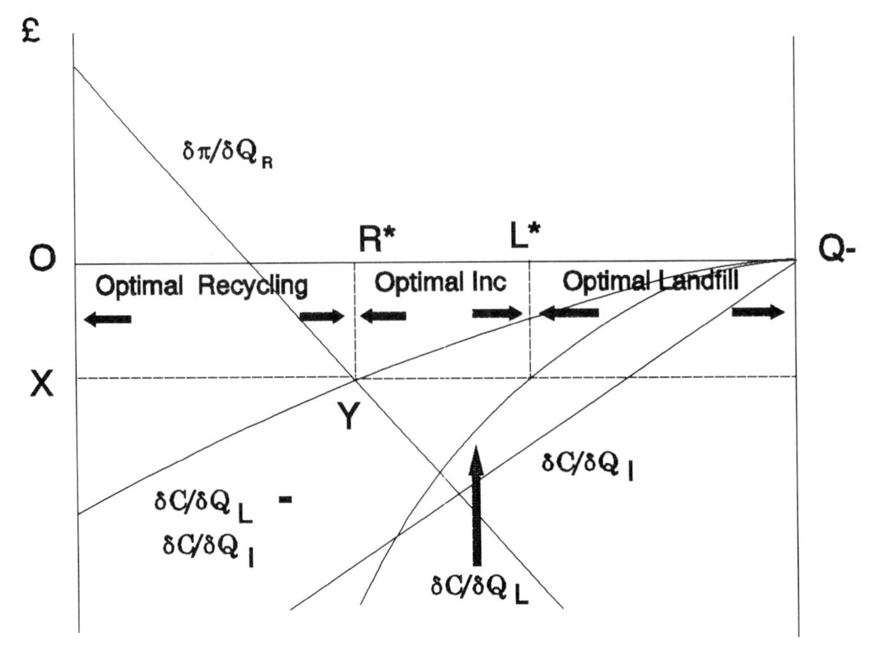

Figure 5.3 Optimal recycling and waste disposal.

Economic instruments have special attractions in the field of solid waste management. Since there are various options for waste disposal and for the reduction of waste at source, changes in the cost of one disposal route should encourage the diversion of waste to other disposal or recycling routes. The following instruments have been, at one time or another, under consideration, or have been implemented:

- recycling credits;
- landfill disposal levy/tax;
- packaging tax (an example of a product charge);
- tax concessions;
- deposit-refund systems;
- levy/tax on virgin raw materials;
- household waste charges;
- non-fossil fuel obligation.

We now take a closer look at the role which selected instruments could play.

5.4.1 Disposal taxes

The objective is to minimise the total social costs (TSC) of disposing of a *given* amount of waste W^* (the given amount is determined by prior policy on the benefits and costs of source reduction), i.e.

$$\min \text{TSC}\ (W^*) \tag{5.1}$$

Let there be three 'disposal routes':

recycling, R
landfill, L
incineration, I

Then

$$W^* = W_L + W_R + W_I \tag{5.2}$$

Forming the Lagrangean, G

$$\min \text{TSC}\ (W^*) =>$$

$$\min G = \text{TSC}\ (W_L) + \text{TSC}\ (W_I) + \text{TSC}\ (W_R) + \lambda\ (W^* - W_L - W_R + W_I)$$

where λ is the Lagrangean multiplier.

$$\frac{\partial G}{\partial W_L} = \frac{\partial \text{TSC}\ (W_L)}{W_L} - \lambda = 0$$

$$\frac{\partial G}{\partial W_R} = \frac{\partial \text{TSC}\ (W_R)}{W_R} - \lambda = 0$$

$$\frac{\partial G}{\partial W_I} = \frac{\partial \text{TSC}\ (W_I)}{W_I} - \lambda = 0$$

Now $\delta \text{TSC}(W_L)/\delta W_L$ is the marginal cost of landfilling an extra tonne of waste and can be written MSC_L. Following the same principle for other notation, we have:

$$\text{MSC}_L = \text{MSC}_I = \text{MSC}_R \tag{5.3}$$

'Decomposing' each element:

$$\text{MSC}_L = \text{MC}_L + \text{MEC}_L - (\text{PEN}_L + \text{MEB}_L) \tag{5.4}$$

$$\text{MSC}_I = \text{MC}_I + \text{MEC}_I - (\text{PEN}_I + \text{MEB}_I) \tag{5.5}$$

$$\text{MSC}_R = \text{MC}_R + \text{MEC}_R - (\text{P}_R + \text{MEB}_R) \tag{5.6}$$

where

PEN = price obtained for any energy generated in landfill and incineration

MEB = external benefits (see below)

MEC = external costs, e.g. CO_2 emissions from incinerators or landfill sites.

In the case of L and I, external benefits will comprise the *pollution displaced* by energy recovery systems. Thus, energy recovered from, say, an incinerator will displace energy produced somewhere else in the electricity grid system. The pollution from this displaced source is thus avoided, and this is a benefit.

Now, the landfill owner charges a price for landfill which should take into account the money he receives for energy recovered. So 'true' MC_L is

$$MC'_L = MC_L - PEN_L$$

Similarly for incineration and recycling

$$MC'_I = MC_I - PEN_I$$

$$MC'_R = MC_R - PEN_R$$

(Note that P_R is the price of recovered materials, not the price charged for recycling activity.) Also writing $MEC' = (MEC - MEB)$, i.e. *net* external cost, we have

$$MSC_L = MC'_L + MEC'_L \quad (5.7)$$

$$MSC_I = MC'_I + MEC'_I \quad (5.8)$$

$$MSC_R = MC'_R + MEC'_R \quad (5.9)$$

and optimality requires that

$$[MC'_L + MEC'_L] = [MC'_L + MEC'_L] = [MC'_R + MEC'_R] \quad (5.10)$$

If markets are in *private* equilibrium,

$$MC'_L = MC'_I = MC'_R$$

So social optimality requires that we *tax* each option by

$$t_L = MEC'_L$$
$$t_I = MEC'_I$$
$$t_R = MEC'_R$$

The nature of these taxes is not straightforward. Thus in the UK a recent report (CSERGE/WSL/EFTEC[8]) suggests that

MEC'_L = £1.3–4.1 per tonne (UK damage only)

MEC'_L = £1.0–4.1 per tonne (Europe damage)

For MEC'_I

MEC'_I = £2.0 to −2.3 per tonne (UK damage only)

MEC'_I = −£4.1 to −4.4 per tonne (Europe damage)

(note that $MEB_I > MEC_I$, which explains why $MEC'_I < 0$).

Prima facie these estimates suggest a tax on landfill of, say, £1–4 per tonne, and a *subsidy* to incineration of £2–4. An alternative would be to combine the tax and subsidy into one tax on landfill of, say, £3–8 per tonne. However, such a tax might not be efficient since the incineration externalities include a benefit item for displaced pollution. It would be perfectly legitimate to argue that this element of externality is best controlled by taxes on the source of the displaced pollution rather than letting it accrue as a subsidy to incineration. In short, there is a further step in the argument, namely that involving the translation of externality costs into workable taxes.

5.4.2 Product taxes

At the product level a pricing rule would be

$$P_{PROD} = MC_{PROD} + MEC_{PROD} \tag{5.11}$$

or

$$P_{PROD} = MC_{PROD} + t_{PROD} \tag{5.12}$$

where P is price, MC is private marginal cost, MEC is marginal external cost, t is a Pigovian tax and 'PROD' reminds us we are dealing with products.

Since the costs of disposal are excluded from P_{PROD} we can decompose MEC_{PROD} to

$$MEC_{PROD} = MDC_{PROD} + MEC_{DISP}$$

where

MDC_{PROD} = marginal disposal costs

MEC_{DISP} = marginal external costs of disposal

all normalised for, say, the weight of the product. To allow for recycling we require the resulting *product tax* t_{PROD} to have the following properties:

t must rise as $(MDC + MEC)_{PROD}$ rises,
t must fall as recycling/reuse rises,
t should encourage source reduction.

Note that in each case we wish focus to be on a tax that encourages the optimal amount of waste generation and recycling.

Consider a beverage container such that

W_i/L_i is the weight of the container per litre of beverage

r_i is the recycling rate

then

$$t_i = 100\,\frac{W_i}{L_i}.\,(1 - r_i).\,(MDC + MEC_{DISP})$$

and

$$r_i = 1 - 1/T_i$$

where T_i is the number of 'trips' a container may make.

For example, consider a PET plastic bottle. Data for the UK suggest:

W_i = 3 kg/100 litres
T_i = 1.05 or r_i = 4.8%
MDC = £20 per tonne = 2p/kg
MEC_{DISP} = £6 per tonne = 0.06p/kg
$\therefore\ t_i = 100.3/100.(1-0.048).(2.06)$
$= 3\ (0.95).(2.06) = \underline{5.9\text{p}/100\text{ litres}}$

Similar product taxes can be calculated for other beverage containers and detailed beverage container product taxes for the UK, Sweden, Japan, and South Africa are set out in Pearce and Turner,[24] Brisson,[21] Pearce and Brisson,[22] and Pearce and Brisson.[23]

5.4.3 Virgin materials tax (V)

Reinspecting the product tax formula and dividing t_i by W_i/L_i gives a virgin materials tax V_i where

$$V_i = 1/T_i\ (MDC_{PROD} + MEC_{DISP})$$

or

$$V_i = 1/1.05\ (2.06) = 1.95\text{p/kg}$$
$$= £19.5 \text{ tonne, for PET plastic.}$$

That is, a tax on PET plastic in its unformed state could replace the product tax.

5.5 Conclusions

Environmental economics offers powerful insights into the problem of pollution. The basic theory is coherent, though not without its problems. It would be unusual if there were no problems. It offers a way of identifying the 'correct' amounts of waste entering the environment, and it offers attractive mechanisms for their control. The cost of regulation is important. If polluters find control costs becoming a heavy burden they will react against them and lobby for their removal (or, more likely, obstruct their introduction). It makes sense, therefore, to look for efficient mechanisms of control that minimise compliance costs whilst maintaining environmental objectives. Market-based instruments have that attraction, and the examples illustrated here show how this approach might be applied in practice to energy pollution and to solid waste disposal problems.

References

1. Pearce, D.W. (1991) Economics and the environment. In: *Companion to Contemporary Economic Thought* (eds D. Greenaway, M. Bleaney and I. Stewart), pp. 316–342.

2. Pearce, D.W. and Turner, R.K. (1990) *Economics of Natural Resources and the Environment*, Harvester Wheatsheaf, Hemel Hempstead.
3. Tietenberg, T. (1992) *Environmental and Natural Resource Economics*, Harper Row, New York, 3rd edn.
4. Pearce, D.W., Bann, C. and Georgiou, S. (1992) *The Social Cost of Fuel Cycles*, HMSO, London.
5. ETSU (Energy Technology Support Unit) (1993) *Report to the European Commission on Social Costs of Energy*, CEC, Brussels.
6. Resources for the Future, and Oak Ridge National Laboratory (1993) *Report to the US Department of Energy on the Social Costs of Energy*, Resources for the Future, Washington DC.
7. Pearce, D.W. (April 1993) *The Economic Value of Externalities from Electricity Sources*, Paper read to Green College, Oxford, Seminar on Environment and British Energy Policy.
8. CSERGE, EFTEC and Warren Spring Laboratory (1994) *Externalities from Landfill and Incineration*, HMSO, London.
9. Norway Central Bureau of Statistics (1991) *Rapporter 91/1A: Natural Resources and the Environment 1990*, Oslo.
10. PACE (1990) *Environmental Costs of Electricity*, Oceana Publications, New York.
11. Brajer, V., Hall, J. and Rowe, R. (April 1991) The value of cleaner air: an integrated approach, *Contemporary Policy Issues*, **IX**, 81–91.
12. Hall, J., Winder, A., Kleinman, M., Lurmann, F., Brajer, V. and Colome, S. (Feb. 1992) Valuing the health benefits of clean air, *Science*, **255**, 812–817.
13. ECOTEC (1992) *A Cost Benefit Analysis of Reduced Acid Deposition: a Revised Approach for Evaluating Buildings and Buildings Materials*, ECOTEC, Birmingham.
14. Nilsson, S. (1992) Economic impacts of forest decline caused by air pollutants in Europe. In *The Economic Impact of Air Pollution on Timber Markets*, US Department of Agriculture, Asheville, NC.
15. NAPAP (National Acid Precipitation Assessment Program), *1990 Integrated Assessment Report*, NAPAP, Washington DC.
16. Eyre, N. (1990) *Gaseous Emissions to Electricity Fuel Cycles in the UK*, Energy Technology Support Unit, Harwell.
17. Fankhauser, S. and Pearce D.W. (June 1993) The social costs of greenhouse emissions, Paper to *OECD/IEA Conference on The Economics of Climate Change*, Paris.
18. Maddison, D. (1994) *The Shadow Price of Greenhouse Gases and Aerosols*, Centre for Social and Economic Research on the Global Environment, University College London.
20. Pearce, D.W. and Brisson, I. (Winter 1993) BATNEEC: The economics of technology-based environmental standards, with a UK case illustration, *Oxford Review of Economic Policy*, 24–40.
21. Brisson, I. (1993) Packaging waste and the environment: economics and policy, *Resources, Conservation and Recycling*, **8**, Nos 3–4.
22. Pearce D.W. and Brisson, I. (June 1993) Economic solutions to the packaging waste problem, Paper to *International Symposium on Packaging Waste*, Budapest, p. 10.
23. Pearce D.W. and Brisson, I. (May 1993) Using economic incentives for the control of packaging waste problem, Paper to *International Solid Wastes and Public Cleansing Association Conference on the German Packaging Directive*, Munich, p. 11.
24. Pearce, D.W. and Turner, R.K. (1992) Packaging waste and the polluter pays principle – a taxation solution, *Journal of Environmental Planning and Management*, **35** (1).

6 Introduction to clean technology

ROLAND CLIFT and ANITA J. LONGLEY

6.1 Clean technology and clean-up technology

There is a useful general distinction between:

1. *Remediation*: repairing damage caused by past human activity (or 'natural disasters').
2. *Clean-up technology*: reducing environmental damage by retrofitting, modifying, or adding 'end-of-pipe' pollution abatement measures to an established plant or process.
3. *Clean technology*: (or sometimes 'cleaner technology'): avoiding the environmental damage at source.

Remediation is considered in Chapter 4. To introduce the second part of this book, we concentrate here on the important distinction between clean technology and clean-up technology.

The concept of clean technology goes beyond 'clean production' which has been defined by the United National Environment Programme (UNEP: see Baas *et al.*[1]) as: 'a conceptual and procedural approach to production that demands that all phases of the life cycle of a product or of a process should be addressed with the objective of prevention or minimisation of short- and long-term risks to human health and to the environment.'

For reasons explored below, the discussion of clean technology throughout this chapter concentrates on the function of providing a human benefit or service, rather than concentrating on products *per se* as in the UNEP definition of cleaner production. However, the concept of the life cycle is central to both clean production and clean technology. Environmental life cycle assessment is introduced below, as a systematic approach to determining all the environmental impacts and resource depletions associated with providing the benefit or service.

To give a simple introduction to the distinction between clean technology and clean-up technology, we assume that the environmental impacts and resource depletions can be aggregated so that they can be represented by a single parameter. This parameter is called here the *environmental load*, but it is also sometimes known as 'environmental stress'. Just as for the environmental load, the financial cost of the activity must be based on the total life cycle cost, including the usual capital and

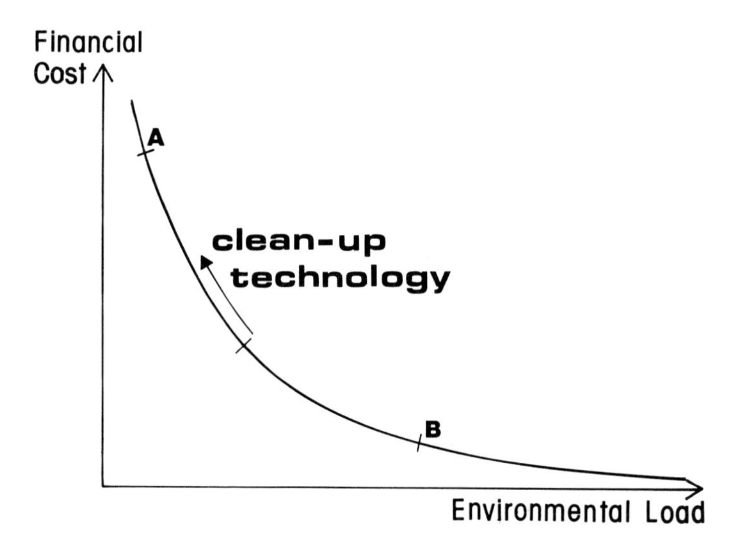

Figure 6.1 Clean-up technology and the efficiency of environmental improvements.

operating costs but also allowing for the cost of decommissioning plant and disposing of waste. For any given activity, the financial cost and environmental load are related by the kind of curve shown in Figure 6.1[2,3] representing the trade-off between financial cost and environmental load. The asymptotic behaviour of the curve at low environmental load arises from the thermodynamic constraint that any human activity involves some environmental impact or resource utilisation, so that the environmental load cannot be eliminated completely. Similarly, the asymptotic behaviour at low financial cost simply reflects the non-existence of free lunches: even the most environmentally profligate activity has a financial cost.

Clean-up technology (or end-of-pipe pollution abatement) by definition involves adding something to the process, or treating or reprocessing the product which provides the benefit or service. Thus it involves moving around the curve in the sense shown in Figure 6.1, increasing the financial cost to reduce the environmental load. The gradient of the curve is inversely proportional to the efficiency of expenditure on environmental improvements. In general, OECD countries will be represented by points around A, corresponding to high cost but relatively low environmental load. By contrast, the Eastern European countries, for example, are represented schematically by points around B, where the environmental load is higher but the efficiency of expenditure on clean-up technology is also much higher. To take the specific case of electricity generation from fossil fuel, the environmental load parameter will be a measure primarily of emissions of acid gases and, for those who subscribe to the view that carbon dioxide causes global warming, of 'greenhouse gases' (see Chapter

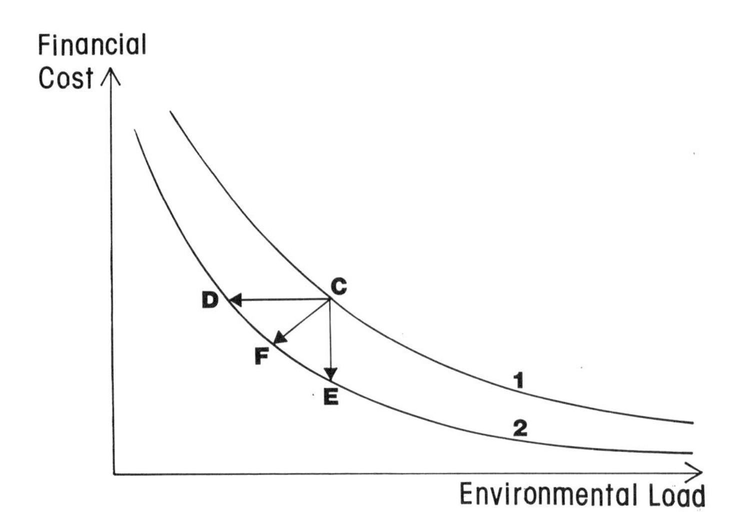

Figure 6.2 Clean-up technology (Curve 1) and clean technology (Curve 2).

2). For example, it has been estimated[4] that expenditure on improving the environmental performance of power generation and use will achieve many times more reduction in environmental load by investing in Eastern rather than Western Europe. In other words, for this particular environmental load, the gradients at points A and B differ by orders of magnitude. This kind of analysis can be developed further, for example as an aid to formulating international policy[5].

Figure 6.2 illustrates the essential difference between clean-up technology and clean technology[3,6]. As in Figure 6.1, Curve 1 represents an established technology to which clean-up can be applied. Curve 2 represents a significantly different technology, which is inherently cleaner in the sense that it is represented by a curve displaced towards the origin. Thus an organisation operating at point C with the established technology can in principle, by adopting the clean technology of Curve 2, reduce its environmental load without increasing costs (point D), or reduce costs while retaining environmental performance (point E), or reduce both cost and environmental load (point F). Whether this is practicable depends, of course, not only on the availability of the clean technology represented by Curve 2 but also on whether there are any barriers to making this change, for example due to existing investment in technology 1 or investment needed to exploit technology 2. Other organisational barriers to the adoption of clean technologies are discussed later in this chapter.

Within this context, general features can be identified which distinguish clean technologies from clean-up technologies. Illustrations of these features will be found below and in the following chapters. For example, in the process industries, a clean-up technology may capture a potential emission from a process and transform it to another state, for example by 'scrubbing' or filtering a gas stream to collect an atmospheric pollutant as liquid or sludge or solid for disposal by other means. A clean technology will avoid producing the pollutant in the first place, or will recycle it within the process.

6.2 How clean is the technology? Environmental life cycle assessment

The UNEP definition of clean production, quoted above, explicitly refers to the 'life cycle of a product or of a process'. Environmental life cycle assessment is a formal approach to defining and evaluating the total environmental load associated with providing a service, by following the associated material and energy flows from their 'cradle' (i.e. primary resources) to their 'grave' (i.e. ultimate resting place, as solid waste or dispersed emissions). Life cycle assessment (LCA) is increasingly being used as a decision support tool in improving environmental performance, primarily by reducing the environmental load associated with specific products,[7,8] with an accepted methodology[9,10] and recognised internationally as the essential basis for awarding 'ecolabels' as a public recognition of products or services with improved environmental performance[11]. LCA aspires to be an objective quantitative approach[12,13]. The idea behind LCA is that, by considering all activities 'from cradle to grave', it is possible to determine whether a product or service genuinely causes reduced environmental load, or whether the environmental load is merely transferred from the immediate supplier to the 'upstream' suppliers or to 'downstream' disposal. LCA must be used to establish the environmental load used as a parameter in Figures 6.1 and 6.2.

Environmental life cycle assessment is a form of environmental system analysis. The basic concept of LCA (and of other forms of analysis, including site-specific environmental impact assessment, EIA) is summarised in Figure 6.3. The productive system which provides the function (i.e. the benefit, service or product) is identified. In order to provide the function, the system will require inputs of materials and energy. It will also generate undesirable outputs in the form of emissions to air and water and solid wastes. For site-specific EIA, the system boundary is drawn around the manufacturing plant, so that the productive system takes the simple form shown in Figure 6.4. However, this level of analysis cannot represent the total environmental load: providing the inputs of energy and materials must involve some environmental loads, while if the output is a tangible product it will eventually become a waste. Environmental life cycle assessment attempts to account for these 'upstream' and 'downstream'

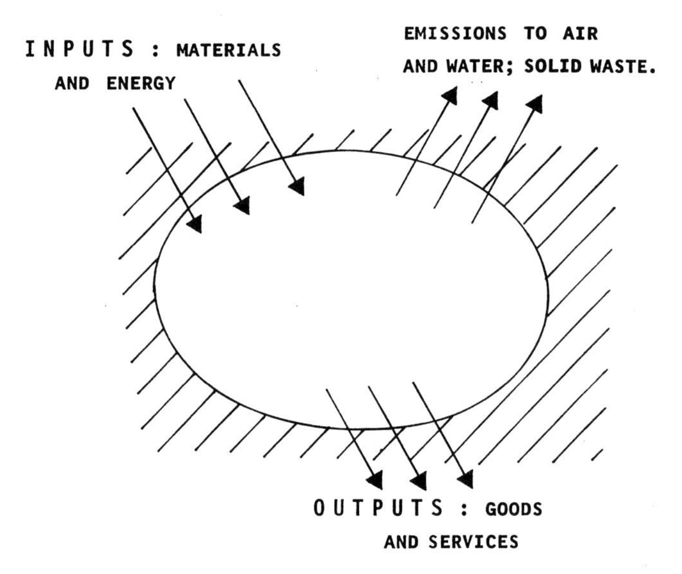

Figure 6.3 Environmental system analysis (after Azapagic and Clift[23]).

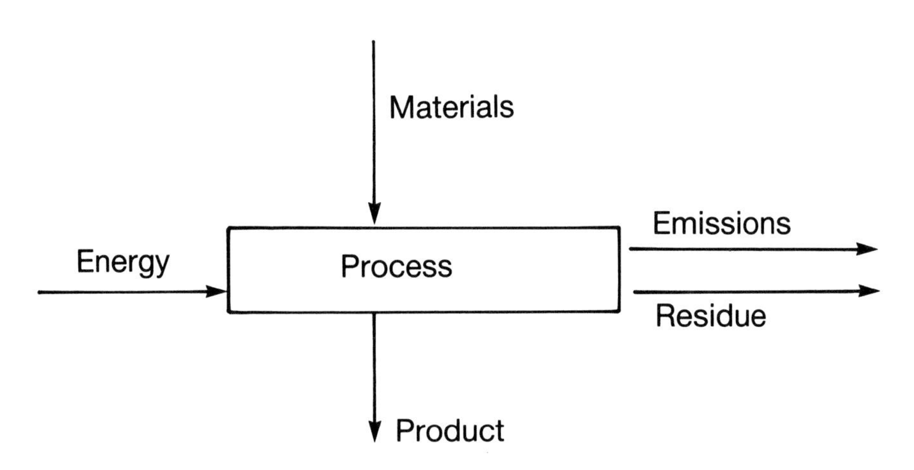

Figure 6.4 Site-specific environmental impact analysis.

environmental effects, as well as those directly associated with manufacture. For life cycle assessment, the system boundary must therefore be drawn around the whole life cycle of the materials and energy flows used to provide the function. The system now takes the general form shown in

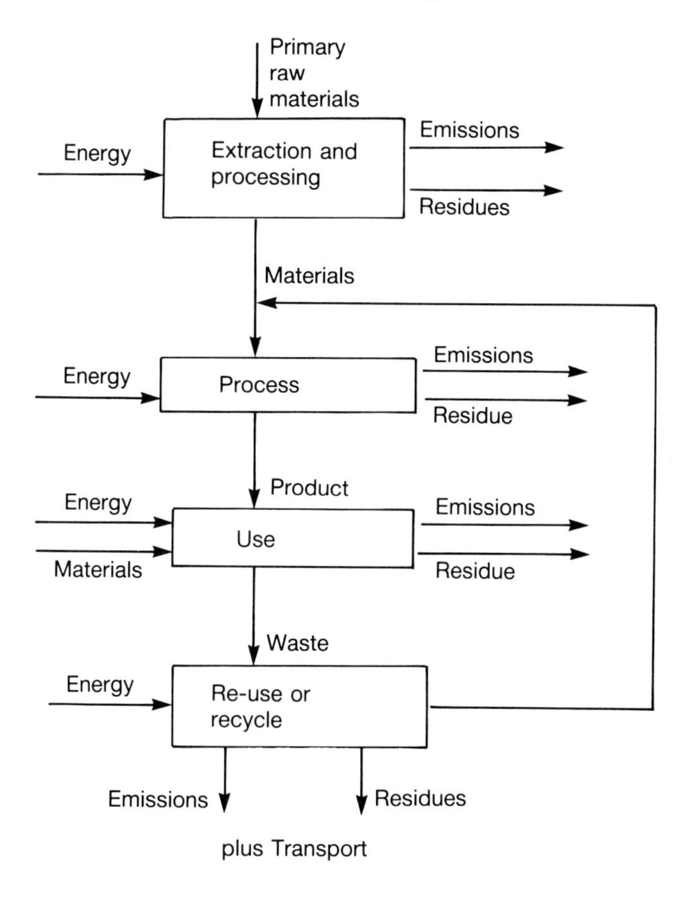

Figure 6.5 Environmental life cycle analysis.

Figure 6.5, and must also include the environmental effects of the transport steps involved.

The formal procedure of carrying out and applying an LCA involves the following steps[b]:

1. Goal definition. In terms of Figures 6.3 to 6.5, the first step is to define the boundaries of the system to be studied. At this stage the *functional unit* must also be defined: the unit of service on which the analysis is to be based. This is not necessarily a quantity of material or a number of manufactured items. For the example of packaging, the functional unit must be defined as a quantity of material packaged, not as a number of packets or a fixed weight of packaging material. When LCA is used to compare completely different ways of providing the same service, then definition of the functional unit can be difficult or contentious.[11,14] In any case, the goal definition step is critical, because it can determine or even prejudge the outcome of the study.

2. Inventory. Next, the inputs of primary resources and the outputs of emissions and solid residues must be defined and quantified. In effect, this amounts to carrying out a material and energy balance over the system defined at the goal definition step, allowing for flows of trace components which may have significant environmental impact. The result of this step is an *inventory table* which quantifies the inputs and outputs per functional unit provided.

3. Classification. Once the inventory table has been compiled, the contributions of the inputs to and environmental outputs from the system (see Figure 6.3) to recognised environmental and health problems are quantified. For example, all atmospheric emissions which could contribute to global warming are weighted according to their 'greenhouse warming potential' (see Chapter 2). By this process, the detailed data in the inventory table are aggregated into a much smaller number of so-called *effect scores* which define the environmental profile of the system.

4. Valuation. Ideally, the effect scores are now weighted to give a simple estimate of the environmental effect of delivering the function; in other words, to determine the environmental load parameter shown in Figures 6.1 and 6.2. This step inevitably requires relative values to be placed on completely disparate environmental effects and resource usages.

5. Improvement analysis. Ideally, when LCA is used within a company, the result is used to identify changes in the product or productive system which will reduce the total environmental load.

Of the five steps above, goal definition and inventory analysis are established processes (although some problems of detail remain to be resolved in inventory analysis; see e.g. Udo de Haes *et al.*[14]). Classification is less well developed, but standard methodologies are emerging (see e.g. Guinée *et al.*[13]). However, serious difficulties arise in valuation. The original aspiration in developing LCA as a decision-support tool in environmental management was that the same relative weightings could be assigned to different environmental effects wherever they occur, i.e. that valuation could be carried out on a 'global' basis. The environmental economics approach set out in Chapter 5 can be regarded as one way to address valuation. However, it is increasingly becoming clear that valuation must be carried out locally rather than globally.[14] Thus the 'environmental load' parameter in Figures 6.1 and 6.2 actually conceals a substantive problem in environmental management.

Overall, environmental life cycle assessment is a complex procedure and is not fully developed. Nevertheless, the discipline of setting out the life

Table 6.1 Distribution of environmental loads through life cycle of a typical machine for washing clothes (UKEB[15])

Categories of environmental load:

1. Energy consumption.
2. Air pollution.
3. Water pollution.
4. Solid waste.
5. Water consumption.

Typical contribution of stages in life cycle (as percentages):

	1	2	3	4	5
Production	4.1	1.5	3.7	7.2	2.1
Distribution	0.3	0.1	0.7	0.6	0.1
Use	95.5	98.4	95.6	87.2	97.8
Disposal	0.1	0.0	0.1	5.0	0.0

Hence 'key criteria' indicating environmental performance:

A. Energy consumption during use.
B. Water consumption during use.
C. Efficiency of use of detergent.
D. Recyclability of materials.

cycle, from cradle to grave, is an important part of identifying where effort should be applied to make a technology cleaner and which stages in the life cycle should be changed or eliminated. A specific example is summarised in Table 6.1: an LCA study carried out to establish ecolabel criteria for washing machines.[15] Four stages in the life of a machine were considered, with the environmental loads 'classified' by aggregating into five broad categories. In this case, it is clear that the great majority of all the environmental loads arise at the use stage, so that valuation to combine the five categories of environmental load is not needed in this case. Furthermore, most of the air pollution and much of the solid waste arise from generating the electric power which the machine uses. Therefore the analysis leads to the four key criteria, A to D. Criteria A to C indicate the environmental efficiency in use, while criterion D recognises that the solid waste associated with manufacturing and disposing of the machine is not negligible. Because the environmental load is dominated by the use stage, it follows that a clean technology approach to cleaning clothes will concentrate on the actual cleaning process rather than on the equipment in which the cleaning is carried out. We return to this example below.

6.3 Services and commodities

The essential distinction between providing a service or human benefit rather than supplying a product or artifact has already been introduced, as the basis for defining the 'functional unit' in environmental life cycle assessment. This distinction is central in developing the idea of clean technology, and arises with remarkable ubiquity. The differences will now be illustrated by some simple examples.

Herbicides. For the first example, we are indebted to Dr Geoff Randall, of Zeneca, for pointing out a simple calculation by Corbett *et al.*[16] Herbicides used to control wild oats, a common weed in cereal crops, typically are recommended for application at rates of the order of kilogrammes per hectare. However, the quantity of herbicide actually needed – in the sense of entering the system of the wild oat seedlings, to kill them – is estimated as a few microgrammes per hectare.[d] In other words, we apply more than 10^9 times as much of the herbicide as reaches its target. If the functional unit is taken as unit mass of herbicide, then the emphasis will be based on manufacturing and distributing the herbicide, i.e. on cleaner *production*. However, the functional unit should be taken as a number of wild oat seedlings killed, or an area of cultivated land treated or, perhaps most appropriately, a quantity of cereal crop produced without damage by wild oats. The calculation is now based on the whole weed control system, and it becomes clear that the limiting step is controlling the weed rather than producing the herbicide. Thus a cleaner *technology* might focus on selective delivery, or on overcoming the wild oats' defences against systemic toxin, or on pesticides whose action is triggered by contact with the pest, but not on production *per se*. Put starkly, there is little benefit in containing emissions during production if the herbicide is subsequently, quite literally, sprayed all over the countryside. One of the most significant environmental developments in agrochemicals over the years has been reduction in application rates rather than cleaner production.

Organic solvents. The idea of providing a service rather than a product can also be illustrated by developments in the use of industrial solvents, for example for degreasing metals. Traditional organic solvents, which are commonly chlorinated, can represent a substantial environmental load. One trend in recent years has therefore been towards use of water-based solvents. However, the environmental load arises not from the use of organic solvents but from their release or escape. An alternative approach to environmental improvement is therefore to concentrate on 'closing the system' to contain the solvent completely and to reprocess it for re-use. This is properly a clean technology, because it reduces environmental load and is economically attractive, even though the degree of technical innovation involved may be relatively small. The innovation lies primarily in the business practice. Rather than being sold to the consumer, the solvent – in this case kerosene (see below) – is leased to the user and then taken back for re-processing, sometimes with the user paying a premium for any material lost. Equipment for using and containing the solvent is also leased to the user. Thus the life cycle has been closed in the commercial as well as the material sense. The supplier retains responsibility for the material from cradle to grave. Sometimes, the solvent residue left after reprocessing may be used as fuel (see below). Thus the shift to what Giarini and Stahel[17] have called 'the new service economy' (see below) is virtually complete: the service or use is traded, not the commodity. This particular case is explored in more detail below, as an example of industrial ecology.

Cleaning clothes. To take a further, partly hypothetical, illustration, we will consider the service used above as an example of LCA: removing soil from human clothing. The basic process is a simple separation, shown schematically in Figure 6.6. The soil is mainly animal fats excreted by the body, plus whatever external dirt the clothes have acquired. If the solid could be kept together, it could be used – for example as food for chickens or to make candles. However, the familiar process for washing clothes does not do this; it disperses the soil into the water system, with the aid of hot water and detergents (see Figure 6.7). It was noted above that the environmental loads associated with washing machines arise primarily in use, and Figure 6.7 shows why: effect chemicals (the components of the detergent) and energy are used once, and then emitted as waste. This represents an example almost as stark as that of pesticides: while it is important to contain releases to the environment from detergent manufacture, there is an inconsistency because the product is subsequently flushed down the drain with hot water. A clean technology approach to cleaning clothes will therefore concentrate on the function of separating soil from fabric, rather than on clean production of detergents.

An alternative approach is suggested by a different technique: dry

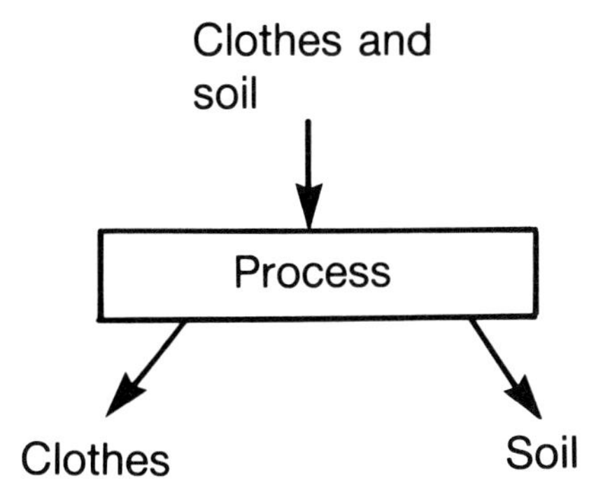

Figure 6.6 Basic operation of cleaning clothes.

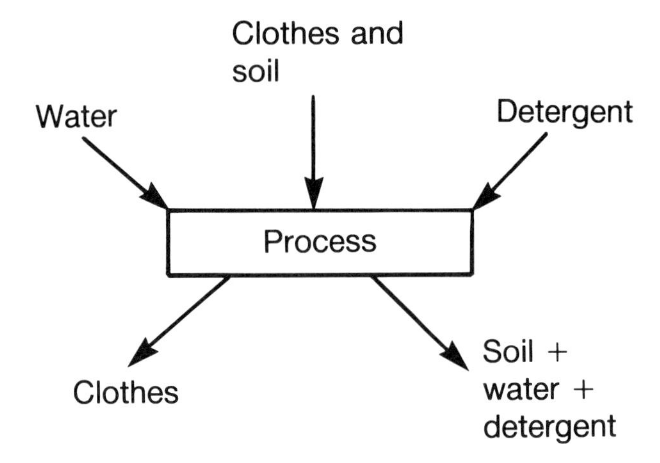

Figure 6.7 Traditional approach to washing clothes (schematic).

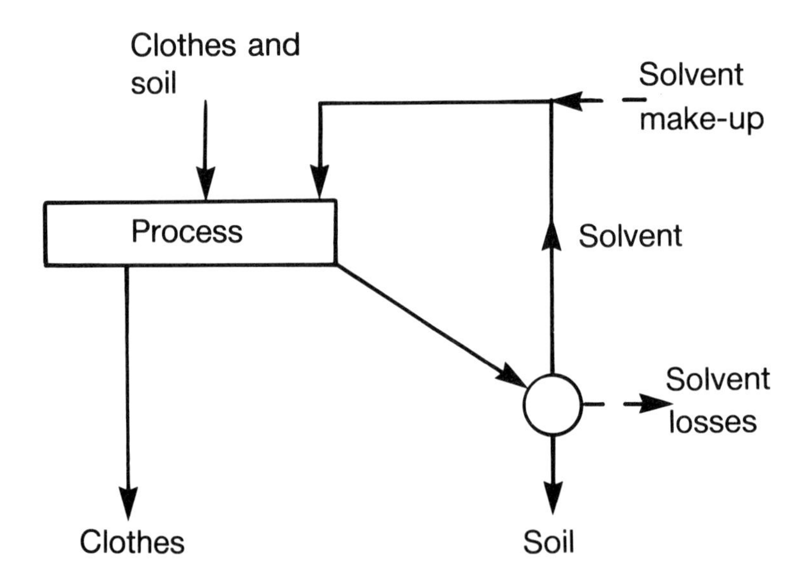

Figure 6.8 Dry cleaning of clothes (schematic).

cleaning. In this case, a solvent is used to detach the soil (see Figure 6.8). Although dry cleaning is not usually seen as environmentally benign, the environmental load arises not because organic (and often chlorinated) solvents are used but because they are emitted. Hence, just as for metal degreasing, dry cleaning can be made a clean technology by containing the solvent, separating it from the soil (for example by evaporation), and reusing it.

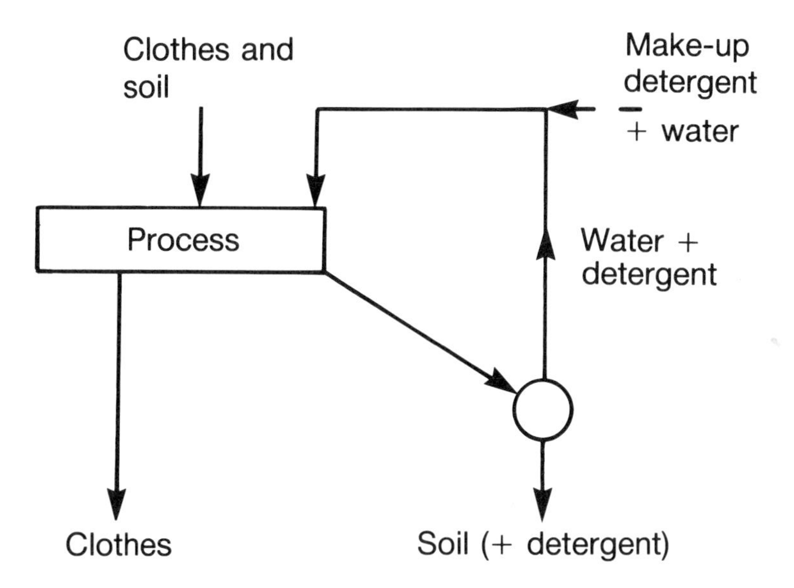

Figure 6.9 Cleaner technology for washing clothes (schematic).

With this example in mind, a cleaner approach to wet washing might be as shown in Figure 6.9. If the soil is separated at source from the water waste, perhaps using a membrane process, then the hot water plus unused detergent can be reused, while the wet effluent keeps the soil concentrated, perhaps with sufficient surfactant to keep it in micellar form. This difference in technology now opens up a number of further possibilities, which go beyond purely technological changes:

1. The detergent chemicals could now be replenished, rather than added in fixed proportions for each wash. The machine would then incorporate dispensers for the main groups of agents making up a detergent, so that their quantities and proportions could be selected for the load according to the nature of the fabric and the soil.
2. If the soil stream is concentrated, then dedicated water treatment at source can be considered. This suggests in turn that fabric cleaning might return to being centralised within a local community, rather than distributed throughout individual households. Signs of this trend can be found in the revival in some countries of 'diaper services' instead of disposable or home-washed diapers.
3. If 'fabric care' becomes centralised, then it becomes possible to consider further approaches to achieving the basic separation of soil from fabric (Figure 6.6). Super-critical fluid solvents would reduce the energy requirement to separate solvent from soil (Figure 6.7), opening

up the possibility of completely different approaches to providing this particular service.

This example has deliberately been pursued beyond current practice, to demonstrate that clean technology depends at least as much on rethinking commercial and social habits as on introducing technological developments. For example, as developed here, a clean technology for cleaning clothes would introduce completely different relationships between the manufacturers of detergents and washing machines: they would combine to provide the service, rather than each selling their own product.

6.4 Materials reuse: the new industrial ecology

One of the goals in clean technology is to provide a service with minimum consumption of energy and primary materials, a goal which has been called *dematerialisation of the economy*[17,18].

Some reductions in consumption can be achieved by *life cycle containment* (with minimal and controlled purge from the cycle of recovery and reuse, as in the examples above of degreasing solvents and Figure 6.8). Prolonging the service life of machines represents an analogous approach for services which are provided by manufactured items. However, given that service life will remain finite, clean technology requires a systematic approach to reuse of materials and components. Like the example of cleaning clothes explored above, this approach implies changes not only in technology but also in commercial relationships. The term *industrial ecology* has been coined to denote a relationship in which systems providing different services and products 'metabolise' successive uses of materials and energy.

Beyond recycling: cascades of use. The 'once-through' use of resources which characterises the profligate behaviour of much human activity is represented by the 'open loop' system of Figure 6.10. An ideal 'dematerialised' system would be represented by the 'closed loop' system of Figure 6.11. However, this popular view of recycling is obviously a simplistic misconception. Very rarely does reuse or recycling entail no resource consumption or environmental load. Even reusable drinks containers (such

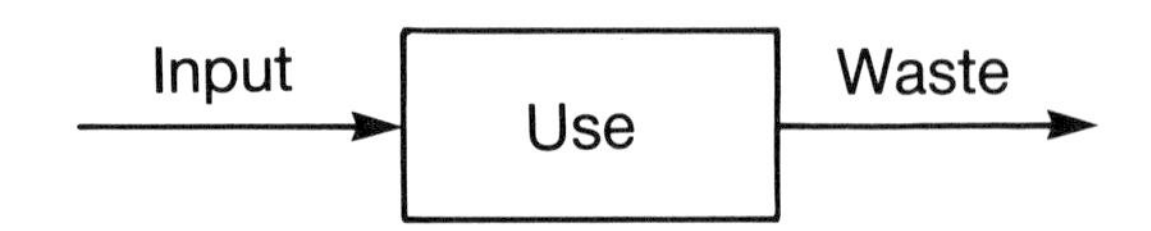

Figure 6.10 'Once through' use of resources: open-loop system.

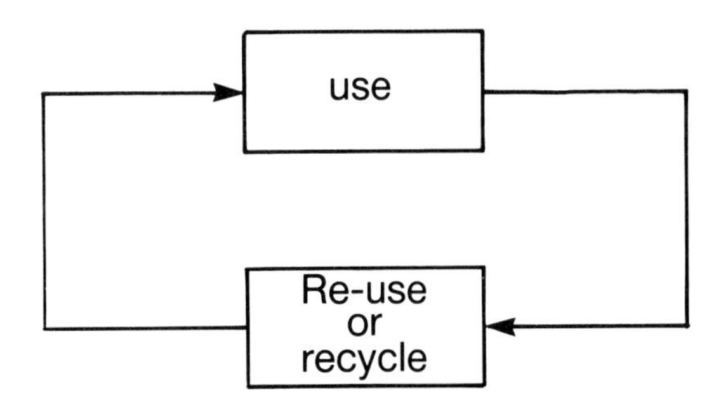

Figure 6.11 Idealised view of recycling: closed-loop system.

as the traditional British milk bottle) must be transported, washed and sterilised before refilling. Furthermore, a completely closed loop is impossible; even in the everyday (at least in the UK) example of the milk bottle, bottles can be broken or chipped. Therefore a real closed-loop recycling system must take the general form shown in Figure 6.12. It is now necessary to enquire whether the resource depletion and environmental load entailed in recycling may offset the benefit of material recycling. In some cases, particularly involving products from renewable resources, recycling may arguably be environmentally damaging; i.e. recycling is not necessarily a clean technology, or even a best environmental option. To take a specific example, it has been argued that paper derived from sustainably farmed forests should be burned as a renewable biofuel, not recycled as a waste material.[3,19,20]

Recognising the practicalities of reuse and recycling leads to the general concept of industrial ecology shown in Figure 6.13. On its passage through the human economy, a material will ideally pass through a series of uses, usually with progressively lower performance specifications (see e.g. Stahel and Jackson[21]). This general concept is familiar in process engineering, exemplified by a counter-current washing or extraction cascade. Industrial ecology extends the concept to encompass more than one process. There may be some recycling around any step in the cascade, and it could be desirable to reprocess the material to raise its specification so that it can be returned to a higher level in the cascade. Ultimately, however, the material will leave the economic system as waste or, if combustible, as fuel. For example, paper products and hydrocarbon-based polymers should be regarded as materials which pass through one or more uses on their way to being used as fuels.[3]

Recognising that a material should pass through such a *cascade of use* introduces another component of clean technology: avoiding contaminating materials with unnecessary additives which will increase the environmental loads associated with subsequent uses. An obvious example is eliminating toxic heavy metals from printing inks, recognising that paper

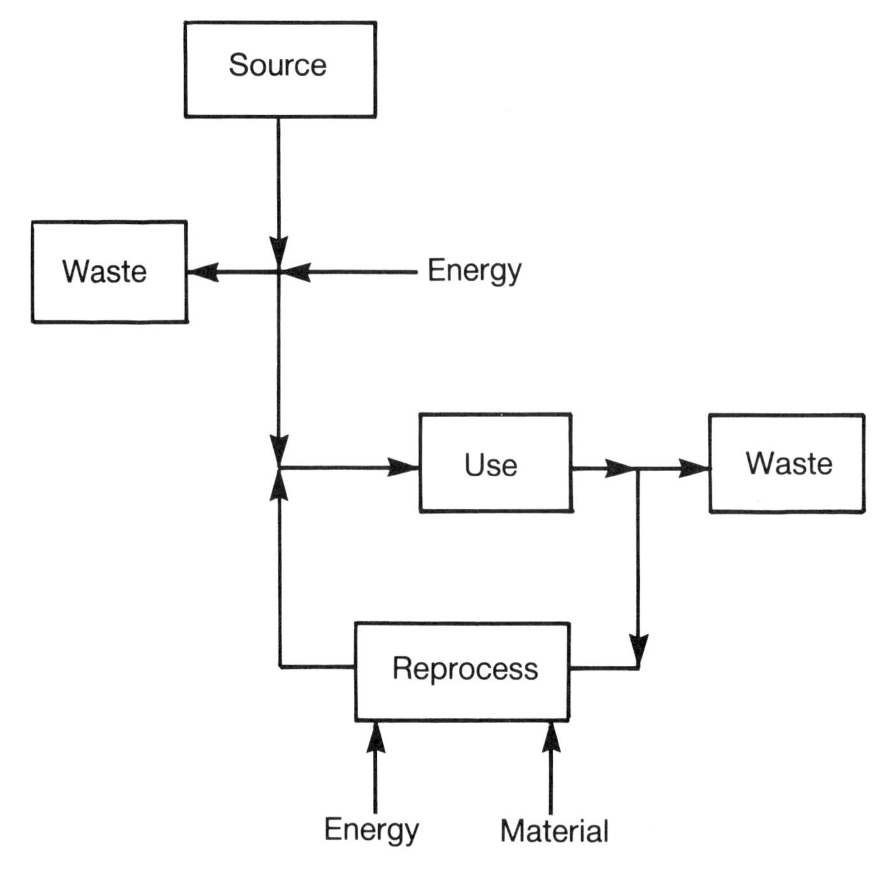

Figure 6.12 Real 'closed loop' recycling system.

should ultimately be burned. *Life cycle design* goes beyond environmental life cycle assessment, to designing the whole use cascade rather than optimising a single use of the material.

Example 1: organic solvents. The example introduced above, of reprocessing degreasing solvents, can be examined in more detail as an example of the development of an industrial ecology and the associated commercial relationships. The following account is a simplified version of the development described by Roberts and Lewis.[22]

The business of leasing and reprocessing kerosene used as degreasing solvent is shown schematically in Figure 6.14(a). The leasing relationship closes the loop in the life cycle, so that the used solvent is returned for reprocessing and subsequent re-use. At this early stage of the commercial relationships, the residue from reprocessing must be incinerated. However, the residue contains a large proportion of kerosene, in addition to

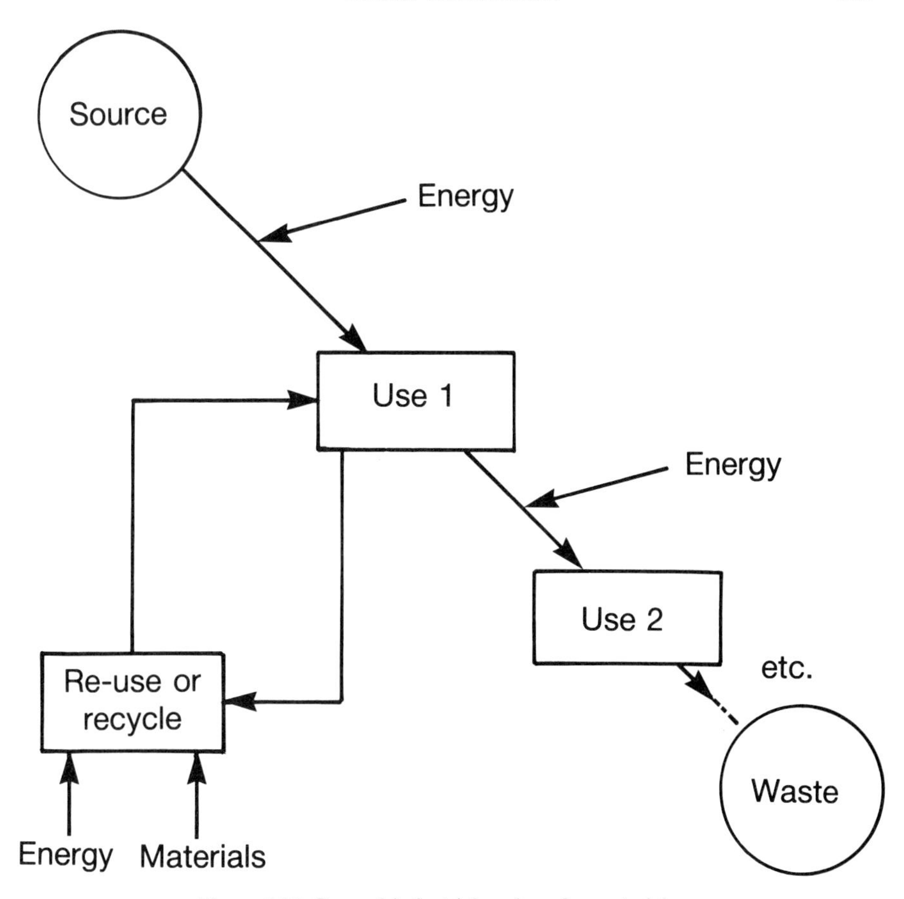

Figure 6.13 General industrial ecology for material use.

aqueous and solid components removed in degreasing, and therefore has a substantial fuel value. As a first development, it can therefore be used as a specification fuel, for example in cement kilns, in this case with the same supplier providing the fuel system to the cement manufacturer – see Figure 6.14(b). Thus the solvent supplier has retained ‘duty of care’ for the organic material throughout its life cycle.

This concept can be expanded further, as shown in Figure 6.14(c). Other organic liquid wastes, such as used lubricating oil or paint thinners, may be brought in and blended with the degreasing solvent residue to increase the quantity of recycled liquid fuel available. The industrial ecology has now drawn in other materials which would otherwise be incinerated or stored, to provide a useful conclusion to their life cycles. Provided that the import of other materials is carefully controlled, the output from this cascade of use is still a recycled liquid fuel produced to specifications. Whether this

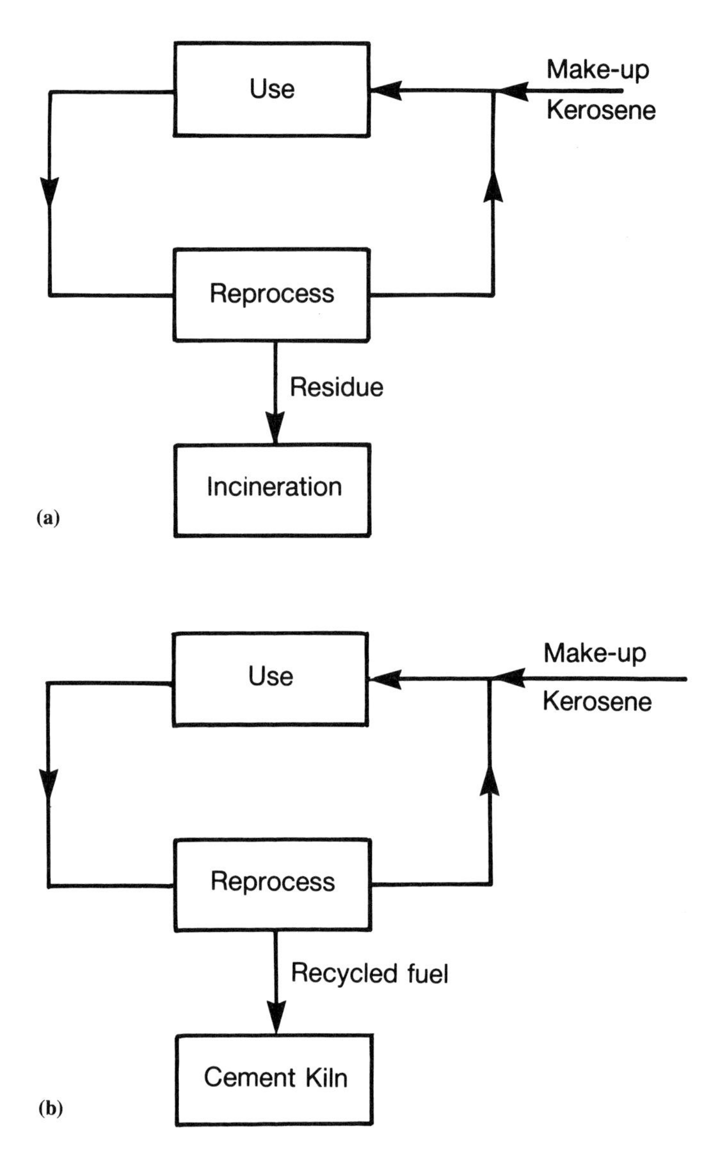

Figure 6.14 'Metabolised' use of organic solvents. (a) Reprocessing of degreasing solvents. (b) Fuel use of solvent residue. (c) Incorporation of other used organic liquids.

fuel is treated as a 'waste' then becomes a question of legal definition only, although the legal status of the material determines the applicable Regulations or Directives and therefore affects its commercial usability. Legal constraints on the practice of clean technology represent a broader area than can be addressed here.

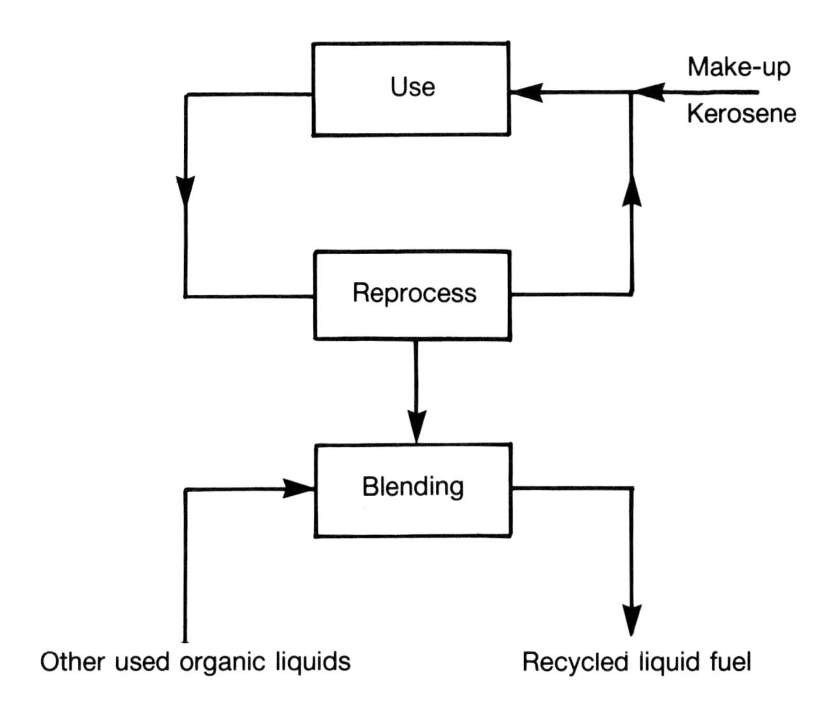

Figure 6.14(c)

Example 2: office equipment. Different technical and organisational changes are necessary for manufactured products. We consider next the specific case of photocopiers. We are indebted to Geene and Kummer[23] for this example, which we have attempted to generalise here as an approach which could be followed for other manufactured products. The user of the equipment self-evidently obtains it for its function: the purpose is to make photocopies, not to have a photocopier. Thus the end-user buys or leases the equipment to obtain a service. When the machine reaches the end of its service life (i.e. it no longer provides the function), it becomes scrap hardware without the service value. Unless it is returned to the supplier or an equivalent agent at this stage, the machine becomes undifferentiated scrap and joins the general waste stream. These broad aspects are common to the example of organic solvents.

If the machine is returned to the supply system as shown in Figure 6.15, then the assembly and components may be re-usable. Machines which have only been subject to light use may be partially dismantled, to remove worn

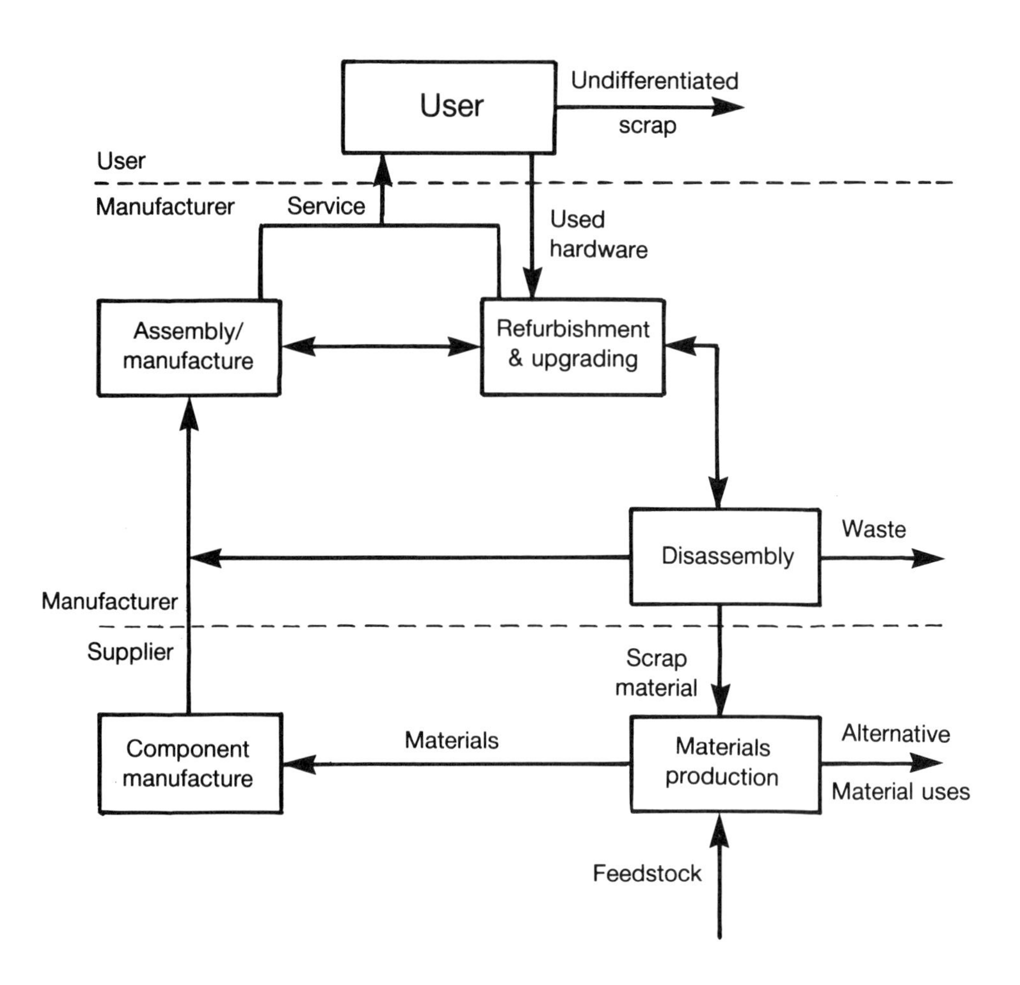

Figure 6.15 Commercial relationships in 'metabolised' use of a manufactured product.

or damaged parts, and reconditioned or returned to the assembly line. However, it requires an uncommon flexibility in the manufacturing process to accept incomplete machines into the assembly line. Machines which have received more use are disassembled completely. Components can then be fed back into the assembly line, preferably with no distinction from new components. Components which cannot be reused, for example because they contain compounds which are no longer used or permitted, leave the system as solid waste. Otherwise, the material is reused or 'cascaded' to another use. In the latter case, the material is preferably returned to the supplier, who can blend it in to fresh material or reprocess it for other applications. The component manufacturer would then be required to accept material which, wholly or partially, has been used before.

Thus Figure 6.15 also shows the business relationships necessary to bring about an industrial ecology. It is essential that the used machine is returned to the supply system. For photocopiers, this step is part of the normal commercial process, because some 80% of photocopiers are leased rather than sold.[23] This is another example of leasing the service rather than selling the product, and it can represent a barrier to adopting clean technology for an industry (such as automobiles) where the accepted commercial relationship involves selling ownership rather than leasing use. It is also important that the manufacturer is able to accept partially dismantled machines and used or reconditioned components into the manufacturing process. This in turn demands quality control procedures in disassembly which equal those in original supply. Furthermore, the material and component suppliers must be involved in the reuse of the material content of components which cannot be reused directly. This requirement demands that individual components contain as few different materials as possible, preferably a single material.

The requirements of designing for easy disassembly and minimising the number of different materials used in any component diverge from conventional design procedures, but emerge as key elements of clean technology in the manufacturing sector.

6.5 Waste reduction at source

The preceding discussion has concentrated on changing to new technologies and new commercial relationships. For an organisation to improve its environmental performance on a time-scale which is shorter than its capital cycle, it is necessary to concentrate on improvements to existing processes and products. Rather than clean technology in the broader sense, this section concentrates on clean production and specifically on reducing waste at source.[18]

The European Union recognises a hierarchy of waste management options: in order of decreasing importance:

1. Prevention;
2. Minimisation;
3. Recycling;
4. Disposal.

Following the discussion earlier in this chapter, *recycling* must be approached in the context of industrial ecology, while *disposal* is considered elsewhere in this book. We concentrate here on *prevention* and *minimisation*. A more closely defined hierarchy, due to Crittenden and

Table 6.2 Hierarchy of waste management practices (from Crittenden and Kolaczkowski[24])

	Top priority
Elimination	Complete elimination of waste
Source reduction	Avoidance, reduction or elimination of waste, generally within the confines of the production unit, through changes in industrial processes or procedures
Recycling	Use, reuse and recycling of wastes for the original or some other purpose such as input material, materials recovery or energy production
Treatment	The destruction, detoxification, neutralisation, etc., of wastes into less harmful substances
Disposal	The discharge of wastes to air, water or land in properly controlled or safe ways such that compliance is achieved; secure land disposal may involve volume reduction, encapsulation, leachate containment and monitoring techniques
	Lower priority

Kolaczkowski,[24] is given in Table 6.2. A classification due to Freeman[25] is shown in Figure 6.16, again making the basic distinction between source reduction and recycling.

In terms of Table 6.2 and Figure 6.16, the clean technology approach concentrates on source reduction (with some attention also to prompt recycling, i.e. recycling within the process itself). Figure 6.16 recognises four components of source reduction.

Input material changes. These can be considered where it is possible to substitute a material which provides the required function but results in reduced environmental load. Examples are the replacement of chlorinated organic solvents by non-chlorinated or aqueous media[f] in cleaning operations or paints; the replacement of chemical biocides or oxidants by ozone or hydrogen peroxide, which decompose to leave no residues or emissions; and the elimination of toxic heavy metals, for example in electrolytic processes and devices. In each case, the whole life cycle of the material must be considered, to ensure that one environmental problem is not simply being exchanged for another elsewhere. Although a substituted material may be more expensive, the increased material cost may be offset by savings in waste treatment and disposal costs. In this case, material substitution constitutes a change to a cleaner technology.

Technology changes. These include the kind of major changes in practice discussed in earlier sections of this chapter. Less radical changes which may be 'retrofitted' to an existing process include:

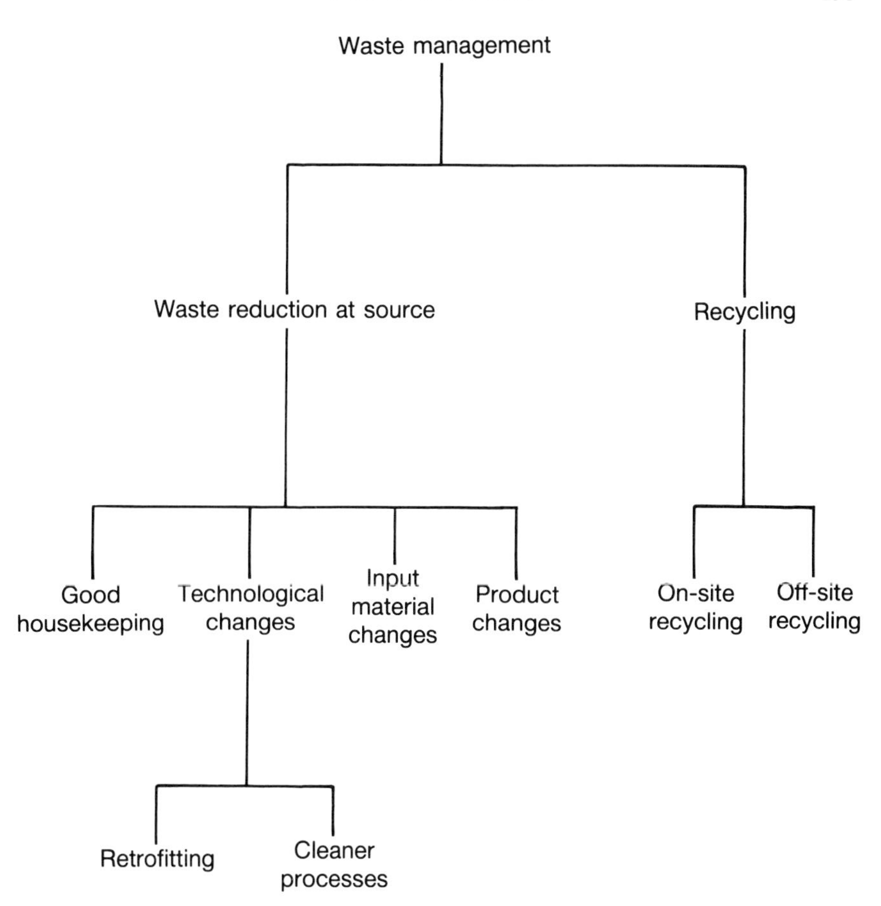

Figure 6.16 Practical techniques for waste minimisation (after Freeman[25]).

1. Improving chemical synthesis, for example by using a more selective catalyst, to reduce by-product formation
2. Improving process control
3. Redesigning or reconfiguring the process to improve heat recovery or to avoid dilution of process streams
4. Improving the selectivity of separation processes.

In addition to avoiding producing waste, technology changes include containing wastes and retaining them in a concentrated form, rather than diluting them for dispersion into the environment or mixing wastes for subsequent treatment. Throughout the manufacturing and process industries, there is a general trend towards treatment of waste at source, recognising that a process to treat a specified waste is likely to be more effective and less expensive than treatment of a mixed waste stream. The move away from the 'dilute and disperse' approach to the strategy of 'concentrate and contain' usually also requires changes in operating practice.

Good operating practice. This includes preventing unnecessary releases, and therefore merges into on-site recycling. Returning to the example of replacing organic by water-based solvents, this kind of material substitution may not represent an environmental improvement if the solvent is simply discharged after use. The discussion of approaches to cleaning clothes earlier in this chapter was a 'domestic' example. If the organic or aqueous medium is used as a cleaning agent, it should be contained and reused; depending on the nature of the cleaning operation, mechanical or membrane filtration may suffice to collect the soil so that the liquid can be reused. A particular environmental problem arises with 'clean-in-place' treatment, for example of food processing or brewing equipment. Emissions of the cleaning fluids can constitute the main environmental load from the equipment. The most urgent issue for improving environmental performance is then management of the cleaning process. Waste stream segregation may represent a minor technology change which yields a disproportionately large improvement in environmental performance.

Product changes. These include changes in final or intermediate products, to reduce waste generation and other environmental loads arising elsewhere in the life cycle. The approach here is usually product-specific and therefore commercially sensitive. Ecolabelling is an attempt to identify consumer products which genuinely have improved environmental performance, without revealing commercially-sensitive information (see Clift[11]).

By this stage, we are reducing clean technology to specific problems. Specific examples of the above approaches are given in later chapters.

6.6 Concluding remarks

The introduction to clean technology in this chapter has invoked examples from several different industrial sectors. However, a number of features characterising clean technology emerge whichever sector is considered:

1. Clean technology is really an approach to providing services and benefits, not a recognisable set of technologies.
2. To ensure that a technology is really clean, it is necessary to consider the whole life cycle of the materials or objects which provide the service or benefit.
3. An organisation intending to apply clean technology must concentrate on the service which it provides, rather than on the products or artefacts

which it sells. It should also retain responsibility for the products or artefacts which provide the service, and either reuse them or pass them on to another organisation for a different use within the industrial ecology.

4. Wherever possible, waste should be avoided at source rather than cleaned up at the end of the pipe. If a waste is unavoidable, it should be concentrated and contained so that it can properly be managed, not diluted and dispersed into the environment.

The following chapters discuss clean technology in different industrial sectors, and serve to illustrate the general points identified here.

References

1. Baas, L., Hoffman, H., Huisingh, D., Huisingh, J., Koppert, P. and Newman, F. (1990) *Protection of the North Sea: Time for Clean Production*, Erasmus Centre for Environmental Studics, Erasmus University, Rotterdam, NL.
2. Warhurst, A. (1992) Environmental management in mining and mineral processing in developing countries, *Natural Resource Forum*, Feb., pp. 39–48.
3. Clift, R. (1993) Pollution and waste management, *Science in Parliament*, **50**, 29–32.
4. CEFIC (1992) *Carbon/Energy Taxation: The position of the European Chemical Industry*, European Chemical Industry Council, Brussels.
5. Löfstedt, R and Clift, R., An Analysis of the Efficiency of International Environmental Aid, to be published.
6. Karlsson, R. (1994) LCA as a Guide for the Improvement of Recycling. In *Proceedings of the European Workshop on Allocation in LCA* (ed. G. Huppes and F. Schneider), SETAC, Brussels, pp. 18–28.
7. Pedersen, B. (ed.) (1993) *Environmental Assessment of Products*, UETP-EEE, Helsinki.
8. Keoleian, G.A. and Menerey, D. (1993) *Life Cycle Design Guidance Manual*, US Environmental Protection Agency, Cincinnati, Ohio.
9. SETAC (1992) *A Conceptual Framework for Life-Cycle Impact Assessment*, Society of Environmental Toxicology and Chemistry, Pensacola, Florida.
10. SETAC (1993) *Guidelines for Life-Cycle Assessment: A 'Code of Practice'*, Society of Environmental Toxicology and Chemistry, Brussels and Pensacola.
11. Clift, R. (1994) Life cycle assessment and ecolabelling, *J. Cleaner Production*, **1**, 155–159.
12. Guinée, J.B., Udo de Haes, H.A. and Huppes, G. (1993) Quantitative life cycle assessment of products – 1: Goal definition and inventory, *J. Cleaner Production*, **1**, 3–13.
13. Guinée, J.B., Heijungs, R., Udo de Haes, H.A. and Huppes, G. (1993) Quantitative life cycle assessment of products – 2: classification, valuation and improvement analysis, *J. Cleaner Production*, **1**, 81–92.
14. Udo de Haes, H.A., Bensahel, J.F., Clift, R., Fussler, C.R., Griesshammer, R. and Jensen, A.A. (1994) Guidelines for the Application of Life Cycle Assessment in the EV Ecolabelling Programme, DGXI of the Commission of the European Communities, Brussels.
15. UKEB (1992) *Ecolabelling Criteria for Washing Machines*, UK Ecolabelling Board, London.
16. Corbett, J.R., Wright, K. and Baillie, A.C. (1984) *The Biochemical Mode of Action of Pesticides*, Academic Press, London and Orlando, p. 343.
17. Glarini, O., and Stahel, W.R. (1989) *The Limits to Certainty*, Kluwer Academic Publishers, Dordrecht.
18. Jackson, T. (ed.) (1993) *Clean Production Strategies*, Lewis Publishers, Boca Raton, Florida.
19. Virtanen, Y. and Nilsson, S. (1993) *Environmental Impact of Waste Paper Recycling*, Earthscan Publications Ltd., London.

20. Daae, E. and Clift, R. (1994) A life cycle assessment of the implications of paper use and recycling, *I. Chem. E. Environmental Protection Bulletin* no. 28, 23–25.
21. Stahel, W.R. and Jackson, T. (1993) Optimal utilisation and durability – towards a new definition of the service economy, Chapter 14 in Jackson, T., ref. 18.
22. Roberts, K. and Lewis, R. (1994) personal communication, on behalf of Safety-Kleen UK Ltd.
23. Geene, F. and Kummer, K. (1994) personal communication, on behalf of Rank Xerox Ltd.
24. Crittenden, B.D. and Kolaczkowski, S.T. (1992) *Waste Minimisation Guide*, Institution of Chemical Engineers, Rugby.
25. Freeman, H.M. (1990) *Hazardous Waste Minimization*, McGraw-Hill, New York.
26. Jackson, T., Costanza, R., Overcash, M. and Rees, W. (1993) The 'biophysical' economy – aspects of the interaction between economy and environment, Chapter 1 in Jackson, T., ref. 18.
27. Azapagic, A. and Clift, R. (1994) Allocation of environmental burdens by whole-system modelling – the use of linear programming. In *Proceedings of the European Workshop on Allocation in LCA* (ed. G. Huppes and F. Schneider), SETAC, Brussels, pp. 54–60.

Notes

a. The environmental impact axis here could be interpreted as the 'external costs' – see Chapter 5 – in which case the parameter is measured in financial terms. However, the idea of clean technology can be developed without introducing the many explicit and implicit assumptions on which this school of environmental economics is based. Therefore the environmental impacts are not introduced here in terms of 'externalities'. The financial cost axis should be interpreted as the conventional 'internal costs' of providing the benefit, including the costs arising specifically from pollution abatement measures but without attempting to 'internalise the external costs'.
b. The steps in carrying out an LCA summarised here are simpler than the formal steps identified in the SETAC Code of Practice[10] but are those suggested by Guinée *et al.*[12,13]
c. The washing machine LCA study[15] has sometimes been criticised for insufficient rigour. Nevertheless, it remains an example of the way LCA can yield useful and sometimes counterintuitive conclusions. One of these is that an environmentally responsible owner of an old washing machine should 'throw it away' and replace it by an energy-efficient machine.
d. The basis for this estimate is as follows[16]

   1. a 3-week-old wild oat plant contains 3×10^6 cells;
   2. assume one molecule per cell gives a toxic response;

   Therefore 3 million molecules need to enter each plant.

   3. assume 10% of the chemical falling on a plant actually gains entry;
   4. only 10% of the chemical gaining entry survives inactivation by metabolism or storage at an inert site;
   5. half of the plant's leaf area is available to chemical sprayed from above;
   6. take representative molecular mass of herbicide as 250 Daltons.

   Then lower limit for herbicidal action is then 2.5×10^{-6} g per hectare.
e. In addition to the practical problems of mechanical damage, there are thermodynamic constraints on reuse and recycling. For a preliminary discussion, see Jackson *et al.*[26]
f. While it is generally considered good practice to replace organic by aqueous solvents, especially if the organic solvents are chlorinated, it was pointed out earlier in the chapter that (consistent with the life cycle approach) it is the escape rather than the existence of the organic material which causes the environmental load. The discussion here therefore refers to cases where it is impractical or uneconomic to contain the solvent.

7 Agricultural and pharmaceutical chemicals

P.A. JOHNSON

7.1 Introduction – agricultural and pharmaceutical chemicals

This chapter outlines the position of the agricultural and pharmaceutical chemical businesses within the chemical industry in terms of annual production, scale of manufacture, product value, product life, type of manufacturing process and environmental impact. The two business areas have been put into one chapter because of the commonality that exists in many of the areas being considered. Where there is a significant difference this is indicated. The concept of clean technology is proposed as a means of maintaining product choice and availability, whilst minimising the effect on the environment of continuation and even expansion of the two business areas. The specific problems and constraints of the agricultural and pharmaceutical chemical industry are indicated. The ways in which clean technology is and can be applied at all stages of product development and manufacture are discussed in detail for agricultural chemicals (both fertilisers and agrochemicals) and pharmaceutical chemicals. The areas of research and development which could give significant improvements but are in the early stages of development are given in outline.

7.2 Why clean technology?

The biological world in which we live uses chemical reactions to enable it to function. It is not therefore surprising that the chemical sciences and technologies are deeply involved in the debate about the ecological future of our planet and that the chemical industry is seen as one of the key industries which is expected to take a leading role in resolving the situation. The products of the chemical industry are integrated into our way of life to a much greater extent than most people realise. Look around a domestic room and what do you see? The ceiling is decorated with a synthetic polymer-based filler which is less likely to crack than the traditional plaster. The walls are decorated with emulsion paint rather than lime wash. The floor is synthetic fibre carpet, or tiles. Many of the objects in the room are made of special plastics. Where wood is used it is often reconstituted (chipboard) using resins. Synthetic materials have taken over from natural materials because the cost of natural materials is much more than the average person could afford or is prepared to pay. The reason that natural materials cost so much is partly because there are not enough of them and nature takes its time to replace them. If you consider carefully the materials you are using and surround you at any time you will start to realise how the chemical industry provides a large proportion of the materials required for manufacturing, building, electronics, etc. In order to meet the demands of our society the chemist/technologist has tried his or her best to mimic nature and has sometimes provided a superior alternative. They have done this by converting an available resource into a synthetic alternative. Agriculture and medicine are not exceptions to this philosophy. The pressure of increasing world population and expectations of higher standards of living have encouraged humans to try to give nature a helping hand. Hence farmers use synthetic fertilisers to increase productivity and pesticides to prevent crop damage. Doctors use drugs to cure or control diseases of both the body and the mind. Whether this is morally right or wrong is beyond the scope of this chapter. In reality we may be asking too much from the world in which we live because our present demands do not appear to be based on a sustainable ecology. What is certain is that the demand for agricultural and pharmaceutical chemicals will not cease overnight, if at all. There is the possibility that agriculture, as a sustainable resource, may be asked to provide certain raw materials as an alternative to oil and the scope of crops used for non-food purposes is expected to increase. There is already some evidence for this trend. Methanol produced from sugar cane is used as a fuel in Brazil, bio-diesel processed from rapeseed oil has been on trials in Europe, and the use of farmed animals to make certain human proteins, namely factor 9 and haemoglobin, is at the development stage. The average human life expectancy is increasing and with recent advances in diagnosis, care and treatment this trend is not likely to change. The manufacture of agricultural and pharmaceutical chemicals uses up resources and inevitably its products and wastes find their way into the environment. Consequently, if we wish to maintain our standard of living and a healthy global ecology, we need to ensure that what resources are available are used efficiently and that the processes that are used to convert one available resource into another are not detrimental to the world in which we live. It is inevitable that during this 'shake down' we may decide that certain materials and processes should discontinue. The phasing out of CFCs is one example of this. Although some people would insist that we need to take a step backwards in time and reduce our dependence on modern technology and accept a reduced standard of living, it is unlikely that this will be acceptable to the population as a whole. The compromise is clean technology.

7.3 Clean technology is not new

The concept of clean technology need not be confined to the process industries. Clean technology is applicable to all aspects of business and living. In the home a new energy-efficient cooking hob is an example of clean technology because it results in a reduction in the use of fossil fuels and consequently less carbon dioxide is generated. In general any technology that saves time and effort is likely to save energy and is therefore a clean technology. Clean technology is not a new philosophy. The chemical industry has been involved with its implementation for many years. The driving force, in most cases, has been based on economics and where it has made economic sense to use cleaner technologies they have been developed. Chemical processes which operate on a large scale (several hundred thousand tonnes per year) and manufacture low cost products (£100 per tonne) cannot afford to waste either materials or energy if they are to remain competitive. Therefore historically there has been an incentive to develop specific processes with complicated recycles for energy and unconverted raw materials, which generate the minimum of by-products. On the other hand chemical processes which only make small quantities (tens of tonnes) of a high value (£100 000 per tonne) patented product are less likely to use technologies which are comparatively material or energy efficient. This is because of (a) the cost of dealing with the quantity of waste coming from the small production is likely to be a relatively small part of the operating cost and (b) the high cost of developing clean process technology compared to the total lifetime value of the business. However, it is difficult to generalise because the major area of cost will depend on the individual process, and although it is assumed that most businesses have identified the high cost areas which can be reduced, it is surprising how many manufacturers are only just studying their manufacturing costs in sufficient detail to enable improvement of medium cost areas to be identified.

This criterion of cost of development against the lifetime value of the business is an important one to understand. Most products with a finite life go through four phases during that product's lifetime. Initially the quantity produced is small whilst the product is finding its place in the market. This is the 'launching' phase. Once it has been accepted as a successful product its market expands. This is the 'growth' phase. It eventually reaches a peak and stays at this level, the 'mature' phase, for a period of time before it is superseded and the product goes into 'decline'. There are three main factors which influence the lifetime value of a business, (a) the quantity of product sold, (b) the value of the product and (c) the product's lifetime. One of the reasons why, over the past decade, we have seen a large number of companies amalgamating and rationalising is to increase the quantity of product sold. The cost of process development for making a change to a chemical manufacturing process is almost the same whether the process makes 100 or 100 000 tonnes of product a year. Thus low value/low tonnage products are less likely to support any significant development compared to low value/high tonnage or high value/low tonnage products. If a company is lucky enough to have a high value/high tonnage product then it should, in theory, be able to fund substantial development to enable utilisation of clean technology. Normally the value of the product is related to the market and is not easy to change significantly.

The remaining important factor in the lifetime value of a business is the time element. The development activity for a company making for example phenol, ammonia or sulphuric acid, all of which are building blocks for which under normal circumstances there is a continuing demand (or infinite lifetime), will be focused on making the product as effectively as possible, using the minimum energy and raw materials. The business is naturally driven towards the use of clean technology. However, if a product has a finite life then its profits must fund the development of its replacement. The willingness of a company to invest in expensive development also depends upon the current phase of the product's lifetime. The launch of a new product is always a problematic period. Many companies will make a new active ingredient in relatively low tonnages during the 'launch' period, often using existing plant and a partly developed process. The period of maximum process development would be during the 'growth' phase so that by the time the product reaches a stable 'mature' phase there is minimum investment and maximum profit. It is during this 'growth' phase that a decision has to be made for the choice of the long-term process and technology. During the steady 'mature' phase development will be maintained for products which are likely to have a long lifetime. It is unlikely, however, that a company will invest in an alternative process during the phase when sales are falling. If put under pressure to make such investment a company is likely to drop the product and concentrate its effort into the development of a new product. For a company that is managing products with a finite life the amount of effort which is put into developing existing products and developing new products has to be carefully judged and can be critical to the long-term future of the business. It can be seen how in some circumstances, with products which have limited life expectancy, the natural driving force is not necessarily towards clean technology. However, the change in public opinion towards caring for our environment, along with the more recent legislation, does encourage the use of cleaner technologies especially for new products once they are established in the market.

7.4 What has caused the move towards clean technology?

Why is there a move towards processes with less environmental impact? The change is the result of public concern about environmental issues. The concerned public has influenced politics as it should in a democracy and this influence has increased considerably since the mid 1980s. Political pressure, as an indication of the 'concern of the public', is being exerted in economic terms by imposition of legislation, taxes and incentives. Consequently because of legislation, typified by the *Control of Pollution Act* in the UK, companies are being set stricter standards to which they have to conform. In order to stay in business they are having either to improve their processes to reduce waste or at the very least reduce the impact of their wastes by using treatment techniques.

Consequently where the cost can be absorbed by the business environmentally based projects are being implemented. Where they cannot it is possible that some manufactures will cease. Many companies, compelled by these factors to make investments to reduce their environmental impact, are actually experiencing a greater pay back than expected because of hitherto unrecognised benefits. The experience of the Aire and Calder programme, initiated by the Centre for Exploitation of Science and Technology with 11 companies in the Aire and Calder catchment area, has identified savings of about £2 m resulting mainly from reductions in use of water and energy and improved effluent treatment. The level of potential savings has been much higher than that expected when the programme was started. The report on the project implies that much of this can be made without significant investment and some is available simply by better use of existing facilities.[1] As a result of these and similar encouraging experiences,[2] companies are now more willing to make investments in environmental projects and change to cleaner technology than they were a decade ago. It should, however, be noted that there are companies and individuals who have been environmentally aware for several decades and who have seen the benefits that waste minimisation can bring.[3]

7.5 The need for an even playing field

Provided the influence in the form of legislation, etc., which drives towards clean technology is even-handed and applied globally, it will result in continued improvement, provided the customer is willing to pay. The danger comes from unilateral strong action, which it could be argued will cause one company/country to be disadvantaged compared to another. This has two effects: (i) it gives an excuse to those inclined to do so to procrastinate whilst contesting the need for 'such stringent measures', whilst (ii) it can disadvantage those who are compliant to such a extent that they can no longer afford to be in business. The net result could be self defeating.

7.6 Position of agricultural and pharmaceutical chemicals businesses

The products of the agrochemical and pharmaceutical chemical industries along with the manufacturers of associated specialised intermediates, in general, fall into the category of 'low to medium tonnage, medium to high value products'. In common with other chemical manufacturers within this category there is scope for the use of clean technology in the future. To what extent this can be justified is impossible to state. Both businesses manufacture products of which the majority have limited life times. In some instances products are sold for many tens of years, whilst in others, for various reasons, the product life may be measured in months.

Both businesses supply materials with the intention that they are released into, and will affect specific aspects of, the environment. The area of effect for agrochemicals is likely to be greater than that for pharmaceuticals. Whereas medicinal products are intended to maintain the health of one species, namely humans, pesticides are required to protect a much wider diversity of species, spread over a much larger physical area. Consequently pesticides appear to have a much higher environmental profile. It should be noted at this point, however, that other businesses which have in the past assumed that their products do not have an environmental impact are having to recognise that they are mistaken. For example, a polythene bowl when it is sold could be considered to be released into the environment but does not, whilst it is usable, have an impact. It has an impact when it is no longer of use and is disposed of into the urban refuse. The automobile manufacturing industry has recently taken steps to address this problem. It is for this reason that moves to recover and recycle such waste from all sources are gradually being implemented. The fact that both agrochemicals and pharmaceutical chemicals are designed to affect living species imparts on them more severe constraints not applied to other chemical businesses. There are both advantages and disadvantages to this situation. The main advantage is that the effects they have on the environment are better researched than other chemicals. Their use is strictly controlled, however, and if there are any shortcomings in the safety/environmental screening it can have extremely serious effects. DDT and thalidomide are examples of this.

7.7 The future for agrochemicals and pharmaceuticals

Whilst accepting the concept of clean technology and its undoubted benefits, we must also accept that it is not possible to apply these principles to every process 'overnight'. In reality we have a choice between no waste, no products and no jobs, which is going to be politically unacceptable, and making a controlled change from the current situation which has developed mostly over the past 50 years. The agricultural industry finds itself in a dilemma. It has responded to the demand for increased food production by doubling its output since the Second World War only to find that in some areas there is over production. The world's population is expected to double by 2025[4] and with an anticipated move to the use of agricultural products as raw material for renewable energy sources and chemical intermediates the overall need is for cropped land to increase by about 1.5% per year. The current rate is about 0.15%.[5]

If we are not to destroy ecosystems which are least suited to cropping then it would seem that we will have to continue with intensive agricultural techniques in the long term. Contributory to the increase in the global population is the improvement in health care. Generally the world's population is living longer and expecting a higher standard of living, which is putting pressure on resources. Consequently it is realistic to expect that both agricultural and pharmaceutical chemicals will be required for several decades to come and the processes used for their manufacture will continue to improve by the use of cleaner technology. In order that we can realise the best environmental improvement in the shortest period of time the chemical manufacturers will need to identify the largest problem areas and prioritise their action. This can be achieved by studying the cost structures and the process mass balances in detail. Such investigation invariably shows where the largest areas of cost and waste are within the process being investigated and makes it easier to prioritise action. This will, however, only enable prioritisation within a company and it is essential that national and even international priorities are established. The environmental legislation will assist in this objective by establishing national and international measurement principles and setting target levels of waste discharge. In reality the process of using cleaner technology will continue indefinitely.

7.8 Importance of agriculture

The wealth of any country is related to its use of available natural resources. These resources are derived from the earth, the air and the sea. Some resources once used are not likely to be replaced easily. For example, oil and coal come into the category of finite resources. Their formation takes a long time and requires certain global conditions. Other resources are renewable albeit on different time scales, grass daily, trees several tens of years. Agriculture is the basis of a vast renewable resource and it is gradually being recognised that with careful development it could help to supplement the use of oil for fuel and non-fuel products. Agricultural turnover in the UK for 1992 amounted to about £12.2 billion which is about 1.3% of GDP.[6] For many countries that do not have any significant mineral wealth, agriculture or fishing may be the only natural resource.

In broad terms the agricultural chemical industry which supports agriculture is divided into two areas: nutrition and control products.

7.9 Plant nutrition

The major products within the scope of plant nutrition are fertilisers. These can be straight, containing only one of the important elements, or compound, containing various ratios of nitrogen (N), phosphorus (P), potassium (K), calcium (Ca) and other trace materials. The chemistry used in their manufacture is inorganic in nature and the raw materials derived from natural gas, air, water and mineral rocks. The scale of the business in terms of tonnage per year is large; about 2.2 million tonnes of plant nutrient (N, P_2O_5, K_2O) were applied by UK farmers in 1992.[7] This corresponds to a gross usage of just over 5 million tonnes. The cost of the product is relatively low, in the region of £80 to £150 per tonne. The products and wastes are mostly inorganic in nature, the majority being generated by the phosphate manufacturing industry where traditionally calcium sulphate is produced as a by-product.

The nitrogen content of the fertiliser is generally in the form of an ammonium or nitrate salt. The processes used are very specific, well developed, continuous and catalytic.[8,9] They are typical of large scale manufacturing processes for low cost bulk chemicals which are highly tuned to reduce energy consumption with low waste of raw materials. Although very efficient processes, the scale of the operation world-wide is likely to encourage continued improvements to increase both material and energy efficiency. If we assume that the spot price for ammonia is about £100 per tonne and the world consumption is about 120 million tonnes then an increase in efficiency of 1% globally is worth £120 million. This represents a worthwhile saving which should encourage those with ingenuity to research and implement the necessary improvements. The process for ammonia is based on the Haber process and uses nitrogen which is easily obtained from the air, and hydrogen, which for the majority of producers is obtained by catalytic oxidation of natural gas. To obtain the hydrogen at the required level of purity makes the process complicated and examination of a typical reaction scheme will confirm that a disproportionate part of the process is devoted to producing the hydrogen. The latter process also produces carbon dioxide as a by-product. Manufacture of hydrogen by electrolysis is normally not energy efficient. Industrially, ammonia is converted to nitric acid by a catalytic oxidation process.[8] Ammonium salts are not used directly by the plant but are converted to nitrate by nitrifying bacteria in the soil. The reaction conditions for the ammonia synthesis and conversion of ammonia to nitric acid in a manufacturing plant; both take place at elevated temperature. In some leguminous plants such as peas, beans and clover, the growth of certain species of bacteria is encouraged which are able to convert nitrogen from the air to nitrate at ambient temperature. This is one of nature's processes which it could be advantageous to mimic and would be a truly clean technology. Alternatively it would be even more beneficial if other plant species can be safely bio-engineered to enable nitrogen fixation to occur in their root systems.[10,11]

Potassium chloride, mined or isolated from brines, is the main source of potassium used in fertilisers and virtually no chemical processing is required to make it available for incorporation into a fertiliser. Although it is the main source of potassium, other potassium salts are used for very specific requirements.

The phosphorus content of fertilisers is normally in the form of phosphate and is derived from phosphate-containing rocks. The manufacture of phosphates for fertilisers has ceased in the UK, where fertiliser manufacturers obtain most of their phosphate from North Africa (Morocco, Tunisia, Algeria) where there are large deposits of suitable ore. The most commonly used process is to convert the calcium phosphate-containing rock to either the mono or dihydrogen calcium phosphate or even phosphoric acid. The more hydrogen and less calcium present the more soluble the phosphate. The process produces calcium sulphate as a by-product. One way in which the calcium sulphate by-product and waste of acid can be avoided is by the use of nitric acid for the digestion process producing nitrophosphates.[12] This process has been adopted by some manufacturers with a subsequent reduction in the amount of calcium sulphate for disposal.

The major area of concern specific to the use of fertilisers is the effect of leaching of nitrate on the environment. The problem of phosphate leaching from farmland is, in fact, virtually non-existent because it is generally immobilised in the soil. In reality the problem is that of maintaining soluble phosphates for the plant to use before they are locked up in the soil as insoluble calcium phosphates. Nitrate is manufactured by bacteria in the soil but with intensive agricultural techniques the rate of manufacture is not sufficient to satisfy the needs of the crop and thus a nitrogen supplement is required. It is the application of nitrogenous fertilisers at the wrong time which is likely to result in unnecessary leaching of nitrate. In some areas of the EU where the ecology is especially sensitive, the use of nitrogen fertilisers is restricted. Although slow release fertilisers are available based on urea formaldehyde condensation products these are expensive. A technique which has been developed to prevent build up of excess nitrate in the soil involves the use of nitrification inhibitors which slow down the conversion of ammonium salts to more easily leachable nitrates. At present, the most effective method of making the best use of the fertilisers appears to be the timing and control of application. Each crop has a period of more rapid growth when nutrients are most needed. The time at which this occurs depends on the crop and the climatic conditions. By understanding the exact nutrient requirements at different times in the life of the crop, and by developing easy-to-use methods of identifying application timing for each crop it should be possible to minimise input whilst at the same time reducing the quantity available for leaching. For example it is now considered bad practice to apply nitrogen in the autumn for winter cereals. At this time growth is not rapid and sufficient nitrogen is normally available due to natural processes taking place within the soil. The recommended time of application is during spring when growth begins to increase. Application of the correct quantity at exactly the correct time depending on soil temperature and soil moisture will ensure that the majority is used and the minimum is leached out. This is one of the key requirements of the Code of Practice for farmers, issued in 1987 by the EEC, aimed at reducing the nitrate leaching problem.[13]

Another approach is that of targeted dosing. In principle this technique finds a use in the application of both fertilisers and herbicides. As indicated earlier, the need for fertilisers is different depending on soil type, temperature, rainfall, season of year and crops, etc. Farmers already make some allowances for the different requirements field by field, but by the use of yield mapping or field mapping along with remotely controlled tractors and application equipment it is possible to selectively apply different quantities of fertiliser to different areas of a field. Yield mapping is the process of identifying the variations in yield over the area of a field. At the moment one technique of achieving this is from historical data gathered during the harvesting operation using specially modified harvesters.[14] Field mapping is the technique of determining the variations in defined parameters likely to influence plant growth over the area of a field. This can be obtained by the use of aerial and satellite photography or even by field walking. By combining the information from yield and field mapping it is possible to regulate the use of fertilisers and other farming activities to obtain optimum yield whilst at the same time minimising input. It can be seen that in wasting less, using less and thus requiring less material to be made, the fundamentals of clean technology are applicable even to the use of fertilisers. There is good evidence that sound integrated management practices by farmers along with regulated minimal use of fertilisers can achieve the required level of agricultural output whilst minimising the impact on the environment.[15]

7.10 Animal health and nutrition

The problems associated with animal health and nutrition products are very closely related to those of the pharmaceutical industry, which are outlined later in this chapter, except that the greatest interest is in the residual material left in the animal at slaughter. The same ethical issues exist about the effect on animals as to the effect on humans.

7.11 Outline of historical development of agrochemicals

Chemicals were used for the control of weed, insect and fungal pests as early as the nineteenth century. For weed control the concept was very basic and total kill was all that was required; such treatment involved the use of crushed arsenical ores, oil wastes, thiocyanates from coal gas washing and creosote. Selective control was obtained by the use of Bordeaux mixture for the treatment of fungi and hydrogen cyanide for the broad control of insect pests. Sodium chlorate and sulphuric acid were used as total weed killers in about 1910 and sulphuric acid was later used on cereal crops to selectively destroy broad leafed weeds. In this latter case, selectivity depended on the growth habit and the user was relying on the physical shape of the leaf to obtain a crude form of selectivity. In 1932 *ortho*-dinitro cresol was used as a contact herbicide and was probably the first organic molecule used for this purpose. 2.4-D ((2,4-dichlorophenoxy) acetic acid) was introduced in 1942 and was the first of the phenoxy herbicide family of products. These were also the first truly selective herbicides in that they had a specific biological effect on certain broad leafed weeds whilst cereal crops were not effected. 2.4-D is very effective at controlling broad leafed weeds in cereals and along with other members of the phenoxy family is still in use today. Subsequently, other active ingredients have been developed to give control of other specific groups of weed species in specific crops. In addition to selective control there are also several sophisticated total kill herbicides which eliminate all green vegetation with or without persistence. The development of fungicides and insecticides has proceeded in a similar manner over about the same time scale. Today there are more than 500 different active materials listed in *The Pesticides Manual*[16] and over 2000 products listed in *Pesticides 93*.[17]

Although considered to be agrochemicals, insecticides are an important element in the control of some of the most widespread and debilitating diseases that are a serious health risk in the Third World. Diseases such as malaria, yellow fever and sleeping sickness are spread by insects and the use of insecticides to eliminate the insect vectors is far more effective than trying to treat the disease once it has been transmitted to humans. In addition many diseases of animals are also transmitted by insects.

The capacity of agrochemical active ingredient manufacturing facilities in the UK varies depending on the active ingredient. For very specialised products or materials just being launched the output could be in the region of 50 tonnes per year, whereas for the more widely used established products the capacity could be in the 10 000 tonnes range. Manufacturers of commodity materials would tend to make and formulate the product in the global area of use, whereas speciality products are likely to be made on one site globally and shipped to other parts of the world. Agrochemical active ingredients made in the UK are sold world-wide. During 1992 the UK exported about £760m worth of pesticides against £413m produced and used in the UK. For the 1991–92 season in the UK a total of 4.5 million hectares of crops were treated during cultivation. In total about 27 million hectares of treatment were applied which is consistent with an average of about six treatments per growing season.[18] These would normally consist of multiple applications at different stages of growth of herbicides, fungicides and insecticides. During the same season about 25 000 tonnes of active ingredient was used,[19] an average use of about 1 kg of active ingredient per hectare per application. The plant capacity and the UK use cannot easily be related because most manufacturers of active ingredients market their products in other countries world-wide. Even taking this into account, however, most manufacturers of agrochemical active ingredients are not considered large. Consequently the agrochemicals manufacturing business is considered to be a low to medium tonnage, medium value market with a limited lifetime for its products. Therefore the drive towards clean technology due to market forces will be high for the larger tonnage manufacturers; for those at the lower end it will need to be assisted by the publicity derived from the good experiences of those companies that have made the move towards clean technology, and legislative rather than market pressure.

The trend has been towards the development of active materials which are more selective, require low dose rates and have reduced relative toxicity. Thus for the early herbicides the dose was about 10 kg or more per hectare for a product with an LD50(rat) of about 300 mg/kg, whereas some of the newer herbicides such as the sulphonyl ureas are applied at only 20–50 g per hectare but have an LD 50(rat) of about 5500 mg/kg.

Where activity is restricted to a particular stereo isomer, techniques are being developed to enable this isomer to be made exclusively. For example, the dose rate for Mecoprop ((*RS*)-2-(4-chloro-*o*-tolyloxy)propionic acid) can be halved by the use of the active isomer and the whole manufacturing process can be scaled down with the consequential reduction of waste.

The 25 000 tonnes of pesticides, expressed as active ingredient, used in 1992 in the UK had a value of about £400m. The quantity used corresponds to a reduction of about 8000 tonnes since 1983, while the total cropped area has not shown the same level of reduction in the same period of time. In terms of global total pesticide treatment by area, Western Europe accounts for about 26%, the Far East accounts for another 25%, with North America accounting for 29% and Eastern Europe, Latin America and others accounting for the remainder. Of the total market about 45% is in herbicides, 30% insecticides, 20% fungicides and 5% other products.[20]

7.12 Outline of problems with pesticides

Pesticides fall into three main classes: herbicides, insecticides and fungicides, with growth control products as a minor group of materials. Pesticides take a very high profile within the environmental lobbies because they are *specifically designed to be released into the environment and are intended to have an effect on specific aspects of that environment.* The perceived problems are that their use is unnatural and that in the past they have achieved more activity to a wider range of species and persisted longer in a wider environment than had been intended. Some pesticides have been detected in drinking water. Consequently maximum levels of contamination by pesticides in drinking water have been proposed. Much discussion, however, revolves around the validity of these levels. Some pesticides residues are found in food and again the question being asked is, What is a safe level?

In some instances pesticides have been detected in the environment because methods of analysis with adequate sensitivity and specificity have only recently become available. The fact that they are now being detected does not mean they have only just arrived in that particular environment. Pesticides have caused the unwanted death of non-target species and this has aroused considerable concern about their use and our lack of knowledge relating to their use. In the ensuing controversy one body of opinion suggests that we should return to organic farming, whilst another opines that this is not possible because of the shortages of food and other related problems that this would cause.[21] It is not the purpose of this chapter to get embroiled in such discussion although it is the author's opinion that having moved to the current position over several tens of years it would be unwise to make any sudden dramatic reversal.

Plants use specific chemicals to protect themselves against disease, and some produce their own insecticides and fungicides.[22] Animal species normally rely on biological methods of protection, such as antibodies, whereas plants rely on chemical protection which may poison the attacker, interfere with digestion or inhibit growth. Numerous natural toxic materials may be present in the food which we eat but generally the level is too low to have any detectable effect. It has been estimated that the average human ingests 10 000 times more natural pesticide than synthetic when partaking of a normal diet.[23]

Over the years numerous people have died as a result of poisoning by mycotoxins resulting from fungal contamination of foodstuff. We probably know less about the effect of natural pesticides on humans than we do about synthetic pesticides. We do know the acute toxic effects of some of nature's most virulent poisons such as strychnine, some snake venoms, fungi (e.g. death cap), etc. What is not known with any certainty is the long-term effects of many toxic materials be they natural or unnatural. Until the no-effect levels are agreed then it is not unreasonable to be cautious in order to limit the possible problems, but being too cautious is also likely to be problematic. The way forward is to make a balanced judgement using the information that is known, and decide on the information that is needed to enable refinement of control and action. This is the process that is at present in operation. There is a lot of work to be done to obtain the required information, mostly in the sphere of ecology, toxicology and analysis.

One of the differences between the use of pesticides made by man and nature is the mode of application. In natural systems the plant makes the pesticide 'on board'. There is no need to use an application system with its associated inefficiency. When a farmer uses a pesticide his method of application is generally much less specific, which results in the need to use more, with the excess finding its way into the environment. It is probable that less than 5% of an active material applied by simple spray techniques reaches the target site and is effective. This immediately highlights an area where enormous improvements can be made.

In some instances the use of pesticides has resulted in unwanted and unexpected changes to the environment. This has occurred because the investigations made, prior to launching the compound, to assess the effect that the material would have on the environment were not in sufficient depth. This is not to say that the investigations were negligent but rather that it was not appreciated many years ago how materials released into the environment could effect the ecosystem. Indeed the analytical tools were not available many years ago to enable the in-depth testing of pesticides that takes place today. As a result of experience with such problems as that posed by DDT (1,1,1-trichloro-2,2-bis(4-chlorophenyl)ethane) the scope of the investigations into use of pesticides has been significantly increased. It would be foolish to say that there will never be another mistake; it might be expected, however, that the number of mistakes will diminish with increased knowledge and experience.

At this time it would appear unlikely that a wholesale return to the agricultural practices of the 1940s will occur. In general it is probably true that where we have used modern technology with care and consideration the experience is positive. The problems have occurred when we have become too confident and made changes that are too large, too quickly. The legislative controls are continually changing in response to our increase in the understanding of or interaction with the environment, and this in turn is setting the standards for the new pesticides of the future. Both DDT and 2,4,5-T ((2,4,5-trichlorophenoxy)acetic acid) are examples of active materials for which the screening was not adequate. In the case of DDT it was the active material itself which was a problem whereas with 2,4,5-T it was the dioxin impurity which was the main problem. Both of these examples will be outlined later in the Chapter.

In addition to the problems specific to agrochemicals, the chemical industry faces the general problem of reducing the quantity of waste generated during manufacture of the active ingredients and formulations.

7.13 The DDT experience

The insecticide DDT was one of the first materials to indicate that the investigations being undertaken were inadequate in showing the effect of using chemical pesticides on the environment. It was first used as an insecticide in 1943 and was found to be exceptionally effective. In addition to controlling insects in crops it was very efficient in controlling insects that spread disease to animals, and it was also used on humans to eradicate certain insect pests. In comparison with previous materials it was persistent yet appeared to have no measurable effect on humans. Eventually its use became widespread and because of its ability to resist biodegradation it accumulated in the environment. Some species of insects actually became resistant because they were able to metabolise DDT to the related compound DDE (1,1-bis(4-chlorophenyl)ethane) which has no insecticidal properties.

Ultimately the persistence of DDT contributed to its downfall. It accumulated in the fatty substances in species with which it came into contact and this caused much concern at a time when new and more sensitive methods of analysis were becoming available. These new techniques, namely GLC (gas liquid chromatography), enabled the problem to be studied in more depth than had previously been possible. The study found that DDT was bioaccumulative and that residues could be found in many species and tissues. DDT was even detected in mother's milk. In some predatory bird species DDE was found to cause the thinning of eggshells and failure to hatch.

The banning of DDT in both the UK and USA took place in the late 1960s and early 1970s, respectively, although restrictions in use were imposed earlier. DDT may, to some extent, have got swept up in the tide of emotion which resulted from the problems related to the use of several chlorinated insecticides, in particular Aldrin and Dieldrin. DDT had during its lifetime prevented the deaths of thousands of the world's human population by preventing disease and crop damage. It had such a low toxicity to humans that it was in effect seen by many users as absolutely safe. Eventually this confidence led to gross misuse and, with growing evidence that some species were becoming resistant, these factors led to its demise. The irresponsible use of DDT was undoubtedly detrimental to the environment but there is a view that it would have been more realistic to place severe controls on its use rather than a complete ban. For the developed nations the use of more expensive, less effective, materials was possible. For the developing countries, however, where control of insects responsible for destruction of badly needed foodstuffs and the spread of some of the most debilitating diseases is required, the most cost effective control was, and still is, the use of DDT. Although DDT is currently banned in most countries it is still used today in some areas of the world, mostly to prevent the spread of malaria. Were DDT to be discovered today it is likely that with more restricted use and careful controls the DDT story might have ended differently. The history of DDT is interesting and well documented.[24] The investigation into the effect of DDT and other pesticides on the environment was probably the start of pesticide ecology as we know it today. Humanity has learnt from this experience, but not without some cost to the environment.

7.14 The 2,4,5-T experience

The environmental impact of impurities in a product can also be more problematic than the active ingredient itself. This is demonstrated by the effect of the dioxin impurity TCDD (2,3,7,8-tetrachloro-dibenzo-*p*-dioxin) in the herbicide 2,4,5-T. Depending on the process conditions used in the manufacture of 2,4,5-T, TCDD can be formed. The problem became apparent when the US army used the herbicide under the code name 'agent orange' to defoliate trees during the Vietnam War. Under the conditions of use many of the United States and Vietnamese troups came into contact with substantial quantities of material. Subsequently problems of infant mortality and deformation along with a higher incidence of cancer were studied and the use of 2,4,5-T has been banned in many western countries[25]. Pure 2,4,5-T does not appear to have the same teratogenic problems as TCDD and despite the fact that it is possible to make 2,4,5-T with exceptionally low levels of TCDD, the indiscriminate use of the product and the emotive issues involved have resulted in the banning of a very useful herbicidal material.

7.15 Legislative control relating to pesticides

It is as a consequence of society's experience with the problems associated with the use of pesticides like DDT and 2,4,5-T and the expansion of our knowledge resulting from the subsequent investigations that legislation has been developed to control the manufacture and use of pesticides world-wide. The British Agrochemicals Association Annual Review and Handbook for 1993 lists 29 different pieces of legislation which control the manufacture, sale and use of pesticides in the UK. Many of these are not specific to agrochemicals and relate to safety, packaging, labelling, transport, weights and measures, etc. Four controls are, however, worthy of special mention.

In the UK the principal health and safety legislation concerned with controlling the workplace during the manufacture, formulation and packing of pesticides is the same as that controlling all other workplace activities. This is the *Health* and *Safety at Work etc. Act 1974* (HAWASA) which aims to protect people at work and the general public against risks resulting from the activities of persons at work. This legislation is not specific to the chemical industry.

The legislation known as the *Control of Substances Hazardous to Health Regulations 1988* (COSHH), made under the HASAWA, is specific to workplaces handling chemicals. The regulations require that employers assess the risk to workers from substances hazardous to health and that they introduce and maintain appropriate controls of that risk.

The *Environmental Protection Act 1990* (EPA) concerns itself with the effect on the environment of workplace practices and again is applicable to the chemical industry or any other industry which could cause pollution. Integrated Pollution Control (IPC) is one part of this legislation. The objective of the legislation is to control and minimise the impact on the environment of all process activities primarily by encouraging the use of processes which minimise the amounts of waste made and defining the standards to be attained using the principle of 'BATNEEC'. The Best Available Technique (BAT) defines the technology/methodology target which should eventually be attained whilst Not Entailing Excessive Cost (NEEC) enables a time element to be introduced. The impact of the IPC legislation is no more or less for the agrochemical industry than for any other part of the chemical manufacturing business to which it is applicable. It is interesting to note, however, that if a process does not have any waste, or has less than a certain minimum quantity of waste, it does not fall within the IPC legislation and does not need the expense of having its processes registered. In general this legislation is a driving force towards clean technology. In common with most other chemical businesses the agricultural and pharmaceutical chemical industry employs the following strategy to help comply with the IPC regulations:

(a) where at all possible avoid making a waste;
(b) where producing a waste is unavoidable make the least possible;
(c) where there is a waste reuse, recover or sell it as a by-product;

(d) where a waste is ultimately generated that has a high calorific value use it as a fuel;
(e) reduce the impact of wastes by suitable treatment.

This strategy is reflected in the methodology for developing processes with the minimum of waste outlined later in this chapter.

The *Control of Pesticides Regulations 1986* (COPR), made under the *Food and Environmental Protection Act 1985* (FAEPA), is specific for pesticide formulation and use. The regulations are in place to (i) protect the health of human beings, creatures and plants; (ii) safeguard the environment; (iii) secure safe, efficient and humane methods of controlling pests; and (iv) make information about pesticides available to the public. Before a pesticide can be sold, stored, supplied, used or advertised it must be approved by the regulating authorities. The regulations do not control the manufacturing process for pesticide active ingredients although they do control its quality in terms of assay and impurities. Thus changes to the manufacturing process, to incorporate improved technologies, are possible without the need to re-register providing the quality constraints are met.

The regulations are designed to ensure that minimal quantities of product are used for effective control, assess the risk to operators, consumers and the environment, and set restrictions on use as necessary. The ultimate restriction is a total ban on use. The net effect of this control procedure is to encourage the development of the ideal pesticide. Such a material will biodegrade at such a rate that it will not bioaccumulate or interfere with subsequent crop uses. This would imply a half life in the environment of not more than 100 days. The toxicity to humans and other relevant species must not have any harmful effect directly, indirectly or as a result of being transferred through the food chain at levels likely to be experienced. This applies to the residues left in the crop and the unused material which finds its way into the environment. The ideal material would not be easily transported through the soil by rain, etc., so that it can contaminate the ground water or rivers. The material needs to be persistent enough to enable it to be effective but not so persistent that it will cause problems due to bioaccumulation. Even if a material is very effective as a pesticide it is unlikely that companies will continue with the development if the product shows either toxicity or persistence problems during the initial screening. Consequently, the new active ingredients will need to be chemically sophisticated, very active, but specific to their target, so that only small quantities need to be applied. The research and development effort required to achieve these objectives is enormous and it is easy to understand why it is so difficult to find a material with the perfect combination of properties and why only the larger companies will be able to afford the development costs. Currently R & D expense accounts for between 6 and 10% of turnover.

It must be noted that currently the registration procedures or standards world-wide are not the same even within member states of the EC, although this may be rectified during the next few years following implementation of the EC Authorisations Directive 911/414/EEC in July 1993. In the UK an agrochemical registration package for a new product takes about eight years to assemble. At present it costs about £20m and approval takes about one to two years; this is expected to become shorter as fewer compounds are submitted for approval.

7.16 Agrochemicals – finding new active ingredients

The concept of clean technology needs be applied at all stages of the development procedure starting with the discovery of the molecular species, through process development, manufacture, formulation and use; it applies also to the disposal of the pack after application and to the subsequent fate of the product in the environment. The discovery of the carly activc ingrcdicnts was madc by thc proccss of 'scrccning'. Largc numbers of pure materials were tested for suitable activity and when that was detected the active species was developed as a product. Eventually the screening process identified fewer and fewer potential products. This technique has been called 'molecular roulette'. In 1956 by screening 1800 potential materials it was possible to find a viable product; by 1970, 7400 had to be screened and by 1977 this had further increased to 12 000.[26] Subsequently the screening process was used to detect 'lead' molecular species. These are materials which themselves have limited pesticidal activity but, with modification by the research chemist, involving the removal or addition of other functional groups, can be converted to very active materials. This has only been possible as a result of the skills and knowledge accumulated by investigating the structure and properties of thousands of prospective materials.

The more stringent specifications required for any new active ingredients in terms of reduced dose rate, increased selectivity and reduced toxicology are making the chances of finding new active materials with the right characteristics significantly smaller; it is for this reason that a new approach to the discovery of new molecular species is being developed. The first step will be to choose the target crops and the problem to control. This is a market research exercise and because the cost of the development programme will be expensive, it is imperative to choose the right target. Having decided on the target it will be necessary to study the biological system and understand the biological processes involved and to decide on the most appropriate way to obtain the required control. For example, if the target is a certain weed species in a certain crop it would be necessary to find some difference in the metabolic processes between weed and crop and identify a method by which it would be possible to inhibit the metabolism of the weed. It may, for example, be possible to attach a molecule to an enzyme in such a way that an important function is impaired. To achieve this one will have to understand the plant biology in order to identify the enzyme, to determine its structure, and be able to create a molecular species that would bind to the enzyme.

The principle of the design process is (a) the understanding of the electronic configuration of the receptor site and (b) the design of a binding molecule with the corresponding electronic configuration that will enable adequate bonding. In addition the material must have the ability to be transported to the target site within the plant where it can have the desired effect. It will be of no use to have a very active material which prevents some important function within the root system if the material can only be applied to the leaf area and remains there until it is degraded.

Consequently future research and development need to be aimed at understanding the metabolic processes which occur in crops and weeds, the way in which materials are transported through the target species, structural analysis of enzymes, etc., together with electronic configuration and structural modelling. This approach has the added advantage that once the mode of action of a particular chemical is understood it will be possible to build up a library of knowledge about different chemicals and predict their unwanted properties. As experience grows, eventually it should be possible to extrapolate and devise many new innovative compounds, possibly based on naturally occurring molecules. The time scale for this will be many tens of years.

At present the discovery process in many companies is an amalgam, the 'lead molecule' approach being combined with the 'design principle'. Lead molecules are being found and during the subsequent development the change in activity is being compared to that which might be predicted from computer modelling if the design principles were correct. With such limited information as is available at present, however, the correlation between actual and expected activity is often confusing. When, and if, the design principle is established it is likely that many of the new products will have their origins in naturally occurring compounds and are likely to have chiral centres requiring increased knowledge in the area of chiral manufacturing techniques.

An alternative strategy to developing new pesticides is to modify a plant species genetically so that it becomes resistant to the activity of a known pesticide. Species resistance may be due to metabolism of the active material to a less active substrate. By studying the mechanism involved it may be possible to identify the gene which is responsible for the manufacture of the enzyme which catalyses the conversion. It may then be possible to insert the relevant gene into the DNA of another plant species in order to make it resistant. Resistance to specific herbicides has already been achieved with tobacco, cotton, rape, rice, flax, tomato, sugarbeet, lucerne, lettuce, melon, chicory, soybean and maize.[27] Although there are several transgenic species at various stages of development, at present, none is commercially available. It is expected that some crop species will be made available during 1994. When considering the use of these techniques one should consider the possibility of a transgenic plant becoming a weed that could be difficult to control in a subsequent crop. Previously it was thought that insects would not become immune to natural insecticides but there is now evidence that this is not the case. Therefore it might be commercially unwise to invest heavily in producing a transgenic crop species with specific insecticidal ability only to find that the insect became resistant.[28] Thus there are many problems associated with releasing such genetically manipulated species into the environment and this approach should be taken with considerable caution.

Identification of an active molecular species is only part of the development, since toxicological and ecotoxicological studies are required as well as the process research to establish a viable industrial manufacturing process and application technology. The time scale for the entire development programme for a new product may be in the region of tens of years and companies may have to contend with the fact that over this timescale the market has changed and the product is no longer required.

One fundamental difference between an agrochemical and a pharmaceutical is that the former may be used for the control of many different problems on many crop species whereas pharmaceuticals are used only on humans.

7.16.1 Agrochemicals – process development

In general agrochemicals made in low tonnages are likely to be made using batch processes. Established stable compounds made in larger tonnages are likely to be either batch or continuous processes depending on their chemistry. Such processes have been evolving for many tens of years and have been adapted to meet the changing needs of the industry. In order to reduce waste on a global basis, the manufacturing process needs to be studied as part of an integrated production package rather than the conversion of a particular raw material into a particular product. The chemist needs to look at the available raw materials as well as the by-products from other processes which would be useful raw materials.

The integration of processes is an important aspect of clean technology and is an advantage to the larger company operating on a large chemical complex. The possibilities of one company using another company's waste as a raw material are minimised due to confidentiality problems. In the UK, however, it is possible that the IPC legislation which requires public disclosure of wastes may help. What is required is a more open attitude towards by-product use and more emphasis during process design on the ability to control and improve the quality of by-products. It is very probable that there are tonnes of materials being treated as waste by one company that would be useful as a raw material by another. Normally companies that manufacture virgin products are not interested in marketing the same materials as by-products. It is probable that they see this as a danger to their virgin product business. Faced with the option of paying to dispose of a potential by-product or modifying their process, many manufacturers could be forced into finding markets for their by-products, causing significant unnecessary disruption to established virgin product businesses. A nationally accepted and widely used database of by-products and 'wastes' could help bring producer and user together with significant savings.

Apart from the traditional chemical process, consideration should be given to biological methods of manufacture by micro-organisms, plants, even animals. It is not unreasonable to assume that the 'designer approach' to identifying a pesticide is more likely to direct attention towards more complex molecules possibly similar to naturally occurring materials. Under these circumstances the use of biological methods of manufacture of the intermediates could be the most efficient while the final modification of the chosen functional groups could be made using conventional chemistry. Where chirality is an important factor, the development of process techniques enabling specific chiral materials to be made needs to be expanded. Again studies of 'natural' processes which enable manufacture of chiral molecules could be beneficial.[29]

In view of the wide variety of conventional chemistry used, the scale of operation and the manufacturing cost for different companies it is impossible to be very specific about the approach to be taken to achieve a cleaner process. It is possible, however, to outline some of the important guidelines and some of the 'tools' available to the development chemist. One of the best tools available is the 'mass balance'. The biggest problem of putting together a mass balance is normally the analysis of the individual process streams and finding an affordable, user-friendly, easily modified mass balance programme. There is no substitute for putting into a mass balance the known information which enables identification of the inadequacies. Examination of the mass balance normally shows up the largest problem areas in the process and identifies possible approaches to take to improve the situation.

Several companies are using factors to indicate the efficiency of their processes in converting input into useful output. Both Rhône Poulenc Agriculture and Zeneca Pharmaceuticals are using a mathematically derived factor to rank the efficiency with which their processes convert raw materials into product.

For a reaction defined as:

$$A + B = P + W$$

The factor is:

$$(A + B - P)/ P$$

where A and B are the weight of raw materials, P weight of product and W weight of waste. Having derived their factors independently, both companies have effectively come to the same conclusion about how to indicate the comparative environmental influences of their processes. A factor of zero indicates that all the materials entering the processes are converted to product. Using the effluent load factor, Zeneca have found, for the processes so far reviewed, a range of 0.25 to 55; similarly RP Agriculture using their process index see their processes in the range 0.2 to 20. Using this technique with care, it should be possible, at an early stage within the process development procedure, to decide the likely environmental acceptability of a proposed process route. Application to existing processes will identify those requiring development or substitution and assist with prioritisation of action plans. The concept of such a factor can be extended to energy or any other resource.

The engineering design of a process is also critical in regard to its ability to convert raw material to product and to avoid loss of material[30], particularly solvent into the atmosphere.

Although not exhaustive the areas worthy of investigation when developing a process in order that waste is minimised are listed below:

1. Use of raw materials of an appropriate quality; too high can be expensive whereas too low can lead to unwanted by-products. The possibility of using by-products or making usable by-products should be investigated.
2. Use the minimum number of solvents; avoid mixing solvents.
3. In a multi-stage process, engineer the reaction so that transfer from one stage to another can be achieved without isolation of the product.
4. Avoid drying stages wherever possible.
5. Optimise concentrations for reactions based on total process; not individual stages.
6. Recycle process washes.
7. Identify the fate of the materials used with a mass balance.
8. Identify the factors influencing the production of impurities so that purification stages can be minimised.
9. Investigate the possible benefits of lower direct yield and recycling against higher yield but more impurities.
10. Do not hesitate to use pressure processes.
11. Avoid compromise on equipment design, mixing, heat transfer, isolation, etc.
12. Discuss with the customer how the product is to be used. There may

be some commonality that will enable elimination of a process operation, e.g. use of the same solvent could save an isolation or drying operation.

13. Investigate the truth of 'myths and legends' of the process at the level of the process operators. Process operators often notice certain correlations whilst operating a process but do not understand the importance of their observations.
14. When using low boiling point solvents consider the effect that this might have on VOC emissions.

Some inherently cleaner technologies exist which are either still being developed or are under-utilised. Electrolysis is often only considered as a solution when other techniques have failed. It is not surprising that under these circumstances it also fails. The point has been made that if, where appropriate, electrolysis were seriously considered as a technique during the early stages of process development it would probably have a larger success factor and be more effective than traditional chemistry. The use of membrane technologies for separation is still in its infancy but is gaining more acceptance. Nature manages to produce some of the most complicated molecules known, mostly in aqueous media at ambient temperature and pressure by the use of catalysts. This fact has not gone unnoticed by the chemical industry and a significant amount of research is being undertaken to find new catalyst systems. Consequently the range of synthetic catalysts available to the process research chemist is expanding by the year and should be considered carefully during route scouting and process development. It has been said that the way forward to cleaner technology is the replacement of 'stochiometric technologies' which often produce inorganic salts as by-products with 'catalytic alternatives'.[31]

The use of natural systems is rarely exploited. Micro-organisms such as bacteria and fungi find niche uses in the brewing and fermentation industries. Raw materials isolated from crops are just beginning to receive recognition. The use of some animal species to manufacture complicated molecules in their milk is becoming of interest and will be described later.

7.16.2 Agrochemicals – waste minimisation case study

A herbicide active ingredient with a market of about 900 tonnes per year was formulated as an aqueous solution of the sodium salt. The chemical process for the manufacture of the active ingredient involved a batch condensation in a solvent followed by transfer into an aqueous medium which produced an aqueous solution of the sodium salt required. The solvent was recycled into the next batch. The product was contaminated, however, with about 15% of co-product and large quantities of sodium carbonate, and before the material could be formulated both impurities had to be eliminated. In the initial process this was achieved by reducing the pH of the aqueous mixture by the addition of hydrochloric acid and filtering off the precipitated product in its acidic form. The acid was then redissolved in sodium hydroxide, giving the formulated product. This had two main disadvantages: (a) there was a loss of product in the filtrate and (b) disposal of the filtrate itself was expensive. In addition it was most frustrating to have what was essentially the formulated product available directly from the process whilst being prevented from using it directly because of the co-product and the carbonate content. A study of the process and formulation were undertaken and it was found that:

(a) if the order of addition of reactants were altered and the reaction temperature more strictly controlled, the amount of carbonate and co-product could be almost halved;
(b) the co-product could be converted back to the product by modifying the pH of the mixture after the reaction was complete. However, this produced more carbonate so it was still important to keep co-product manufacture at a minimum;
(c) the quantity of an expensive raw material could be substantially reduced if the moisture content of the raw materials was reduced from about 0.5% to 0.1%;
(d) provided the carbonate content was maintained at a sufficiently low level the carbonate could be decomposed by reducing the pH of the mixture using a small quantity of hydrochloric acid. The extra chloride presented no problem provided it was kept below a specific level.

The process development used about 1.5 chemist years of resource and the equipment modifications cost about £250 000. The process has operated using the modified procedure for several years with the following results.

- in the first three years of operation 2800 tonnes of product were made;
- 731 fewer tonnes of organic chemicals were required;
- 966 fewer tonnes of inorganic chemicals were required;
- 6200 tonnes of liquid waste contaminated with organic chemicals with a COD equivalent to about 1950 tonnes of oxygen were eliminated;
- 3250 tonnes of water were saved;
- solvent losses were reduced by 567 tonnes;
- in addition to the energy consumption, labour requirement and other services related to the purification step were eliminated;
- the estimated savings during a three-year period were in excess of £1 million.

This case study is indicative of the opportunities that are available for improving existing processes. This is typical of a product which was initially launched for a market of about 50 tonnes per year, for which the initial process was adequate, but which increased to 900 tonnes per year over a 7-year period; at this level of output the initial process was inadequate. During that time no review of the process and its development opportunities was made due to lack of resources. When the review was made, using the techniques outlined earlier in the chapter, the development potential and strategy became obvious and in this case was realised with significant advantage.

7.16.3 Agrochemicals – formulation and application development

The active ingredient must be presented to the retailer/customer in such a form that it can be stored, transported and used safely and effectively. This process is known as formulation. Two aspects of formulation for agrochemical products are becoming more important. First, facilitating easy measurement of the quantity of material used while minimising the risk of contamination of the application operative and, second, enabling the product to be targeted more specifically thus minimising the use of material and any waste. The type of formulation and application technique used is largely influenced by the pesticidal activity and mode of action. The most common methods of application are as a spray, granule or dust; paints, pastes and gases are sometimes used for the control of fungi and insects. With the exception of spray techniques most formulations are used as supplied by the distributor. For some materials it would be advantageous if the formulation enabled the active material to be released slowly, thus reducing the number of applications of materials which are easily biodegraded; such slow release formulations minimise overdosing and wastage.

Many techniques are used to obtain slow release and almost all rely on the slow degradation or dissolution of a substrate into which the active material is incorporated or contained. The primary function of the formulation is often to provide a medium for enabling a small amount of active ingredient to be distributed over a large area. In addition, the appropriate use of fillers, solvents and surfactants can often be made to optimise the efficiency of retention, reaction and distribution of the active ingredient: they can make it showerproof, enable the mixing of two normally incompatible materials and numerous other functions.[32]

Products which are applied by spraying are supplied in several different forms, including water- or oil-based liquids which contain the active ingredient in solution or as a very fine suspended solid. For spray applications the concentrated formulation of the active ingredients is normally mixed with water as a diluent. Recently much attention has been focused on this mixing process because it is at this time that the operator is most likely to come into contact with the concentrated product. Once formulated the material has to be packed in such a way that it can be easily used by the farmer, whilst minimising the chance of the user becoming contaminated. The use of smaller quantities of more active materials along with increased concern about user contamination when preparing mixtures has prompted the introduction of formulation/packing combinations which minimise user contact and facilitate measurement of the quantity to be used. Earlier developments aimed at solving this problem relied on more elaborate containers which minimised drips and enabled splash-free action whilst providing improved measuring ability; more recent innovations have utilised water-soluble sachets and soluble tablets. Both systems enable the provision of packs of multiple small doses which along with more accurate measurement of the spray diluent enables more accurate mixtures to be made, providing the experienced operator with the opportunity to minimise pesticide use; in addition, risk of contamination from empty packs is minimised. The introduction of these new forms of spray concentrate is likely to be a serious competitor to simple liquid formulations.

During application the user must be careful to avoid contamination of himself and others. The formulation is generally diluted substantially for application, reducing the possibility of problems arising from accidental contamination.

The concept of clean technology has been reflected in formulations and applications technology, where the overall objective is to enable use of less active product by improved targeting of the active ingredient. In addition, the use of volatile organics in formulations is avoided to reduce emissions of volatile organic compounds (VOC). Thus where possible oil-based formulations are being replaced with equivalent water-based systems: similar trends are evident in the paint industry. As a result of the promotion of more responsible use of pesticides, the amount of material used per hectare in the UK has now fallen by about 32% since 1984 to about 5.5 kg per hectare.[33] This has been achieved mostly by more economical use of materials and it is therefore possible that further reductions may be achieved by the use of improved applications equipment and technologies.

The application of synthetic pesticides may be relatively inefficient in respect to target delivery. When a farmer sprays a field with, for example, a contact herbicide only part of the applied dose actually makes contact with the target plants. Normally blanket spraying is carried out despite the fact that only part of the field may have a weed problem; only part of the material applied is able to be effective. Taking account of the possibly low levels of absorbance into the plant and distribution to the target enzymes upon which the herbicide is active, it is possible that only a few percent of the applied material is actually effective. Even partial resolution of this problem would help reduce the quantity of material manufactured and used.[34] In recent years much attention has been given to various aspects of this problem. The size and uniformity of spray droplets is important in the control of spray drift and ensuring adequate foliage cover. Numerous techniques have been developed to optimise droplet size; pneumatic nozzles, electrostatics, spinning discs and hydrodynamic nozzles. It is considered that by the correct combination of uniform sized droplets and the use of an air curtain to direct the spray the use of blanket spray techniques can be optimised.

The ideal system would be one in which the target can be identified, and then given a minimal dose of pesticide. Sophisticated computer control systems are being developed. Such a system, which is termed 'spatially selective application', depends on identification of areas of weed infestation, and the use of a controlled delivery system for application. Weed detection can be derived either from historical data or from a real time detection system. Such systems have the potential of considerable reduction in use of product especially on larger farm units.[35, 36]

Taking this concept even further, if it became possible to differentiate between weed species it would be possible to have multiple pesticide concentrates in a delivery system which could then apply the most appropriate active ingredient. There is much scope for the development of reliable, easily managed delivery systems that can help the farmer to minimise the use of pesticide whilst achieving the desired degree of pest control. The best technique will depend on the crop, the target pest, the active material and the likely climatic conditions. The long-term objective must be to develop formulation and application systems which enable the use of less material, using targeted rather than blanket application techniques.

7.17 The pharmaceutical chemicals industry

The development of the pharmaceuticals industry has taken a very similar course to that of the agrochemicals industry. In the case of pharmaceutical materials the toxicity to humans and possible side effects are the paramount criteria, whereas on the whole the effect that a material might have on other species and the environment has not normally been considered of great importance. As with pesticides, some problems have occurred because of shortcomings in the breadth of screening procedures for some pharmaceutical materials. Thalidomide is perhaps the best known example. The following sections outline the interaction between the pharmaceutical chemicals industry and clean technology; only those areas where the problems and solutions are not the same as those already discussed for agricultural chemicals are discussed in detail.

7.18 Outline of historical development of pharmaceutical chemicals

The trend in pharmaceutical chemicals virtually parallels that of agrochemicals. Chemicals have been in medicinal use for centuries. The concept of using materials which at first were quite crude and in some instances very toxic to give nature a hand has subsequently progressed to the use of very pure materials which are specific and require very low dose rates to achieve results. In the very early days naturally occurring materials, found to have curative properties, were often used directly or perhaps as an aqueous extract. An example of this would be the application of tree bark or mosses to abrasions, cuts, etc., which by experience were found to have beneficial effects. Over the centuries it was established that individual chemicals were responsible for the curative action and gradually the means whereby these chemicals could be produced have been established. A revolutionary step involved the development of antibiotics. The first true antibiotic was penicillin, a chemical which was produced by the fungus penicillium to protect against bacterial attack. The concept of improving on natural materials evolved and eventually novel synthetic materials were being produced. An early synthetic antibiotic was sulphapyridine which contributed to saving many lives during the Second World War. The range of medicinal products is now extensive, extending from prophylactic to curative and covering anti-infectives, dermatologicals, and products for action on muscular, skeletal, genito-urinary, reproductive, alimentary, cardiovascular and central nervous systems.

More recently the mode of action at the molecular level of some beneficial materials has been elucidated and the understanding of animal and human biochemistry is progressing. Eventually the combination of this knowledge may enable the chemist to design molecules which have specific activity. Recent understanding of the mechanism by which DNA influences the development of living creatures and the nature of genetic modification has opened the door to genetic engineering along with the attendant ethical problems.

The size of manufacture for a pharmaceutical chemical depends on the relationship between dose rate and quantity of material required. For a dose rate of about 10 mg/day for a two-week period, 1 tonne of active ingredient is sufficient to give one treatment to over 7 million people. Assuming a population of about 55 million for the UK, one treatment for the entire UK would require 7.7 tonnes. For the USA with a population of about 700 million the demand would be only 100 tonnes. The amount of material required for pharmaceutical use is normally small. Though there are exceptions to this, for instance production of active ingredients for OTC (over the counter) analgesics, cough remedies, etc., could be in the thousand tonne range. In general the cost of pharmaceutical chemicals is high because they are sophisticated and because the development costs are high; the large tonnage active ingredients used in OTC products, however, are likely to be much less expensive. Typically 100 tablets of 500 mg paracetamol cost about £1.50 whereas 100 tablets of 100 mg Retrovir (AZT) cost about £120.

The UK pharmaceutical industry is the fifth largest manufacturer of medicinal products in the world, meets about 80% of the home requirements and is the second largest contributor to Britain's balance of trade. The value of the pharmaceutical business world-wide is thought to be about £120 billion. The largest manufacturer is the USA (30%) followed by Japan (22%), Germany (7%), France (7%), with the UK at 6%. Of the global research effort about 8% takes place in Britain, indicating the high regard of the rest of the world for the UK's innovative ability. At this level research and development accounts for about 17% of turnover. The UK trend would appear to be upwards in terms of output value of sales and R & D expenditure.[37]

7.19 Outline of problems with pharmaceutical chemicals

The perceived problem with pharmaceutical chemicals is much less than that with agrochemicals. When used they are normally administered in very small doses and targeted much more accurately than agricultural chemicals and therefore seldom affect species other than humans. The results can be devastating, however, when the procedures designed to detect unwanted side effects are inadequate. Thalidomide is a good example. As a result of these experiences, the screening procedures for pharmaceutical chemicals are continually being reviewed and the requirements are reflected in the controlling legislation. As for agrochemicals, these controls are not the same world-wide.

The relatively small manufacturing tonnages for the majority of pharmaceutical chemicals along with their limited lifetime and high value are all factors which are likely to check the natural progression towards the use of clean technology. Consequently the move towards clean technology is likely to need the assistance of legislative encouragement and the publicity derived from good experiences of those companies which have achieved progress in that direction.

Within larger companies the use of manufacturing sites for multiple business activities will increase the probability of the transfer of innovative clean technology from one business area to another, without breach of confidentiality. In some instances where a business resides as a satellite on another business area's site, the policy for that site may compel the use of cleaner technology.

7.20 The thalidomide experience

Thalidomide was first used in Germany in 1956 and was considered a very effective sleeping aid which also relieved the effects of morning sickness. It was not until 1961, by which time nearly 500 cases of possible teratogenicity had been reported, that investigations into the cause of the abnormalities began. In the UK a similar number of problem cases were reported in the 1959–61 period. The marketed product was a racemic mixture of the two optically active isomers. Investigations revealed that the (−) isomer had teratogenic properties whilst the (+) isomer did not.[38] This is a good example of how the effects of active materials can be dependent on the specific isomer of a chiral molecule. Subsequently thalidomide was only used for certain conditions, being administered only when the possibility of pregnancy was very low.

7.21 Legislative control relating to pharmaceutical chemicals

In common with the agrochemicals industry, the manufacture of pharmaceutical chemicals is controlled by general legislation such as the *Health and Safety at Work Act* (HAWASA) and the EPA as previously described for agrochemicals. Legislation which is more specific to the pharmaceutical industry is incorporated into the *Medicines Act 1968*, which:

(a) regulates the development during clinical trials;
(b) approves processes for the manufacture of materials;
(c) licenses products;
(d) licenses wholesale dealers;
(e) controls the labelling and packaging;
(f) controls the retail supply;
(g) monitors the adverse reactions to a marketed product.

Some form of control of drugs has been in operation since 1316 when a code of quality was established by the Ordinances of Guild of Pepperers of Soper Lane. Until the introduction of the Medicines Act in 1968, there was, however, no legislation actually controlling the introduction of new drugs in the UK which related to **their safety in use**. Until that time most of the legislation was related to quality of therapeutic materials and the control of substances known to be either poisonous or narcotic. The thalidomide problem resulted in the English and Scottish Standing Medical Advisory Committees recommending future legislation and the immediate formation of the Committee on Safety of Drugs in 1963. This had no legal powers but worked on the basis of voluntary co-operation with the pharmaceutical industry in the UK as an interim measure until the Medicines Act became law in 1968.[39] Subsequently moves have been taken by the EEC to harmonise legislation for registration as set out in Directive 87/22/EEC of 22 December 1986.

Currently the time from discovery of a new molecule to the launch as a new product is very product dependent. Materials showing activity in the anti-cancer and anti-AIDS areas are likely to be developed in a shorter time frame, extending from about 5 years up to 10 years. Products requiring extended development times are not likely to be pursued because of the higher cost of development and the reduced life of patent protection. The cost of launching a new material is normally in excess of £100 m.

For a pharmaceutical product the manufacturing process is part of the registration package and is controlled by the Medicines Act. Consequently any changes to the production process have to be notified to the registration authorities before implementation. It is possible that the process changes may require that extensive and expensive toxicology studies are repeated before re-registration can be approved. Consequently most pharmaceutical manufacturers are reluctant to change their manufacturing processes unless the modified process has a significant advantage.

7.22 Finding new pharmaceutical actives

The screening procedures previously described in the agrochemicals section have also been used for the discovery of new pharmaceuticals; it is probable that the same chemicals have been screened for both functions. Currently the success rate for pharmaceutically active compounds is very low since it is unlikely that registration will be given for a new active material unless it has a significant advantage over any existing material; the probability is lower than that for a new pesticide. The game of molecular roulette is now more like Russian roulette! Consequently some of the larger companies are developing the 'designer concept' or rational design.[40] The conventional screening technique is similar to building a house with doorways of any convenient shape and size and then asking a carpenter to build a door without telling him the shape or size. You would be very lucky if you found one that actually fitted. The 'designer concept' is similar to measuring the doorway before making the door. The chances are that the door will fit. The exact fit will ultimately depend on the skill and experience of the carpenter and the reliability of the measurement. Continuing with the similarity, the current state of technology with the 'designer concept' is that we are trying to find the doorways and are also trying to develop the ruler.

Because the pharmaceutical business is market driven, at this time most large chemical companies are focusing their attention on the same market areas, research into the prevention of AIDS, cancer and heart disease being high on the list of priorities. The major effort is therefore directed towards those diseases which are prevalent in the developed countries. Consequently the problems of the under-developed countries are not being given the same level of attention.

New products brought onto the market will be expected to be more effective at lower dose rates, selective and easier to administer. Environmental persistence has not been considered as a problem but it would be foolish to find that a very useful widely used drug was not biodegradable and was contaminating drinking water to such a level that some members of the population were receiving, involuntarily, a detectable dose. Again, the experiences associated with thalidomide have highlighted how important it is to recognise the toxicity and efficacy of the individual isomeric species; the question of chirality is exceptionally important when developing a new pharmaceutical compound.[41]

7.23 Pharmaceutical process development

In general, the principles used in the development of manufacturing processes for pharmaceutical chemicals are similar to those used for agrochemicals. The product volume ranging from one to one thousand tonnes is likely to be an order of magnitude smaller than its agrochemical counterpart; the types of process used are, however, similar. The size of manufacture, high value and the possible limited lifetime of these products, however, have reduced the drive towards clean technology. The cost of development and conforming with the regulatory requirements is considerable, whereas the cost of waste in the manufacturing process is normally less important. Changes to an existing registered manufacturing process, although simple from a chemical or engineering viewpoint, can be very complex when the problems of registration are concerned. It could be more difficult for the pharmaceutical industry to adopt cleaner technology for existing processes, as required by IPC, whilst not incurring expensive re-registration costs. It is important, therefore, that the most advantageous process is developed during the process development stage. There is a time limitation, however, on this development because of the current patent legislation. The life of a patent starts when the application is made, normally at the discovery stage; every year taken to bring the product to the market loses a year of patent protection. It is, therefore, very important to get the correct balance between the development time and the number of years of protected sales.

Assuming that new active ingredients developed by 'design' are likely to mimic nature, they are expected to be complex and many could be expensive to produce by conventional chemistry. It is thus probable that many of the raw materials and intermediates will be obtained from animals, plants or micro-organisms, with the final stages of manufacture being carried out using conventional chemistry in order to introduce the required functional groups. As an example the raw material for 'Taxoterre', which is obtained from European yew tree leaves, is modified by traditional chemical processes to give the final product. The expansion of the available techniques for the manufacture of chiral materials will be required.[42] It would seem probable that, initially, this approach will result in an expensive product which will be more suited to the pharmaceutical market. Once sufficient experience has been obtained and the procedures become established and less expensive, the techniques are likely to be used for the manufacture of other products, particularly agrochemicals which are also likely to become more 'designer' orientated.

Although currently controversial, the use of bioengineering is likely to play an ever increasing part in our ability to manufacture materials, treat certain diseases and improve agricultural efficiency in an environmentally sensitive manner. Research into genetics will enable us to monitor and prevent the occurrence of certain problem diseases. There are now numerous examples appearing in the literature relating to missing genes, wrong genes or extra genes causing problems for certain individuals, families and ethnic groups. This information should enable many problems to be resolved in the future. Genetic engineering will enable us to transfer beneficial functions from one species to another. This has already been accomplished, under controlled conditions, with many species. Transgenic sheep have been developed to enable ewes to produce milk containing a large quantity of the human blood clotting agent, factor 9.[43] Plants are being modified genetically to enable them to produce materials which are expensive to manufacture by conventional synthetic chemistry. Experimentation has also been made using micro-organisms. What of the dangers of these new technologies? Are we about to make the same mistakes again that we made years ago during the rapid development of the use of pesticides and drugs?

At this time, considering the benefits that this technology could achieve, it would probably not be so worrying to the general public that large animals, which were the result of genetic engineering, were grazing in fields providing of course that they were not suffering as a result. However, when the implications of the use of genetically engineered plants or micro-organisms are considered, there is increasing public concern regarding guarantees of the necessary degree of control. It is not surprising that there is much debate about the controls and screening required before we can make use of these advanced techniques. However, the advantages of these technologies are potentially so great and so much resource has been put into their development that there will be a lot of pressure on legislative bodies to approve such techniques as quickly as possible.

7.24 Pharmaceutical formulation development

Formulation development for pharmaceutical materials is focused on enabling the active ingredient to be readily applied or transported to that part of the body requiring treatment. In many respects this is the same objective as for agrochemicals. By targeting the material it is possible to reduce the dose rate and obtain a more effective treatment with fewer side effects. This continues to be the concept of clean technology. It enables less waste because fewer materials are required, manufactured and released eventually into the environment. Much consideration has been given to making the use of the product as easy and pleasant as possible. For instance, it is much easier to take one tablet a day that releases its active constituents slowly during that day than four tablets that release their active material in four bursts. It is far more acceptable and reliable to wear a patch which slowly releases a material which is absorbed into the skin and then to the rest of the body, than to take a tablet once a day or even having to have an injection. Drugs which are targeted at the respiratory system can be more effective if delivered in the form of an oral or nasal spray. The different techniques of achieving the targeting required and being 'user friendly' are too numerous to be elaborated in this text. The concept of the 'golden bullet', where a material is administered to the patient and becomes localised in the required area is, normally, exceptionally difficult to develop.

7.25 Conclusion

Clean technology is a compromise which enables the public to retain the benefits of the products of both the agrichemical and pharmaceutical industries while minimising the impact on the environment of their related activities. The concept encompasses the identification of new active materials, the process and formulation development, the manufacture, use and after care. For the majority of such manufacturers the cost of waste in the past has not been the prime concern and they have not been naturally driven towards the use of clean technology. There are, of course, exceptions to this generality. The three determining factors of importance which influence the choice of technology include the value, the expected life and the scale of manufacture of the product. Taking these factors into account, the manufacture of agrochemicals is more likely to use clean technology than the manufacture of pharmaceutical chemicals.

The cost of waste is now being given a much higher profile and companies are beginning to see that waste minimisation programmes are having a positive effect on their profitability. Often there is little or no capital expenditure required to achieve some of these savings. The overall tightening of controls for disposal of waste and the ramifications of IPC are increasing the cost of waste disposal, thus giving impetus for investment in technologies that enable manufacture without waste. There is a need for research and development in many areas so that we understand the biology of the systems we wish to control, can identify materials that will give us the desired control, can improve the technologies for manufacturing these materials and the techniques for their use by the consumer. The quality demanded of new products is superior to those developed in the past in terms of activity, specificity, toxicity, manufacturing process and method of use. In order to meet these demands the scientific community has to be exceptionally innovative, responsible and work crazy. Good hunting!

Acknowledgements

Acknowledgement is made to my colleagues in R.P. Agriculture, R.P. Rorer, Zeneca Pharmaceuticals, Horstine Farmery, and The Fertiliser Manufacturers Association who have helped me with this chapter. Also for assistance received from Association of British Pharmaceuticals Industries, British Agrochemicals Association, British Crop Protection Council.

Bibliography

Hassall, K.A. (1990) *The Biochemistry and Uses of Pesticides*, 2nd edn, Macmillan, London.
Cremlyn, R.J. (1991) *Agrochemicals: Preparation and Mode of Action*, John Wiley & Sons, Chichester.
Barrington Cross (1988) *Pesticide Formulations: Innovations & Developments*, American Chemical Society.

References

1. CEST (1993) Press release – Aire and Calder Project, Centre for Exploitation of Science and Technology, London.
2. Dorfman, M.H., Muir, W.R. and Miller, C.G. (1993) *Environmental Dividends*, INFORM, New York.
3. Stryker, P. and Lone, P. (eds) (1995) *Cutting Chemical Wastes*, INFORM, New York.
4. Borlaug, L.E. (1990) The challenge of feeding 8 billion people, *Farm Chemical International*, summer issue, 10–12.
5. Urban, F. and Dommen, A.J. (1989) The world food situation in perspective, World agricultural situation and outlook report, Washington, DC, Dept. of Agriculture, *WAS* **55**, 8–16.
6. MAFF (1993) *Agriculture in the UK*, MAFF, Cambridge, p 4, 57.
7. FMA (1993) *Fertiliser Review 1993*, Fertiliser Manufacturers Association, Peterborough.
8. Kirk-Othmer (1985) *Concise Encyclopaedia of Chemical Technology*, Vol 6, 23nd edn, John Wiley, New York, pp 92–94, 465–468, 786, 869–871.
9. Blanken, J.M. (1987) *The Ammonia Industry: thoughts about the past, present and future*, The Fertiliser Society, London.
10. Chatt, J. (1976) *Nitrogen Fixation – future prospects*, The Fertiliser Society, London.
11. Beringer, J.E. and Day, J.M. (1981) *The Role of Biological Nitrogen Fixation in UK Agriculture*, The Fertiliser Society, London.
12. Diehl, L., Kummer, K.F. and Oertel, H. (1986) *Nitrophosphates with Variable Water Solubility: Preparation and properties*, The Fertiliser Society, London.
13. FMA (1993) *Fertiliser Review* (1993), Fertiliser Manufacturers Association, Peterborough.

14. Stafford, J.V. and Ambler, D. (1992) *Mapping Grain Yield for Spatially Selective Field Operations*, AFRC Silsoe Research Institute, Wrest Park, Silsoe, Bedford.
15. Domangue, M. (1993) Pesticide, nitrate survey reinforces findings of original Cape Cod study, *Euro-turf and landscape management*, June 1993.
16. Worthing, C.R. and Hance, R.S. (1991) *A World Compendium – The pesticides manual*, 9th edition, The British Crop Protection Council, Farmham, Surrey.
17. MAFF, *Pesticides 93*, HMSO, London.
18. British Agrochemicals Association, (1993) *Annual Review and Handbook*, BAA, Peterborough, UK, p. 28.
19. British Agrochemicals Association, (1993) *Annual Review and Handbook*, BAA, Peterborough, UK, p. 26.
20. British Agrochemicals Association, (1993) *Annual Review and Handbook*, BAA, Peterborough, p. 30.
21. Anon. (1993) Low inputs don't make it green, *Farming News*, 29 January, 20.
22. Felix, J.P., Duffus, C.M. and Duffus, J.H. (1991) *Toxic Substances in Crop Plants*, Royal Society of Chemistry, Cambridge.
23. Ames, B.N. (1983) Dietary carcinogens and anti carcinogens, oxygen radicals in degenerative diseases, *Science*, **221**, 1256–1264.
24. Mellanby, K., (1992) *The DDT Story*, British Crop Protection Council, Farnham, Surrey.
25. Sterling, T.D. and Arundel, A. (1986) Review of recent Vietnamese studies on the carcinogenic and teratogenic effects of phenoxy herbicide exposure. *International Journal of Health Services*, **16**, (ISS2), 265–278.
26. Corbett, J.R. (1979) *Chem. Ind.*, London.
27. Marshall, G. and Walters, D., (eds) (1994), *Molecular Perspective in Crop Protection*, Ch. 6, Chapman & Hall, London.
28. Holmes, R. (1993) The perils of planting pesticides, *New Scientist*, 29 August, 34–37.
29. Tambo, G.M.R. and Bellus, D., (1991) Chirality and crop protection, *Angewandte Chemie* [English version], **30**(10), 1193–1386.
30. Smith, R. and Patela, E. (1993) Waste minimisation in the process industries, reprint of extracts from the *Chemical Engineer*.
31. Sheldon, R.A. (1992) Organic synthesis – past, present and future, *Chemistry and Industry*, 7 December, 903–906.
32. Wilkins, R.M. (ed), (1990) *Controlled Delivery of Crop-Protection Agents*, Taylor and Francis, London.
33. Anon. (1992) *Agricultural Supply Industry*, 25 September, 22–23.
34. Johnson, N. (1993) *Cleaner Farming*, Centre for the Exploitation of Science and Technology, London.
35. Duff, P. (1993) Detecta spray, *Abst. Meet. Weed Sci. Soc. Am., Armidale, Australia*, **33**, 45.
36. Stafford, J.V. and Miller, P.C.H. (1993) Spatial selective application of herbicide to cereal crops. In *Computers and Electronics in Agriculture*, Elsevier Science, Amsterdam.
37. Association of British Pharmaceutical Industries (1993) *PHARMA Facts and Figures*, A B P I, London.
38. Mason, S. (1984) The left hand of nature, *New Scientist*, 19 January.
39. Griffin, J.P. (1989) *Medicines Regulations Research and Risk*, Queen's University, Belfast.
40. Anon. (1993) Taking a short cut to drug design, *New Scientist*, 29 May, 35–38.
41. Hunt, A.J. (1991) Drug chirality, *Chirality*, **3**, 161–164.
42. Collins, A.N., Sheldrake, G.N. and Crosby, J. (1993) *Chirality in Industry*, John Wiley and Sons, Chichester.
43. James, R. (1993) Human therapeutic proteins generated in animals, *Genetic Engineer and Biotechnologist*, **13**(3).

8 Plastics

R. MACKISON

'We do not inherit the land from our ancestors, we borrow it from our children'[1]

8.1 Plastics today

Plastics are an integral part of our lives. At the corner shop, in the supermarket or in the shopping mall, we see a large proportion of consumable goods packaged in plastic. At home, at work or in the car, if we stop for a moment and look around we will see examples of plastics in use. The 'dustbin' is made of plastic as is the waste bin in the kitchen. Washing up liquid is contained in a plastic bottle that serves to store and dispense the liquid. This also applies to shampoo and liquid detergents that we use to wash our clothes. In some cases, we purchase consumable materials based on the way they look and are packaged. This may be an aesthetic choice based on presentation alone, or perhaps the purchaser considers a particular package is superior from an ecological point of view. If this is the case, what are the reasons and is this a balanced judgement based on all the facts? The iron, the radio, the television, the video, the tape recorder and CD player all have plastic components. The consumables that we use with the tape recorder, video and CD are all made of and packaged in plastic. Children's toys contain more plastic today than when we were children. At work and in the office, the telephone, the word processor, a ruler, card index boxes, the floppy discs store and the waste paper bin are all items that are totally or partially manufactured from plastic.

This chapter addresses the following questions. What are plastics and how are they used? Where do they come from? How can cleaner technology be used to avoid environmental damage at source? How can waste from primary and secondary manufacture be minimised/eliminated? What is the environmental cost and how does the user compare this with alternative materials? Can plastics be reused/recycled? What are the options and how do these compare? How can the user best recycle plastic materials? Is there a case for having biodegradable plastics? What legislation is in place now and for the future? What is planned in the short term (up to the year 2000) and in the longer term into the next millennium? What research is taking place now? Are there other avenues that need to be addressed for the future? Where can one go to keep up to date and obtain a balanced view of the issues that will affect our lives in the future?

The packaging, car and domestic waste industries will be looked at in some detail to examine the issues involved within each area.

8.2 Source and nature of plastics

Today feedstocks for plastics come from one of the earth's finite limited resources – crude oil. Seven percent from a barrel of crude oil goes to make petrochemicals. A little over half of this (4% of the barrel) goes to make plastics. These figures should not be considered in isolation but rather alongside the other uses for crude oil. Table 8.1 shows how the crude oil is used in Europe. It can be seen that 29% is used for transport and 22% is used for power generation. Chapter 11 deals with energy and a part of this is power generation. The European Centre for Plastics in the Environment (PWMI) suggests that the efficiency of electricity generation in the UK is less than 30%.[2] This compares with a reported efficiency that is greater than 75% in Norway. With the main goal of cleaner technology being to avoid environmental damage at source, one wonders why there is not more pressure to have more efficient power generation.

Oil refineries produce a naphtha fraction and it is this material that is one of two hydrocarbon feedstocks that is fed to an ethylene cracker. Another hydrocarbon feedstock is ethane and this comes from natural gas. The products from the ethylene plant, ethene (ethylene), propene (propylene), butenes, butadiene, and aromatics, are the basic building blocks that are the starting point for the production of petrochemicals and plastics.[3–6] An individual compound that is the basic building block for the production of a particular plastic is called a monomer. The monomer for polyethylene is ethene and is dealt with in the chapter on clean synthesis. All manufacturers aim to optimise the individual stages in the process using all the tools available – modelling, on-line analysis, computer control, energy saving

Table 8.1 Uses of crude oil in Europe

Use	%
Heating	35
Transport	29
Energy	22
Miscellaneous	7
Petrochemicals*	7

* 57% of the petrochemicals production is used to produce plastics.

Source: Adapted from *Plastics in Perspective* APME.

interstage heat exchange. A very real part of this process is to minimise/eliminate fugitive emissions and reduce stationary source emissions. A further goal is to minimise the discharge of organics to aqueous effluent. Safety is paramount in the oil and petrochemical/plastics business; there must be no compromise as far as safety is concerned. This applies equally to the plant, the staff and the population that live adjacent to the site. There are two main types of plastic, the thermoplastics and the thermosetting plastics (thermosets). Thermoplastics can be put through a melting and solidification cycle numerous times while the thermosets, due to their chemical structure, behave in a completely different way.

Thermoplastics that are in common use are polyethylene (PE), polypropylene (PP), polyvinyl chloride (PVC), and polystyrene (PS). The starting materials for making these plastics are called monomers. The monomers that form the above polymers on polymerisation are ethene (ethylene), propene (propylene), chloroethene (vinyl chloride) and vinyl benzene (styrene) respectively. Ethylene and propylene are obtained directly as products from the ethylene cracker while vinyl chloride and styrene monomer are obtained as a result of further chemistry on products from the ethylene cracker. Styrene, for example, is produced by first reacting benzene and ethylene to produce ethyl benzene. This is then dehydrogenated to form the styrene monomer. Individual processes and process options are described in references 3 to 6.

It is interesting to consider the production of polystyrene, in that there are options available at each reaction stage. Figure 8.1 shows an outline for a chemical process in general. The principles can be applied equally to the monomer and polymer processes. Starting with the feedstocks, ethylene

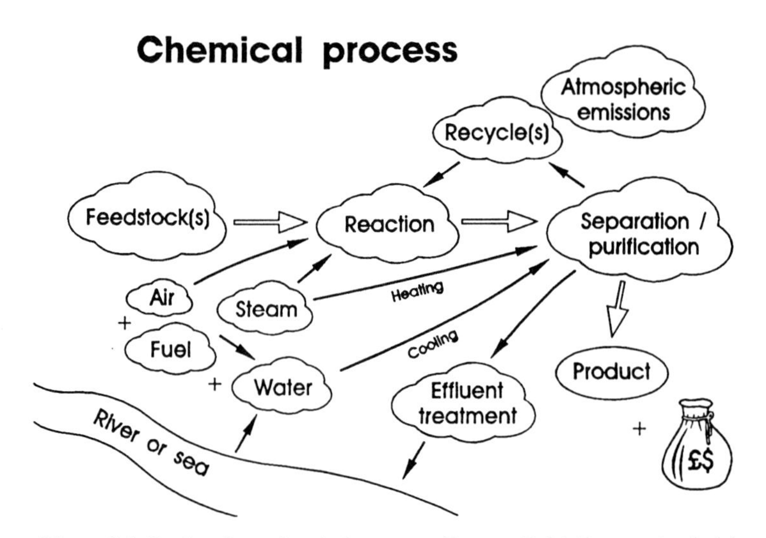

Figure 8.1 Outline for a chemical process. (Source: R.M. Process Analysis)

and benzene, obtained from an ethylene cracker, these react in a reactor to produce ethyl benzene and by-products. There are a number of routes available for the manufacture of ethyl benzene. Today one would be looking to optimise the ethyl benzene plant from a production and environmental point of view. This includes optimal utilisation of energy, minimising by-products and recycling, while at the same time minimising discharges to the atmosphere and aqueous effluent. The same requirements also apply to the dehydrogenation step. The majority of the petrochemical and plastics producers today employ total quality[7,8] techniques to assist in maximising their performance. The simplest way to describe this is to say that every operation must be on the basis of *right first time*; this means exactly what it says. Every operation must be performed correctly first time and this applies to everyone in the company from the doorman to the chairman. It includes office and production functions. The production of an off-specification product is not acceptable or permitted. This is achieved by a number of routes that all involve minute attention to detail. On-line real time analysis enables the plant to operate in a more optimal way, for example, by operating with minimum reflux ratio to provide on-specification product. As a result, the amount of steam required to heat the individual distillation columns is reduced, which again saves money and reduces emissions to the environment. The styrene monomer may be polymerised in a number of ways to produce a number of products. These include different types of polystyrene as well as copolymers with acrylonitrile–butadiene rubber to give acrylonitrile/butadiene/styrene (ABS) copolymer, which is known as toughened polystyrene.

Thermal polymerisation is one route for the polymerisation of styrene. In this case styrene monomer together with an initiator are fed into a series of polymerisation vessels. The temperature and contact time are controlled to obtain a polystyrene having particular molecular weight range. Within this process every effort will be made to conserve heat. This will be achieved by having adequate insulation on the vessels and distillation columns. Where possible energy-saving schemes will have been implemented. If vacuum is required, and this is obtained using steam ejectors, the steam may also be used to preheat the feed styrene monomer. Another way of polymerising styrene is emulsion polymerisation. Here small droplets of styrene monomer are suspended in water; this suspension is assisted by having an emulsifier and an emulsion stabiliser present. Further details of the polymerisation methods available can be obtained from Kirk-Othmer.[9,10]

Thermosets include such polymers as phenol formaldehyde resins, urea or melamine formaldehyde resins, unsaturated polyesters, alkyd resins, epoxy resins and polyurethanes. It is not possible with all these materials to melt and solidify numerous times. Once the material has been shaped and moulded, it is not possible to remelt and reshape it. This is because the thermoset resins polymerise and crosslink in such a way that makes remelting impossible.[4,5] As with the thermoplastics, the thermosets are manufactured to optimise the return for the producer while minimising waste, environmental emissions and discharge to aqueous effluents.

8.3 Common plastics and their uses

Plastics have become an indispensable part of modern life especially in Western countries where a high proportion of consumption relates to packaging (Table 8.2).

Some common plastics are outlined below together with their primary and possible secondary uses after they have become redundant in their primary use.

8.3.1 Primary and secondary uses of thermoplastics

Polyethylenes. Low density polyethylene (LDPE), high density polyethylene (HDPE) and linear low density polyethylene (LLDPE).

Primary uses. Cling film, bags, bin liners, toys, flexible containers, general film, bottles for food and household products including detergents and cosmetics, bottle caps, containers, housewares, fuel tanks and industrial wrapping and film, sheets, gas, waste and irrigation pipes.

Secondary uses (i.e. uses after recycling). Waste disposal bags, industrial bags, flexible bottles, pipes, containers, flexible membranes, agricultural film and wood substitutes, e.g. animal flooring, fencing, detergent bottles, pipes and containers. A method has been developed to incorporate recycled HDPE with virgin polymers.

Table 8.2 Percent plastics consumption by market, 1991

	West Europe	Japan	USA
Building	22	10	22
Packaging	32	28	30
Teletronic	9	14	6
Automotive	4	9	3
Other transport	3	1	1
Other markets	30	38	38

Source: EuPC Monograph, 1992.

Polyethylene terephthalate (PET)

Primary uses. Carbonated drinks bottles, food packaging, carpets, cords for vehicle tyres.

Secondary uses. Textiles for bags, webbing and sails, jackets and pillows, cushions, sleeping bags, rope, string and carpets.

Polypropylene (PP)

Primary uses. Packaging such as yoghurt and margarine pots, sweet and snack wrappers, vehicle battery cases, cereal packet linings, microwave proof containers, medical packaging, milk and beer crates, automotive parts, carpets and fibres and electrical components.

Secondary uses (i.e. after recycling). Milk and beer crates, timber substitutes, automotive components, tool boxes, auto batteries, chairs and textiles.

Polystyrene (PS)

Primary uses. Packaging, dairy product containers, electrical appliances, thermal insulation, tape cassettes, cups and plates.

Secondary uses (i.e. after recycling). Thermal insulation, trays, office accessories, rubbish bins.

Polyvinyl chloride (PVC)

Primary uses. Window frames, rigid pipes, flooring, wallpaper, bottles, packaging film, guttering, cable insulation, credit cards, medical products including plasma bags.

Secondary uses (i.e. after recycling). Pipe fittings, conduits, floor tiles, fencing rails, containers, footwear, garden furniture.

A number of specialist plastics have not been detailed above. Some of these include polyamides (PA), polycarbonates (PC), polyvinylidene chloride (PVDC), polymethylmethacrylate (PMMA) and styrene copolymers (e.g. ABS: see Section 8.2).

8.3.2 Primary uses of thermosets

Epoxy resins. Adhesives, automotive components, electrical/electronic components, sports equipment, boats.

Phenolics. Adhesives, bonding wood laminates, ovens, toasters, plugs, handles for pots and cutlery. Automotive parts and electrical components such as circuit boards.

Polyurethane (PU). Coatings, finishes, additives to improve rubber's resistance to chemicals and ozone. As elastomers they form bumpers, gears, diaphragms, gaskets and seals. PU foams are found in cushions, mattresses and vehicle seats. The technical aspects relating to recycling epoxys and phenolics are still under development. PUs can be recycled for carpet underlay and shoe soles.

Source for above uses: *Plastics in Perspective*, published by the Association of Plastics Manufacturers in Europe (APME).

8.4 Cleaner technology in plastics production

Figures 8.1 and 8.2 suggest minimisation of waste at each stage of production. Within each plant in the chain of processes that go to make the monomer and polymer, efforts will be made to reduce and eventually eliminate fugitive organic emissions to atmosphere. The same will apply to the aqueous effluents from these processes. Efforts will also be made at each stage to reduce the energy requirement while at the same time

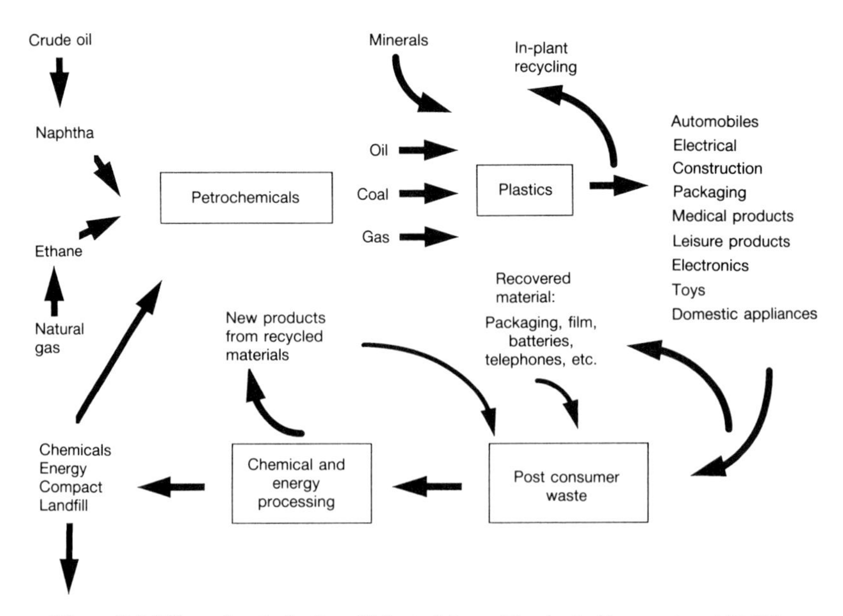

Figure 8.2 Life cycle of plastics. (Adapted from *Plastics in Perspective*, APME)

maintaining or improving quality. Within the polymer production and processing all efforts will be used to reduce the polymer waste.

This may be summarised as follows:

Reduction of waste at source. In packaging applications the weight of plastic used to make vessels for food and chemical products will continue to be reduced by innovative use of design and materials. For example, an item such as a 125 g yoghurt container made from a polystyrene resin would have weighed 6.5 g in 1965; today an equivalent article weighs only 3.5 g.

Primary recycling. This means returning clean plastic manufacturing waste to the process for reuse. Efforts should be made to keep this to a minimum as recycling costs money in terms of energy and effort. It is advantageous to design items with minimal waste.

Post-consumer recycling. One of the problems with plastics is that there are numerous different types and if one wants to recycle these the solution must be environmentally sound. In the USA a numbering system for the identification of different plastics is in place; this is now being introduced in Europe. Figure 8.3 shows the numbering system devised by the American plastics manufacturers.

It makes no sense to recycle a material if it is going to cost more in energy than the value of the new recycled material. If the goal is to recycle individual polymers there must be an economic way of dealing with the identification/separation problem from domestic waste and from the other plastic items that are present. This operation has to be on a large enough scale to permit one to gain from economies of scale. It has been suggested that the size of current single solid waste disposal operations that include composting and incineration is an order too small to take advantage of all

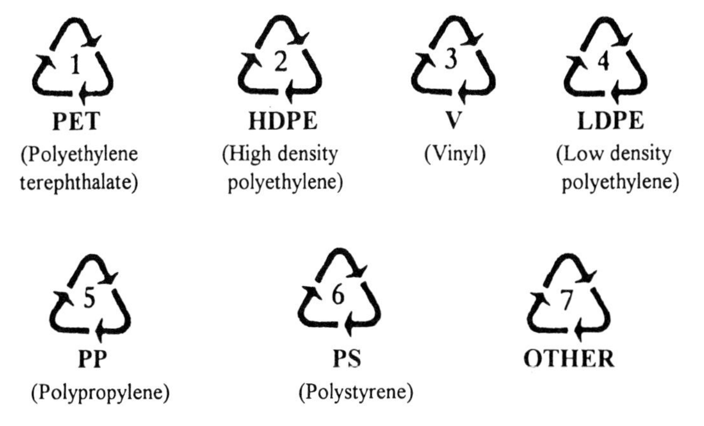

Figure 8.3 Numbering system devised by American plastics manufacturers for the identification of plastic materials.

the possible economic benefits. Incineration with energy recovery appears to be a good option; however, an operator who gets paid a price per tonne at the gate of the incineration plant would say that plastics are bad news because the price paid at the gate is a fifth of that paid for domestic garbage when one takes the calorific value into account. Today the economics of the incineration plant are worked out on the high weight/low calorific value figure. The incineration of plastics containing PVC results in hydrogen chloride being evolved. This has to be removed from the waste gas effluent prior to being discharged to the atmosphere to comply with HMIP requirements.

8.5 Life cycle analysis

8.5.1 Life cycle analysis – standardisation bodies

Life cycle analysis and or assessment is actively applied by both the plastics manufacturers and the plastics users (i.e. the producers of the item that is sold to industry and/or the public).

Life cycle assessment is discussed in detail in Chapter 5. For plastic products this is a very important technique and it is worth reiterating the relevant definitions. Life cycle analysis or, more strictly speaking, environmental life cycle analysis (LCA_{env}) has been defined by the European Committee for Standardisation (CEN/TC261/SC4/WG1/TG2) as a method used to quantify environmental burden based on an inventory of environmental factors for a product, process or activity from the extraction of the raw materials to their final disposal. Life cycle analysis is the collection and calculation of data which produces an inventory. Life cycle assessment goes further and adds evaluation to the inventory. The Canadian Standards Association has published a new standard Z760, Environmental Life Cycle Assessment – Part 11, Life Cycle Assessment.

The International Organisation for Standardisation (ISO) has formed a Technical Committee (TC), ISO/TC 207, on Environmental Management. ISO/TC 207 was formed as a result of recommendations of an ISO/IEC (International Electrotechnical Commission) group called the Strategic Advisory Group on the Environment (SAGE) formed in 1992 to investigate the need for international standards for the environment. Six sub-committees (SCs) and one Technical Committee Work Group have been formed within ISO/TC 207. SC 5 is for Life Cycle Analysis and within this SC four working groups have been formed as follows:

1. SC5/WG 1 on Life Cycle Code of Practice;
2. SC5/WG 2 on Life Cycle Inventory;
3. SC5/WG 3 on Life Cycle Impact Analysis;
4. SC5/WG 4 on Life Cycle Evaluation and Improvement Analysis.

Life cycle analyses are quantitative and objective, life cycle assessments take into account qualitative issues and are subjective.[11] Life cycle assessment has been defined as 'an objective to evaluate the environmental burdens associated with a product process or activity by identifying and quantifying energy and materials used and wastes released to the environment, and to evaluate and implement opportunities to effect environmental improvements' (SETAC Code of Practice for LCA, June 1993).[12] Life cycle analysis and assessment are actively applied by the plastics manufacturers and the plastics users (i.e. the packaging users).

Figure 8.2 provides a simplified view of a life cycle for plastics. There are numerous factors that have to be considered and any life cycle assessment requires a clear definition of the goal(s) before attempting to obtain information for the life cycle analysis. All systems impact on the environment. Life cycle assessment is a tool to help ensure that impact is minimised in the most objective manner.[13]

The Environmental Unit within PIRA International has developed a life cycle assessment computer model known as PEMS. The model is the result of a UK Government and industry-funded research project with a budget of £500 000. PEMS provides a 'transparent' methodology using best practice recommendations as specified in the SETAC Code of Practice for life cycle environmental studies. The PIRA LCA team is led by Dr Neil Kirkpatrick, who is the British Standards nominated UK expert in this field. Dr Kirkpatrick provides the UK with a link into the appropriate committees within CEN (European Committee for Standardisation), ISO (International Standards Organisation), SETAC – Europe (Society of Environmental Toxicology and Chemistry – Europe) and IAPRI (International Association of Packaging Research Institutes).

8.5.2 Life cycle assessment – plastic packaging examples

Franklin and Boguski in their paper presented at Recycle '94 used a life cycle inventory (LCI) to compare HDPE reusable containers and recycled corrugated boxes.[14] One conclusion was that both systems were dominated by transportation and a high usage of petroleum-derived fuel. It was possible to show that improvements could be made in each system with respect to energy use and solid waste.

A Swiss study compared one litre sized containers for the supply of milk to the consumer. The favoured package on environmental grounds was a lightweight plastic pouch which has the lowest energy and waste emission. This study concluded that the most convenient package for transport, stacking, portion delivery and losses was the traditional paper carton.[15]

A Norwegian study[16] carried out on the same subject considered the following four options:

1. Gable top cartons (current situation) (28.4 g/l of milk).
2. Refillable glass bottle (396 g/l of milk).
3. Refillable polycarbonate bottle (75 g/l of milk).
4. Polyethylene pouch (7 g/l of milk).

The overall conclusion was that the pouch was the most favourable from an environmental point of view, but with rather large spillage of both packaging and milk being the most negative factor. For the carton packaging option the most negative factor is the volume going to landfill. The study did not conclude that from an environmental point of view refillable containers were superior to one way packaging systems. The author understands that the pouch system for milk was in use in Germany in 1972.

Kerbside plastic collection schemes operate in Europe and the USA. In one scheme Coca Cola receive used 2 l PET bottles from an area in the South of England. This is centred on the Adur multi-material home recycling project sponsored by the European Recovery and Recycling Association (ERRA) of which Coca Cola is a founder member. Bottles are baled and returned to the USA where they are depolymerised (chemically unzipped), to recover the original starting materials, terephthalic acid and ethylene glycol, using technology developed by the Hoechst Celanese Corporation in partnership with Coca Cola. These materials are then polymerised again to make 'virgin' polyethylene terephthalate and this is mixed with new polymer up to a concentration of 25% to make new bottles.

8.5.3 Life cycle assessment – solid domestic waste

Every effort must be made at each stage of the manufacturing process to minimise/eliminate waste and to recycle material back to the process. A number of options are provided in Figure 8.2 at each production stage. It must be remembered that plastics are different from other materials being considered in this book, in that a large proportion are used for short term packaging. After domestic use, these materials are going to be discarded and it is not sufficient to rely on householders' good will to return used packaging materials. It needs more than this. In Germany a plastics recycling facility was planned for 100 000 tonnes and they were presented with 400 000 tonnes which totally swamped the system. This resulted in the export of plastic waste from Germany, which compromised the recycling plans within the receiving countries throughout Europe. Evidence for the dumping of this waste has been found on rubbish tips in France.[17] Once a plastic item is used and placed in a general garbage bin, the municipal waste authorities have to dispose of the material as best they can. Today in the UK the probability is that the plastic will be discarded to landfill. If it did find its way to a solid waste incineration plant it would be burnt and ideally energy recovered from incineration will be exported in the form of electricity to the national grid. Andrew Bond, writing in the *Process Industry Journal*, presents the following view: 'Recycling of mixed consumer plastic waste is highly labour intensive and of questionable overall economic benefit, given that the recycled feedstock is frequently more expensive than the original raw material. Incineration, by contrast, reduces pressure on the original feedstock and is least likely to have unforeseen disruptive effects elsewhere in the product cycle.'[18]

The APME Plastics-Resource Optimisation document entitled 'Waste to Energy' states that the Saint-Ouen incinerator turns the waste of 2 million inhabitants of Paris into energy. This energy is converted into heat supplying a district of 70 000 dwellings and produces 15 400 megawatts of electricity per year. This can be compared with the Edmonton solid waste incineration plant in North London, where in the year 1993–94 503 399 tonnes of waste were combusted and 231 917 megawatts of electricity obtained.

8.5.4 *Life cycle assessment – automobile industry*

As one would expect, motor manufacturers are involved in recycling. In the USA the American Automobile Manufacturers Association (AAMA) goals include the following:

1. Design: Integrate parts. Reduce the number of materials and parts within an assembly. Facilitate disassembly. Select fastener systems that permit disassembly by various methods, including destruction at the end of useful life. Reduce fasteners. Reduce the number and type of fasteners used. Mark plastic parts. Use SAE/ISO recommended practice for marking parts.
2. Materials selection. Where possible, only recyclable materials to be used, and also materials having recycled content. Minimise substances of concern. Control or eliminate substances that are potentially hazardous in the manufacture and recycling of vehicles.

In the USA the automobile shredder residue (ASR) of a car is 24%, of which 34% is plastic components.

In 1992 Toyota presented their 'Earth Charter'. The waste reduction policy is based on their 5R Programme. 5R includes the first letter of the following words:

Refine, Reduce, Reuse, Recycle and Recover to Energy

Toyota are developing plastic recycling technology within their plant, initially to deal with in-plant waste, and then they will apply the technology to post-use vehicles.[19,20]

Elf Atochem are also actively engaged in exploring ways of standardising plastic parts and increasing the amount used per vehicle to reduce the weight of the vehicle and thus save gasoline.[21]

8.6 Processes available for recycling plastics

From the primary use of virgin plastic there are a number of options available for recycling, recovery and disposal. The following possibilities are available for post-consumer plastic materials:

(a) Reuse in the same role as the primary use.
(b) Recycle to secondary use.
(c) Treatment (e.g. thermal cracking) to provide material that can be fed back to the refinery and or ethylene cracker.
(d) Incineration with energy recovery (i.e. generation of electricity).
(e) Incineration with no energy recovery.
(f) Landfill.

A number of different systems are either in operation or are being built to recycle plastic waste back to the refinery and/or ethylene cracker. Some of these processes are described in outline below.

8.6.1 Polymer cracking processes

BP Chemicals started to develop a polymer cracking process in 1989; this is now funded and managed by the ‘Plastics to Feedstock Recycling Consortium’. Members of this Consortium are DSM, Elf Atochem, EniChem, Petrofina and BP Chemicals.

A description of this process was presented to the Recycle ’94 Conference.[22] The process uses a mixed plastics waste feedstock, separated from municipal solid waste and post-consumer industrial waste (e.g. drums, shrinkhoods, crates, etc.). The mixed used plastics extracted from the waste stream are shredded to medium sized pellets and can cope with up to 5% of impurities.

The plastics particles are ‘decomposed’ (depolymerised) at a temperature between 400 and 600°C in a fluidised bed reactor containing sand. Chlorine and heavy metals are removed from the gaseous reactor product, which is partially condensed to produce a wax. Up to 20% of the wax is mixed with naphtha and can be used as a feedstock for steam crackers or refineries. The steam cracker produces the monomers for making plastics. A small scale plant is due to be commissioned at BP Chemicals, Grangemouth, later in 1994.

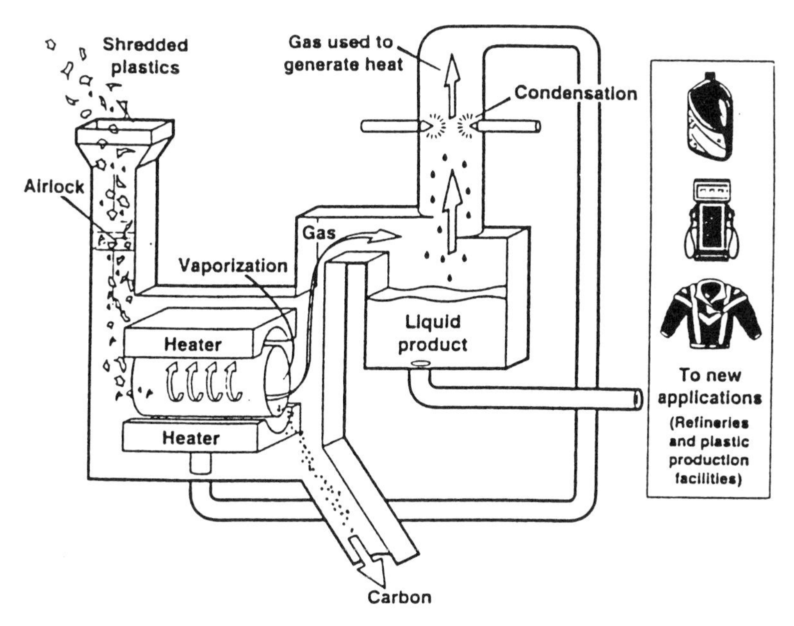

Figure 8.4 Conrad recycling system.

Mark Meszaros of Amoco presented a paper at the Recycle '94 Conference entitled Conrad Advanced Recycling Project.[23] The Conrad Project is funded by the American Plastics Council. This uses an auger kiln reactor to recycle plastics back to liquid hydrocarbons. This stream can be used in conjunction with other petrochemical feedstocks. Heat is applied to the plastics in the absence of oxygen to produce liquid petroleum, carbon and gas products (Figure 8.4). Shredded plastic is fed into the system through an air lock to prevent air entering the system. The plastics then move via an auger, through a heated tube that reaches around 1000°F (538°C) to 1450°F. At these temperatures the plastic depolymerises to a gaseous product with a small amount of carbon. The carbon exits the system and is sold. The gases are cooled and a liquid product is collected. Any gases that do not condense are used to heat the reactor. The liquid product is shipped to refineries and plastic production facilities for use as a feedstock for producing more plastics (70–80% of the output ends up as liquid product). The economics of the Conrad recycling unit are still unclear; further studies are being undertaken to better understand the economics of the process. (Further information can be obtained on the Conrad Advanced Recycling Process from Bill Conrad, Conrad Advanced Industries Inc., 121 Melhart Road, Chehalis, Washington, USA or Conrad Advanced Recycling Project, Bailey Condrey, American Plastics Council,

1275 K Street, NW, Washington, DC 20005. This paper deals with such problems as the evolution of HCl from PVC and possible production problems with terephthalic acid from PET bottles.)

Gebauer[24] in his paper at Recycle '94 described three processes that have been developed to reduce the molecular weight of the polymer to enable reprocessing to take place: breaking, pyrolysis and decompositional extrusion of plastics. Leuna-Werke used a twin spool extruder (ZSK 53, Werner & Pileiderer). With thermal decomposition in the extruder dehydrochlorination results in chlorine levels of 0.02% being obtained.

Veba Oel have developed gasification processes that produce synthesis gas and heating gas. These processes operate at 40 bar. This compares with the cracking processes above which operate at or near ambient pressure. All the major oil companies will be involved in the development of processes for depolymerising plastic material to lower boiling components that can be fed back to the refinery and/or ethylene cracker.

What are the environmental economics of these processes; are they really sound from an environmental point of view? As an informed customer I am unable to make up my mind. This is because the complete picture in terms of conversion from plastic to feedstock is not presented – one needs to have a complete materials and energy mass balance and this information needs to be audited by an independent body with nothing to gain or lose from analysing the data.

8.7 Biodegradable plastics

What is the nature of biodegradable plastic? Is there a case for having biodegradable plastics? The biodegradable bags that are on the market are normal plastic mixed with starch. If buried in a landfill situation the part that is made from starch will degrade and leave the original plastic polymer. In this case the polymer is not biodegradable. ICI have developed a truly biodegradable plastic. However, it is very brittle and further research is required to obtain a fully developed product. ICI have developed polyhydroxy butyrate which is biodegradeable but at the moment it is orders of magnitude dearer than conventional polyethylene.

In some quarters there has been opposition to biodegradable plastics simply based on the fact that, if the biodegradable polymer contaminates an ordinary polymer during the recycling process, this could ruin other recycled plastic material.

Narayan, from the Laboratory of Renewable Resources Engineering at Purdue University, Indiana, has combined the strength, flexibility and light weight of petroleum based polymers with the biodegradability of starch based natural polymers. This was achieved using a graft copolymer.[25] Biodegradable polymers are, and will continue to be, required in medical applications. This area of the plastics business will continue to be researched for specialist applications and will grow in the future.

8.8 Legislation

8.8.1 European Packaging Directive – political position

The targets adopted at a meeting of the Environment Ministers on 16 December 1993 were as follows:

- five-year target only;
- 50–60% by weight of packaging to be valorised;
- Within this, 25–45% of total packaging waste to be recycled as material (not energy?);
- Of the 25–45% total recycling, each material must achieve at least 15% recycling.

Additional modifications:

- Articles relating to prevention and hierarchy to be watered down.
- A five-year derogation on time to be introduced for islands and mountainous zones (i.e. targets to be reached in 10 years).
- Germany, The Netherlands and Denmark allowed to continue to impose national targets exceeding those above, provided they can prove they have the capacity required to achieve them and will not impede free trade. These countries opposed this position but did not have enough combined votes to block a qualified majority.

The next stages towards adoption:

- Greek Presidency to progress to 'Common Position' – bureaucratic procedure requiring no negotiation.
- European Parliament second reading (3 months). Could throw out Directive under recently acquired powers.
- Germany takes Presidency July 1994.

8.8.2 Switzerland

In Switzerland since 1991 specific regulations have been imposed to stop the trend to throwaway packaging. In 1992 this resulted in 80% of drinks being bottled in reusable containers; 3300 t of PET or 53% of the material used for beverage containers was recycled. This introduces its own special problems as PET cannot be heated to the same temperature as glass in the washing sterilisation procedure.[25]

8.9 Conclusions

Without plastics the volume of domestic waste would rise by up to 150%, due to the greater amount of materials required to do the same job achieved by a minimum quantity of plastics. If this were to happen it would cause more problems than it apparently solved.

Waste can be a valuable source of secondary resources and can assist in conserving raw materials. However, the case for recycling plastics and using them to generate further feedstock material has not been proven to the author's satisfaction (this might be because of commercial confidentiality) and further information is required.

As users we all have a right to be presented with the options that are available and the relative environmental 'cost' in terms of energy and emissions compared to the alternatives. This is of course difficult due to commercial constraints.

Acknowledgements

The author would like to thank the following for taking the time to discuss the various issues involved as well as providing the author with information and their views at the research stage for this text: Dr John Brophy and Steve Hardman, BP Chemicals Research; Jeff Cooper, Waste Reduction Officer, London Waste Recycling Authority; Dr Alan J. Griffiths, Director of Communications, Association of Plastics Manufacturers in Europe; Martin Jones, General Manager and Engineer, North London Waste Authority, Edmonton Solid Waste Incineration Plant.

8.10 Information Sources

1. Association of Plastics Manufacturers in Europe (APME), Avenue E. Van Nieuwenhuyse 4, Box 3 B-1160 Brussels, Telephone (32–2) 672 82 59 Fax (32–2) 675 39 35.
2. European Centre for Plastics in the Environment (PWMI), Avenue E. Van Nieuwenhuyse 4, Box 5 B-1160 Brussels. Telephone (32–2) 675 32 58 Fax (32–2) 675 40 02.
3. Environmental Information: *A Guide to Sources*, Nigel Lees and Helen Woolston, ISBN 0–7123–0784–4. The British Library, Science Reference and Information Service, 25 Southampton Buildings, London WC2A 1AW. Telephone 071–323 7485.
4. National Recycling Forum, 6/8 Great George Street, Leeds LS1 3DW.
5. British Federation of Plastics, 6 Bath Place, Rivington Street, London EC2A 3JE.
6. Waste Watch, 68 Grafton Way, London W1 5LG. Telephone 071–245 9998.
7. Department of Trade and Industry, Environment Unit, 151 Buckingham Palace Road, London SW1W 9SS. Telephone 071–215 5000. Recycling Advisory Unit, Warren Spring Laboratory, Stevenage. Telephone 0438 741122.
8. Save Waste & Prosper Limited, PO Box 19, 6/8 Great George Street, Leeds LS1 6FT. Telephone 0532 438777.
9. Lin-Pac Plastics International Ltd, A1 Business Park, Knottingley, West Yorkshire WF11 0BS. Telephone 0977 671111/607766.
10. ENDS. Environmental Data Services, Finsbury Business Centre, 40 Bowling Green Lane, London EC1R ONE. Telephone 071 278 4745.

11. Environmental Business. Information for Industry Ltd, 521 Old York Road, London SW18 1TG. Telephone 081–877 9130. Fax. 081–877 9938.
12. SETAC-Europe, Society of Environmental Toxicology and Chemistry, Avenue E. Mournier 83, Box 1, Brussels. Telephone 32 2 772 7281. Fax. 32 2 770 5386.
13. PIRA International, Randalls Road, Leatherhead, Surrey KT22 7RU. Telephone +44 (0) 372 376161. Fax. +44 (0) 372 377526.

References

1. Dow Chemical Company Annual Report (1989), quotation assigned as a 'Native American Adage', p. 8.
2. Boustead, I. (May 1993), Eco-profiles of the European Plastics Industry, Report 4: Polystyrene, European Centre for Plastics in the Environment, Brussels.
3. Wiseman, P. (1986) *Petrochemicals*, Ellis Horwood, Chichester.
4. Stephens, H.K. and Hollis, C.E. (1976) Petrochemicals and Polymers (Plastics, Rubbers and Fibres) in Our Industry Petroleum, British Petroleum Company Limited, London.
5. Candlin, J.P. (1986) Polymers, In *The Chemical Industry* (ed. Heaton, C.A.), Blackie and Son Ltd., Ch.1.
6. Kirk-Othmer (1984) *Encyclopedia of Chemical Technology* 3rd edn (Exec. Ed. Grayson, M.) John Wiley and Sons, vols 1–24.
7. Crosby, P.B. (January 1980), *Quality is Free. The art of making quality certain*, New American Library.
8. Spenley, P., Pera International, UK (September 1992), *World Class Performance Through Total Quality – A Practical Guide to Implementation*, Chapman & Hall, London.
9. Kirk-Othmer (1984), *Encyclopedia of Chemical Technology* 3rd edn (Exec. ed. Grayson, M.), John Wiley and Sons, Vol. 21.
10. Kirk-Othmer, *Polymer Science and Engineering*, Wiley Interscience.
11. Kirkpatrick, N. (July 1992) Life cycle analysis and ecolabelling, PIRA International.
12. Kirkpatrick, N. (Jan. 1994) The contribution of life cycle assessment (LCA) to responsible waste management, PIRA International.
13. Kirkpatrick, N. (Jan. 1994) Life cycle assessment (LCA) – a tool for managing environmental performance, PIRA International.
14. Franklin, E.W. and Boguski, T.K. (1994) HDPE Reusable containers and corrugated boxes, a life cycle inventory and comparison of reuse and recycle. *Recycle '94, 7th Annual Forum*, Davos, Switzerland.
15. Source (October 1991), Central Union of Swiss Milk Producers.
16. Kildal, S. (March 1994), Life cycle analysis for milk packaging, *Recycle '94*, Davos, Switzerland.
17. Gandy, M. (1994), *Recycling and the Politics of Urban Waste*, Earthscan Publications, Chs 2 and 5.
18. Bond, A. (1994), The Germans have a word for it, *Process Industry Journal*, The Hemming Group, March, 11.
19. Charlton, C. (1994) The environment – learning by example, *BSI News*, April.
20. Kawamura, N., Toyota Motor Corp. Japan, and Ichikawa, S., Toyota Motor Europe Marketing and Engineering – Belgium (March 1994), Effective re-utilization of material – Toyota's progress in the field of plastic automotive parts recycling, *Recycle '94*, Davos, Switzerland.
21. Jean, A., and Burkle, D. (March 1994), Plastics and their contribution to the development of environmentally friendly cars, *Recycle '94*, Davos, Switzerland.
22. Troussier, C., BP Chemicals (March 1994), Raw material recycling – a solution for plastics waste, *Recycle '94*, Davos, Switzerland.
23. Meszaros, M., Amoco Chemical Company (March 1994), Conrad advanced recycling project, *Recycle '94*, Davos, Switzerland.
24. Gebauer, M. (March 1994), Options for primary recycling of plastic recyclables, *Recycle '94*, Davos, Switzerland.
25. Fahrni, H.-P. (March 1994) Recycling in Switzerland, Swiss Environmental Protection Agency, *Recycle '94*, Davos, Switzerland.

9 Clean technology in the food industry

P. FRYER

9.1 Introduction

9.1.1 The food industry and the environment

The aim of the food industry is the efficient production of safe food. The scope of the industry is wide, and extends from the farm, where food is grown, to the factory where it is processed, and then to the shops where it is sold. Each of these stages involves an environmental cost. Other parts in this book consider the environmental impact of farming; here, the use of clean technology in food production will be examined.

The food processing industry uses many of the techniques of the chemical industry; it is designed to process liquids and solids in ways which often involve chemical reactions and physical and chemical separations. However, in environmental terms it differs significantly. The major requirement of the chemical industry, which rarely makes safe products, is process safety. In contrast, the food industry requires both product and process safety; foods must be shown not to be harmful to the consumer. The avoidance of chemical and microbiological contamination is thus central to food production. This imposes a significant need to control and minimise the interactions between food and the environment on the designers and operators of food processing plant.

In addition to possible contamination from the environment, the food industry produces a significant amount of waste. Waste streams produced by the chemical industry are often toxic, while, in contrast, those produced by the food industry may not be toxic in themselves, although they may become contaminated with bacteria; their environmental hazard lies in their high oxygen demand. In some cases waste streams can be used directly as food; for example, solid wastes can be sent as animal feed and liquid wastes discharged onto farmland as irrigation. In many cases, however, materials are treated or discarded.

The food industry has evolved efficient strategies for waste management, but has not generally considered ways of minimising effluent. The aim of this review is to outline some of the areas in the food industry where environmental concerns are important, to discuss the approach of the food industry to product safety and food hygiene, and to seek areas where research in clean technology may be appropriate to the industry.

9.1.2 *Consumer perception of environmental issues*

The structure of the food industry varies significantly from country to country. In the UK, the food retailing market is dominated by the major supermarkets, who have a close relationship with their suppliers, and which sell both branded and own-brand goods. The industry is consumer rather than technology driven; the requirements of the consumer determine what the industry can sell. Historically, the only food available in the shops was that which was in season or which had been preserved by technologies which reduce the quality of the food, such as canning. A number of social and technological trends have changed the character of the industry, including:

- the reduced number of people who stay at home with time to cook, which increases the demand for 'ready-meals' which require minimal processing within the home;
- the increasing use of the home freezer and the microwave oven, which allow storage of food and a rapid way of reheating it;
- the changing pattern of the population, which includes an increasing number of older people who, for health reasons, require higher protein and lower fat levels in food;
- efficient transport and storage systems which allow a wide range of material to be sold, such as all-year-round strawberries.

The consumer has become increasingly concerned about food quality and safety. The less processing a food has received, the more like home-cooked food it can taste. To extend the shelf-life of foods it is possible to add chemicals which enhance stability. Several years ago consumer concerns about the levels of additives resulted in the removal of many preservatives, artificial colours and flavours from foods. In many cases this gives the product a shorter shelf-life; other foods, such as 'cook-chill' meals, have also been introduced which have received minimal processing. To keep such foods safe and saleable, a highly efficient distribution network, together with precise control of the conditions under which food is kept, is required to ensure product safety. In the UK, the approach of retailers to quality assurance differs; some demand formal adoption of quality standards such as BS 5750, in conjunction with systematic procedures for determining areas of possible microbiological contamination. One such procedure is HACCP (Hazard Analysis of Critical Control Points). Here, the process is analysed to identify areas where microbial contamination can occur; remedial action can then be taken to remove the hazard, or procedures devised to ensure contamination is prevented.[1]

In addition to worries about the composition of food ingredients, the consumer is becoming concerned about the environment. In UK retailing, this has been most obvious in sales of detergent and washing powders, with the increasing sale of concentrated powders, but concerns are extending to foods. The implications of consumer environmental concerns for the food industry in the US are discussed by Sloan;[2] key environmental issues were identified by consumer surveys as:

- Recycling: two-thirds of a sample indicated that it was important for packages to be recycled. However, the ability of a package to preserve freshness and quality was preferred, when the respondents to the survey were given a list of possible uses.
- Over-packaging: about 80% of consumers think foods are over-packaged.
- Solid waste: 75% of Americans believe that the disposal of packaging and foods is a serious problem.

However, Sloan notes that the desire for convenience, price and quality in foods ranks far ahead of environmental concerns. The most desirable factor is that any tampering with the food should be evident; however, this can require extra packaging; for example, strips of plastic around the lids of jars which show whether the lid has been removed. Whilst products which declare themselves to be 'green' will sell, they must compete effectively in the marketplace with products which may be cheaper. Any use of clean technology within the food industry must be considered against this fact; extra expense which increases the price of a product will reduce its sales, however 'green' its concept or labelling.

Without a strong market for food products which have been prepared in an environmentally sound way, the industry will not respond readily. However, even without strong consumer demand, there are a number of areas in which the environment affects food processing, and where process modifications which benefit the environment may reduce the costs of manufacture. These will be considered below under five headings:

1. ensuring minimal contamination of food ingredients;
2. ensuring minimal environmental contamination of food processes;
3. efficient cleaning of food plant;
4. minimising the waste from such processes;
5. designing environmentally efficient packaging.

9.2 Ingredients and processing: environmental contamination of foods

Food processing involves taking natural ingredients and processing them to produce safe and palatable food. If the raw materials are contaminated, either with chemicals or biological materials, then this can render the final material unsaleable. It is important to understand interactions between the environment in which the food is grown or from which its ingredients are sourced, and the final food product. As well as contamination of solid food ingredients, water – a vital component in most food processes – can also be contaminated; for example, by nitrate residues from fertilisers.

The possibility of contamination from environmental sources, especially from heavy metals such as cadmium, arsenic and lead, has been noted for some time, and is discussed at length in Nriagu and Simmons.[3] Many agrochemicals and metals are persistent in the environment, and can be highly toxic. Possible contaminants included the following.

Heavy metals. In the pre-DDT era, arsenic compounds were used widely as pesticides, especially within the USA. Widespread use of arsenic was encouraged, using advertising techniques such as:[4]

Spray your roses, for the slug,
Spray the fat potato bug;
Spray your canteloupes, spray them thin,
You must fight if you would win.

Heavy metals can accumulate in the food chain and thus eventually transfer into human food ingredients. Other metals which contaminate foods include lead,[5] although the amount of lead in the environment is now decreasing, as a result of the decreasing use of leaded petrols, and mercury.[6]

Pesticides. Organic pesticides that do not contain metals, but are still potentially hazardous, are also found in foods.[7] The key event in the public perception of pesticide contamination was the publication of *Silent Spring* by Rachel Carson. Since then, increasing care has been taken in the use of pesticides which can accumulate in the environment. The most common residues in food are (i) organochlorines such as DDT which can be very persistent within the environment, (ii) organophosphates, and (iii) carbamate materials. Some of these materials can be concentrated during processing; for example, non-volatile materials can be concentrated into foods by evaporation. Concerns about pesticides have led to the increased popularity of 'organic' foods, which have been prepared without 'chemicals', despite their higher price and often lower quality.

Contamination of the above sort is difficult to overcome save by careful selection of sources. Food producers are now increasingly concerned to minimise the amount of pesticide in the food chain. Equally, they are concerned to reduce the use of fertilisers, which can also contaminate food feedstocks.

The contamination resulting from organic chemicals and metals has been extensively documented by food scientists. Contamination can also arise from the presence of biological organisms, such as bacteria, in the material, or materials such as fungal toxins, including aflatoxins from grains and nuts. Microbiologists distinguish between *intoxication*: illness caused by the presence in the food of a toxic substance produced by a micro-organism, and *infection*: illness caused by the entrance of bacteria into the body through ingestion of contaminated food. Food poisoning agents are widespread and can cause a variety of symptoms. In canned or other types of 'sterilised' foods which have received significant thermal processing, the major potential danger is the presence of anaerobic organisms such as *Clostridium botulinum*, which produce spores which are highly thermally resistant. A variety of contamination agents are possible; some of these are listed below.

Fungi and moulds. Although special moulds are involved in the production of some types of food, such as Brie cheese and soy sauce, the majority of fungal infestations of food are unwanted. In some cases, toxic by-products are produced, such as (i) ergotism – lysergic acid derivatives produced from the growth of mould on wet corn (*Claviceps purpurea*) which can produce gangrene as well as hallucinogenic effects, and (ii) aflatoxins, from the growth of *Aspergillus flavus* and similar species on grains, especially peanuts.

Bacterial. Two main types of food intoxications are (i) botulism, caused by the presence of toxin generated by *Clostridium botulinum*; generally fatal in >30% of cases and found in home-canned food in the US, preserved meats and fish (tinned salmon) in the UK; and (ii) staphylococcal intoxication, caused by *Staphyloccus aureus*. Symptoms are nausea and vomiting after 2–4 h, with an outbreak duration <48 h, and the intoxication is caused by letting meat/fish/milk be too warm for too long.

Food infections can be divided into two types: those in which the food does not ordinarily support growth of pathogens but can carry them, i.e. pathogens that cause diphtheria, cholera, hepatitis; those where the food can serve as a culture medium, such as *Salmonella*, *E. coli*. *Salmonella* infections take 12–36 h to develop, then cause nausea and vomiting; coliforms give diarrhoea in 'traveller's tummy'; gastroenteritis from *Campylobacter*.

Some contaminations are mixtures of the two; for example, toxins are generated within the body to cause gastroenteritis by *Bacillus cereus* and *Clostridium perfringens*.

Listeriosis: from *Listeria monocytogenes*, which can grow at 4°C, i.e. in chill cabinets, in coleslaw, milk, soft cheeses.

Viruses. The viruses which cause polio and hepatitis can be transferred in uncooked meats or milk. More recent concerns have been viruses with longer term effects such as bovine spongiform encephalopathy (BSE), a virus similar to that which causes scrapie in sheep. The difficulty of detecting such viruses makes it difficult to confirm their destruction.

Other problems. Many contaminations can result. Some of the most serious are: (i) parasites: such as trichinosis from infected pork, where nematodes take up residence in muscle tissue; and (ii) shellfish poisoning: in addition to the susceptibility of shellfish to contamination from bacteria, they can also be contaminated by toxic algae, such as *Gonyaulax catenella*.

Introduced organic contamination must be removed during processing, usually by destruction in some form of heat treatment. One excellent study of the processing implications of environmental contamination is that of Phillips and Griffiths,[8] who examined the microbiological hazards associated with the processing chain of pasteurised milk. Milk is pasteurised by heating to about 75°C for a few seconds to reduce the bacterial load and increase the shelf life of the material. Contamination before thermal processing can arise from a number of sources:

- from the cow, for example from animals with mastitis, or from material on the outer surface of the udder, such as dung and bedding materials, which can be contaminated with spore-forming bacteria;
- from milking equipment which has been poorly disinfected;
- during refrigerated storage, where new bacteria can be introduced and existing ones can proliferate;
- during transportation to creameries;
- during storage at the processing site.

Pasteurisation and associated processes can remove much of this contamination by giving a thermal treatment sufficient to kill most of the bacteria. Further contamination can, however, arise after the heat treatment, which is generally carried out within equipment such as a plate heat exchanger.[9] A number of workers have identified contamination within the heating unit; thermotolerant bacteria can build up on the surface of the heat exchanger, or within gaskets and dead ends of the heating unit. Leakage between processed and unprocessed milk can occur within the heat recovery section of heat exchangers, but modern designs of heat exchanger minimise this possibility. Similar problems can arise during holding of the milk and during packaging. HACCP analysis is designed to detect and quantify such possibilities.[10]

Although the above example describes the processing of milk, the principle is a general one; to study the influence of the environment on food processing it is necessary to consider each stage of the production process. This is more important still in processes such as canning and UHT treatment, where the food is to be sterilised to have a shelf life significantly longer than that of pasteurised food.[11,12] Any contamination which persists through to the final product can cause significant food poisoning.

In food manufacture, it is thus important to separate the food from its environment. This ensures that food plants are, in some ways, designed to be inherently clean. However, the requirement for process and product sterility means that it is necessary to clean process equipment frequently. In liquid food processing, this cleaning is highly automated, but, as shown in the next section, may not be optimal, leading to excessive production of effluent, a significant source of environmental contamination. These problems will be discussed further in the next two sections.

9.3 Cleaning of food process plant

9.3.1 The need for cleaning

It is vital that food process plant be clean and hygienic. During operation, food plant gradually becomes contaminated. Many of the components of foods, such as starches and milk proteins, are highly temperature-sensitive, and thus tend to form solid fouling deposits on heat transfer surfaces during thermal processing. These deposits require removal if they are not to become the focus for corrosion and microbial contamination. The removal of solid deposits, together with the requirements of product sterility and safety, means that food process plant requires frequent and expensive cleaning. The cleaning of food surfaces is a familiar process, but one which is not well understood, and the fouling and cleaning of surfaces in contact with foods remains one of the major processing problems in the food industry. In many industries, such as in dairy processing, complex cleaning-in-place (CIP) techniques have been developed.[13] The primary aim of CIP is the removal of residues from a manufacturing process without significant dismantling of the processing equipment, and the establishment of an environment which will not contaminate the next process operation. This requires highly complex and expensive pipework, pumps and vessels, which allow cleaning fluids to be circulated automatically through the plant at set intervals. CIP requires a series of stages; a pre-rinse step to remove product and poorly bound deposit, followed by cleaning steps, each followed by a rinsing stage. Industrially, two types of cleaning treatment are used to remove deposits from dairy plant:

1. *Two-stage* cleaning using both acid and alkali, commonly sodium hydroxide, followed, after rinsing, with nitric or phosphoric acid.
2. *Single-stage* cleaning with formulated detergents, which contain compounds to enhance cleaning, such as surface active and chelating agents.

Two-stage cleaning is more complex in practice, requiring extra dosing equipment and more rinsing steps. Sodium hydroxide and acid alone may be insufficient to achieve a clean surface, so that single stage cleaners were thus developed to provide a clean surface in a short time. Although the chemicals needed are more expensive than two stage-systems, they offer the opportunity of reducing total cleaning costs by more economical use of chemicals, wash water, labour and down time.

Although cleaning is ubiquitous in the food industry, the kinetics and processes involved in cleaning are not well understood. The duration of the operating and cleaning cycles on a given plant are often set by trial-and-error during the commissioning stages of the plant, and it is not known whether the combination of time, temperature and cleaning agent used are optimal, and whether it might be possible to save money by changing the cleaning strategy in a given plant.

The costs of cleaning are significantly greater than the cost of the chemicals. Extra costs arise in a number of areas:

- *capital costs* of the CIP system and for overdesign of valves and pipework needed to resist fouling;
- *energy costs* for the operation of the plant during cleaning;
- *loss of production* as result of down time; in the dairy industry, up to 42% of available production time can be taken up with cleaning and sterilisation;[14]
- *labour costs* for the operation and set-up of the clearing operation;
- *effluent treatment* costs will be significant. Cleaning effluent is often voluminous, contains high BOD waste and will often be at alkaline pH. They are thus potentially hazardous and expensive to dispose of.

As environmental concerns become more important to the public and this is reflected in the costs of disposal, it will become increasingly important to minimise the waste produced by any process plant. It is thus important to understand the processes which are taking place during cleaning so that the environmental impact of the effluent is minimised and the time which plant is not operating is minimised.

Cleaning in the food industry has traditionally been an art rather than a science. Rules of thumb are still in use industrially;[15] for example, it widely quoted that a minimum cleaning velocity of 1.5 m/s is needed. This approach can be attributed to the engineering need to develop cleaning chemicals and cleaning-in-place procedures before the principles of fouling and cleaning were understood. Little is known about the kinetic processes which underpin the design of cleaning systems.

Limited current understanding of the cleaning process is partly due to the complexity of fouling. Fouling is a complex transient process, frequently involving a lengthy induction period during which no change in heat transfer occurs, then followed by rapid deposition and consequent decrease in performance. Fouling occurs from many food fluids, but the one which has been best studied is milk, reflecting its importance as an industrial fluid. Fouling is a function of factors such as process temperature and flow rate, in addition to the chemistry of the system and the surface temperature. The chemical composition of milk deposits, and the ways in which deposit forms are now well understood.[12,16] The deposit is a mixture of proteins and mineral salts; deposition of proteins appears to result from a sequence of stages:[17] denaturation and aggregation of proteins in the hot region of the fluid is followed by mass transfer to the surface, and in a further reaction, protein is incorporated into the surface of the deposit. The fully-developed deposit consists of a thin mineral-rich layer adjacent to the metal surface, underneath a protein-rich layer; however, the first layer of deposit which forms on the surface is proteinaceous.[18] The structure of the deposit influences the design of the cleaning system; in two-stage cleaners, the alkaline wash removes the protein layer, and then the acid removes the mineral. Less is known about the structure of deposits from other types of food, such as those from starch-based systems: without an understanding of the nature of the deposit it is difficult to devise an efficient cleaning technique.

9.3.2 Process variables and cleaning rates

Many studies of cleaning have been carried out on pilot scale or commercial scale plant (e.g. ref. 19); whilst these studies are useful for determining the effect of fouling and cleaning of such plant, they are of limited use in finding the basic kinetics of the process. In any real food plant, a wide range of process conditions is encountered, such as changes in local temperature and flow rate in a plate heat exchanger. This will result in significant variations in the fouling rate and thus the amount of deposit. When such a deposit is cleaned, different local cleaning rates will result; data produced by analysing the output of the plant will be impossible to interpret simply. In order to understand the kinetics of cleaning a uniform fouling deposit must be produced.

Limited data are available on the kinetics of cleaning under uniform conditions. Graßhoff[20] showed that removal of protein-based deposits occurred in lumps rather than uniformly. Bird and Fryer[21,22] described a study of milk fouling deposit cleaned by sodium hydroxide solution and demonstrated that the cleaning rate (i) increases with temperature and velocity, and there appears to be no minimum velocity below which the cleaning rate is zero; and (ii) is a very complex function of the cleaning agent concentration. Figure 9.1 shows the variation of cleaning time; a clear *minimum* in the cleaning time is found for sodium hydroxide concentrations in the region of 0.5% NaOH. Visual observation, and electron micrographs of the surface of the deposit (Figure 9.2) suggest that the action of the NaOH is to transform the surface into a spongy gel; at high concentrations this gel resists removal, whilst at the optimal concentration the structure of the gel is such that it can be readily removed by the action of fluid shear.[23]

This type of result is important; it suggests that the intuitive approach, of increasing the cleaning agent concentration when faced by an intractable

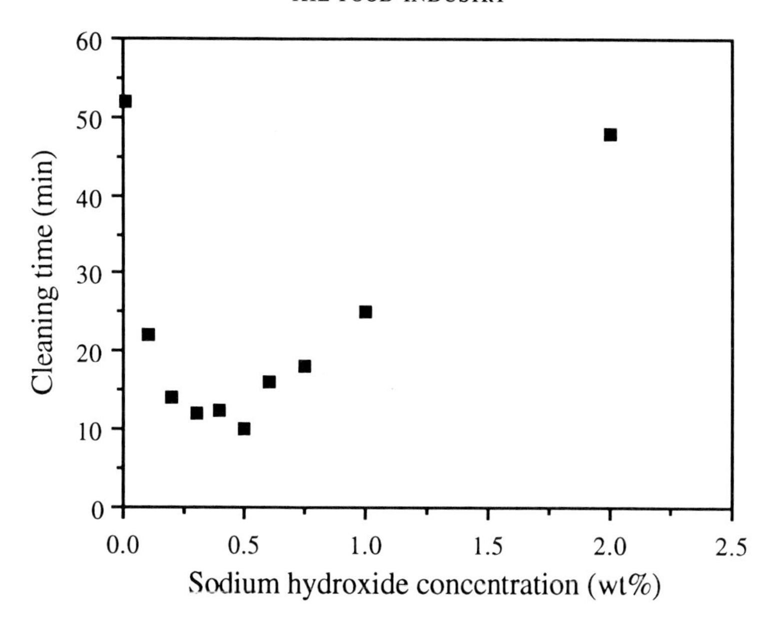

Figure 9.1 Variation in cleaning time for a whey protein deposit as a function of concentration of sodium hydroxide for experiments at 50°C and 0.175 m/s.

deposit, may be incorrect in some cases. The environmental consequences of this are also clear; it may be possible to increase the cleaning rate, and thus the profitability of the plant, by decreasing the potency of the cleaning material.

The data quoted above are from a laboratory scale experiment, and more work will be needed to identify the savings which are possible. There is some evidence that the conclusion is a general one; however, de Goederen *et al.*[14] demonstrated that concentration optima have been noted in plant operation, whilst recent work at NIZO in the Netherlands has found similar results. Some work on other systems and solids such as starch shows similar effects.[24,25] The optimisation of cleaning cycles also requires the correct management of process plant; computational techniques are available[26] by which this might be achieved, but more work will be needed to produce methods which can be used by industry.

9.3.3. Hygienic design

In addition to optimising the cleaning chemical dosage, it is possible to increase the cleanability of process plant by its *hygienic design*. Such design involves, for example, the selection of valves which can be cleaned easily, and the design of systems which eliminate dead spaces in pipes and crevices in equipment where contaminant may remain during cleaning. Reviews of the method are given in ref. 27 and in a series of papers produced by a

(a)
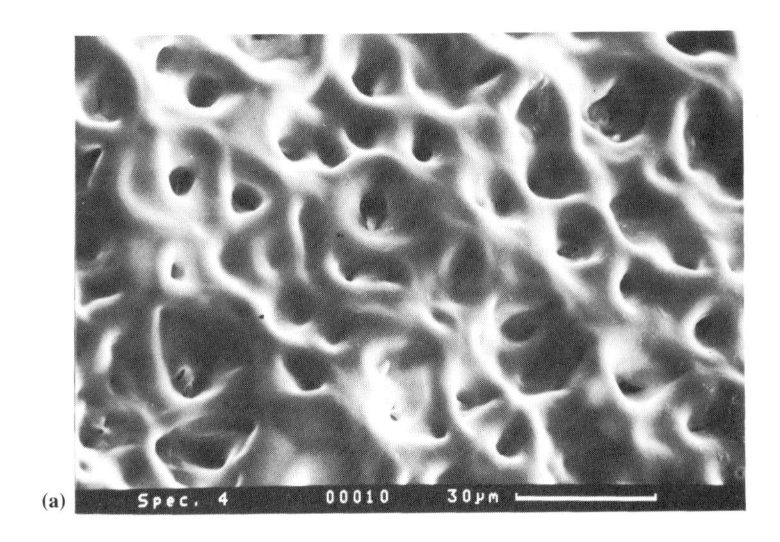

(b)
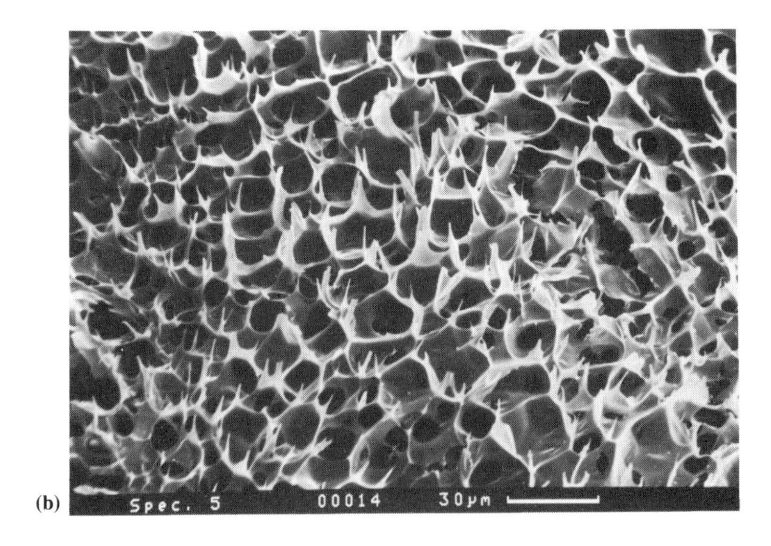

Figure 9.2 Scanning electron microscope photographs of whey protein deposits contacted with sodium hydroxide of various concentrations[23] at 50°C for 2 min: (a) 0.1%, showing a closed surface, (b) 0.5%, showing an open structure, (c) 2%, showing a denser structure than (b).

European committee from the EHEDG (European Hygienic Equipment Design Group).[28] Holdsworth[11] lists the following necessary design characteristics:

(a) surfaces in contact with food must be inert to the food;
(b) surfaces must be smooth;

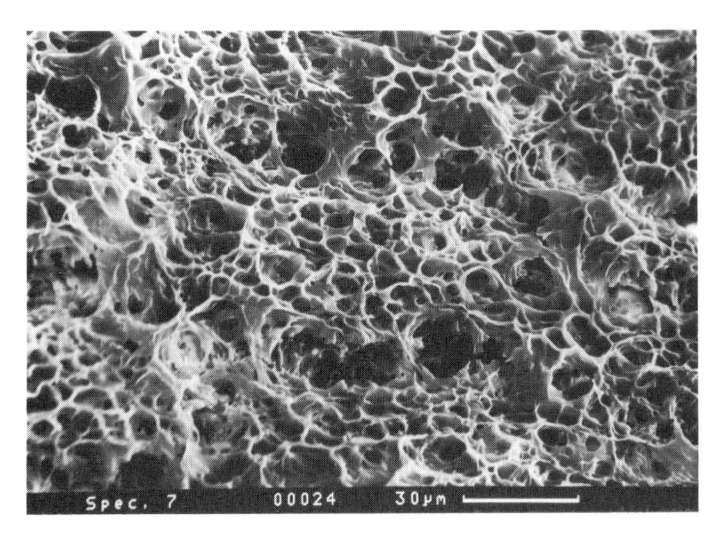

Figure 9.2(c)

(c) it must be possible to clean and then disassemble the plant to demonstrate the efficiency of cleaning protocols;
(d) equipment must be self-draining;
(e) the plant must protect the contents from contamination;
(f) exterior surfaces must not act as a source for contamination.

The primary purpose of such design is to minimise the possibility of microbial contamination; however, correct design will reduce the cleaning load on the plant, and thus the environmental discharge. The physical, chemical and microbiological cleanliness of the plant must be considered. This is an important area of future research: minimisation of the cleaning times and of the resulting amount of effluent produced would both increase process efficiency and minimise environmental damage.

9.4 Liquid and solid wastes

The need for cleanliness and hygiene at all stages of the process leads to heavy demand for cleaning water and consequent effluent production. Despite attempts to minimise effluent, food process plants can produce very high volumes of liquid waste. The amount of liquid waste produced depends on the type of process, and can vary from thousands of litres per day from a bakery to several million litres per day from a cannery. The criteria under which this waste is examined are the standard ones for waste waters:

- BOD and COD and total oxygen demand: in some cases, the waste stream contains valuable components;
- oil and grease: these can form insoluble layers or be present as emulsions;
- suspended and dissolved solids;
- nitrogen and phosphorus contents;
- pH: it is important that waste reaching the treatment plant has a uniform pH, as variations may well upset the biology of the plant;
- heavy metal and specific organic contaminants: the food industry should not generate this type of waste.

Three types of waste water are produced by any food process:

1. *Process wastes*. Produced by leaks, product washing and equipment cleaning. The different types of process waters are discussed below.
2. *Cooling and heating waters*. Thermal processing requires cooling; in the case of canning, very large amounts of water are involved, as cans are cooled from 125°C to ambient temperatures. These waters are often recycled through cooling towers, as they contain corrosion and bacterial inhibitors: significant blowdown is, however, needed to prevent buildup of salts and additives.
3. *Domestic wastes*. These are produced by the plant operators; it is vital that they are kept separate from other plant wastes, as mixing would make it impossible to ensure hygienic conditions for water reuse and/or product recovery.

A number of specific types of food effluent can be identified as causing particular problems, as follows.

Meat and poultry. Slaughtering and subsequent processing release considerable amounts of organic waste, ca. 5 m^3/kg of carcase produced during meat processing and 8 m^3/kg of carcase produced in poultry processing. This waste has a very high BOD, which can, however, be reduced significantly by blood collection. The waste may also have a high bacterial count, including faecal coliforms.

Dairy. Most (up to 90%) of the waste produced from a dairy plant arises due to dilution of the milk. The BOD of raw milk is ca. 100 000 mg l^{-1}, approximately 250 times that of sewage, so dilution will still produce a highly active effluent. Wide ranges in the characteristics of the effluent can be found; as discussed above, the use of highly alkaline cleaning solutions is commonplace, and so high pH waste can be produced. This can obviously cause problems in the process control of waste treatment plants.

Brewing. The drinks industry is highly complex and difficult to generalise. However, a number of waste components, such as steep liquor and fermenter effluent, have a high BOD. These components may be a small part of the total effluent, as significant waste waters are also generated in cooling and washing; this is one area where it is important to keep the waste streams separate. Recovery of solids such as excess yeast is common; these can be sold as animal feeds.

Fruit and vegetable processing. Liquid waste can arise during vegetable processing, such as 15–20 m^3 of liquid/tonne processed generated during washing and peeling of foods such as potatoes; this waste can have a high pH and suspended solids content. The preparation of fruit juices also involves the generation of both liquid and solid waste, and their thermal processing may involve a heavy cleaning duty.

Solids processing. Industries such as cereals and baking involve the processing of solids rather than liquids. These create special cleaning problems; it may be acceptable to hose down a dairy to remove waste, but it is much less acceptable to wash a bakery. By contrast with liquids processing, solids waste handling is poorly understood; whilst it is known how to keep a surface clean by washing with chemicals, to keep a dry surface clean is much more difficult. It is, however, imperative to keep liquid and solids waste apart to minimise subsequent problems.

Food waste tends to be treated either by (i) direct discharge, (ii) land application, or (iii) treatment, usually by some type of activated sludge process. In some areas, the waste stream has a high content of valuable material which can be used as food; solid wastes to pig feed, liquid wastes to irrigation. Some areas which produce waste of very high BOD, e.g. meat factories, receive special treatment, which may involve separation of a high-value protein stream, as discussed below. Treatment of food wastes may be complicated by the presence of a variety of components, such as proteins and lipids.[29] However, the majority of wastes are combined and then treated together.

9.5 Opportunities for waste minimisation

Few studies have been made of ways in which losses from food processes can be minimised.[30] As noted above, some of the 'waste' streams produced by the industry, such as blood-rich streams from abattoirs or milk-rich streams from dairies, contain valuable materials. The approach of the food industry is to combine waste streams and to treat them either in-house or dispose of them. Mixing valuable and non-valuable wastes in a single treatment plant is the type of 'end-of-pipe' solution which clean technology seeks to overcome. However, except in areas where the value of a waste stream is obvious, the segregation of waste is not common in food processing. This may lead to problems, as discussed below.

Modern techniques of waste management, which seek to minimise the losses from the material and the use of water, have tended to be adopted by the food industry on only an ad-hoc basis. In the author's experience of the food industry, waste disposal tends to be an issue only when it becomes a problem, whether or not it was optimal before. Zaror[31] discusses the application of modern techniques to waste management to the food industry, with specific reference to water management. In general, minimising the water usage of a plant requires careful management: (i) waste audits should be carried out to find areas where savings can be made, and (ii) a continuous effort to detect waste is needed.

Excessive production of waste arises due to:

(a) Inefficient technology, such as poorly maintained plant or control processes;
(b) inadequate processing, resulting from a poor knowledge of the process by management or operators;
(c) incorrect production planning, such as unsuitable scheduling of process operations. The correct scheduling of operations in the food industry is highly complex; for example, a plant consisting of two production canning lines may be required to produce ten products. Incorrect management of changeovers may lower productivity and produce waste. Some progress in the design of efficient process scheduling has been made.[32]

Many of the modifications required to reduce effluent are straightforward, such as improving management and supervision of the existing plant with a view to making savings; in a management sense, this can be easy to carry out, since the workforce will wish to behave in a 'green' manner. It is more complex to consider process modifications, such as the introduction of waste recycles. However, where streams of significantly different character are produced in a plant, the possibility of segregation to facilitate recovery or treatment should be considered. For example, once a food diluent stream has become contaminated with sodium hydroxide from a cleaning residue then it will not be possible to recover any material. Equally, if water is to be recycled to the process it should not contain chemicals which are either toxic or lower the product quality.

The key process is one of waste separation. First, liquid waste should be kept apart from solid wastes: for example, a high volume of water with a low organic load of solids can be reused after passing through filter screens and washing. Second, low volume effluents with high organic load should be kept separate from high volume loads with lower loads, either for waste processing or solids recovery.

As noted by Zaror, and discussed above, the meat and poultry industries in the UK generate very large amounts of waste, equivalent to 24 000 tonnes of fat and protein; if it were possible to recover some of this it could

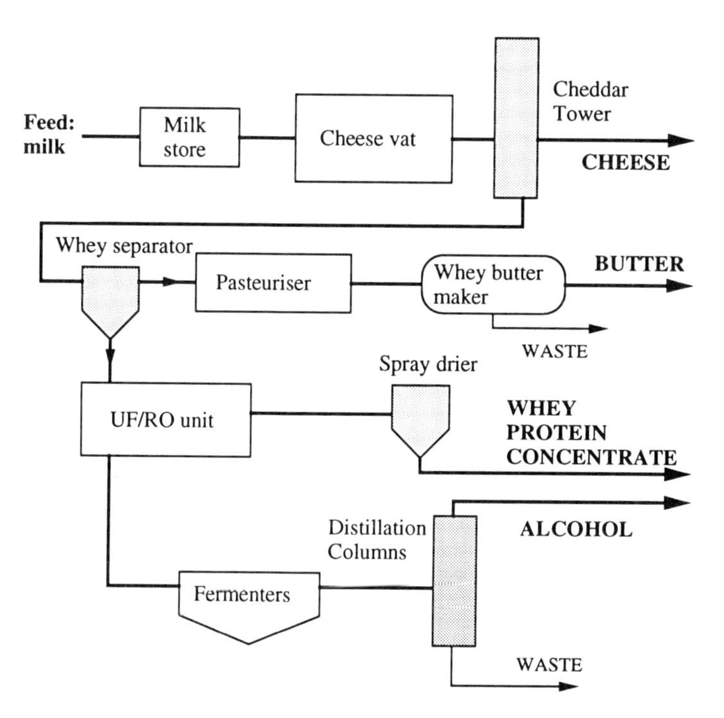

Figure 9.3 Schematic flowsheet of an efficient milk process.

be sold and reduce treatment costs.[33] Similar problems are found in the dairy industries, but some plants have been made highly efficient. For example, the simple flowsheet of Figure 9.3 shows a milk plant in which all the components of the milk are used, either for the production of cheddar cheese or for the making of butter, whey protein concentrate or alcohol, which can then be combined with the whey to produce cream drinks. The production of whey protein concentrate is a good example of the processing of a former waste stream; historically, the waste from cheddar production was discarded, thus letting streams of high value go to waste. Whey protein concentrate is used for animal feed and – more lucratively – as an food additive which can impart texture to a product.[34,35]

Whey recovery is an example of a 'waste' stream which was discovered to have considerable value. Similarly, a number of proteins from wastes can be marketed as food supplements, biological products or food additives. Proteins can be obtained by:

- recovery of existing proteins in effluent streams, i.e from whey, potato, fish and poultry;
- synthesis from other nutrients, e.g. fermentation of sugars;
- biomass generation as a result of waste treatment.

Table 9.1 Protein recovery from food waste streams (from refs 31, 36, 37, 38)

Protein product	Original material	Precipitation agent
Food grade proteins	Potato effluents	Heat
Food grade proteins	Meat effluents	Dicalcium phosphate
Food grade proteins	Food effluents	Lignosulphonic acid
Food grade proteins	Food effluents	Sulphuric acid (isoelectric)
Lactalbumin	Whey	Heat
Glycinin	Soya waste	Acid precipitation
GP dehydrogenase	Muscle extract	Ammonium sulphate
Various proteins	Blood plasma	Ethanol
Alcohol oxidase	Mycelial extract	Polyethylene glycol
Various proteins	Pig liver extract	Polyethylene glycol
Lysosyme	Egg white	Polyethylene acid

Several separation techniques are available for the recovery of protein from waste streams (see ref. 36), but their implementation is limited by the costs of the process. Precipitation techniques are probably the easiest for product recovery.[31] The use of heat is appropriate where the biological function of the protein is not important. Protein can also be precipitated by changing its pH to the isoelectric point, by the addition of salts such as sodium lignosulphonate, or by the addition of organic solvents. Table 9.1 lists proteins recovered by a variety of extraction techniques. Other separations have also been used, based on adsorption techniques such as ion exchange and membranes. A number of products are also made from 'waste', such as gelatin from crushed bone.

In the chemical industry, process integration, in which process flowsheets are designed as a unity, rather than as a series of individual operations, is commonplace.[39] The minimisation of effluent is an integration operation, and thus well suited to process integration techniques.[40,41] The techniques require recycling of both energy and waste. The amount of recycling done by the chemical industry is much greater than for the food industry, in part because of hygiene, in part because streams in the chemical industry can usually be recycled whereas in the food industry it is more difficult (for example, to remake cakes from broken cakes) and in part because these technologies have not yet been applied within the industry. Research into the application of process integration into the food industry would thus be a useful way of introducing clean production.

9.6 The environmental impact of food packaging

The previous sections have considered the impact of the processing of foods. However, as noted in the introduction, to the food consumer it is the packaging, rather than the food itself, which raises environmental concerns. Whether or not the recycling of materials is cost-effective, to the consumer this is the area which has become identified with environmental issues, and thus one in which the producers and retailers have concentrated.

Most food is sold in some type of package. For raw foods this is commonly a plastic or paper bag, to keep the food intact until it is brought home; for processed foods, the package serves a much more complex purpose and is consequently much more carefully designed. Until the last few years, packages were made from either paper, glass or metal-based materials, i.e. packets, bottles, jars and cans. The widespread introduction of plastic packaging has widened the range of products which can be sold; for example, the introduction of plastic trays has made it possible to market 'ready-meal' products. The role of packaging is three-fold; presentation, protection and preservation. In the first category, it is important that the material is attractive and of the required strength and stiffness. For protection and preservation, the package should resist contamination, by providing a barrier against natural contamination and defences against deliberate tampering and accidental punctures and tears, as well as being impervious to the gradual diffusion of gases and liquids which occurs over the length of time that the food is designed to stay in the pack. One other concern is the interaction between packaging and the food material; limits on the possible migration of components of the packaging into the food are set by EC directives and implemented by the UK government.[42]

The food sector dominates the packaging sector; in the UK, food packaging forms two thirds of the market.[43] The environmental impact of such packaging must be considered by the food industry, influenced strongly by the willingness of the consumer to consider environmental issues. Brown[44] reviews progress in this area; much of the rest of the section has been drawn from that paper. Brown notes that the environmental impact of packaging can be studied in several ways: (i) litter, (ii) the usage of raw materials, (iii) the ease of disposal, (iv) the consequences of careless disposal, (v) the feasibility of reuse, and (vi) the energy content of the material.

A series of methodologies are available for the reduction of the environmental impact of packaging. These are as follows.

Reducing packaging. By designing packages which use the minimum amount of material, both costs and environmental impact are reduced. ICI[45] note that the typical weight of a PET (poly(ethylene tetraphthalate)) bottle fell by 33% between 1975 and 1990. Such design often requires the use of complex computer programs to calculate the strength of packages, but can be a highly efficient way of reducing waste.

Reusing packaging. This is restricted by both safety and economic factors. Unless an efficient collection system – such as for milk bottles – is available, this may not be efficient. Once packages have been collected, it must be certain that they are clean and safe before reuse. Few plastics can resist the high temperatures needed for steam sterilisation.

Recycling packaging. Mechanical, chemical and biological recycling techniques are available. The direct mechanical recycling of packaging of materials is limited both by the transportation costs and by the safety and hygiene implications. In addition, without some way of recognising different types of plastic, recycling of packaging is as limited in food processing as in other industries. It is unlikely, given the need for hygiene, that food packaging will be directly recycled; however, it may be possible to use the material in other industries.

Chemical recycling, in which plastics are depolymerised and then reused, is a possible solution which would allow food plastics to be reused within the industry. PET prepared this way has been manufactured by Hoechst and approved by the US FDA for food contact use. One option for the industry is the use of biodegradable plastics, which break down slowly under microbial attack to harmless products. Essentially, these materials, such as Zeneca's 'Biopol', can be composted rather than dispatched to landfill; however, their efficient use will require separation of the materials which can break down from those which cannot. The materials are also likely to remain more expensive than other packaging materials. Degradation of this sort is difficult to control; there is no point in 'environment-friendly' packaging which disintegrates in the supermarket or the home.

Energy recovery. Incineration of plastic wastes is a possible way of disposing of material and releasing the chemical energy within the material. This is the probable solution for much food packaging, as the volume of waste can be reduced by a factor of ten.

In terms of the concepts of clean technology, the logical solution is to minimise the amount of waste, such as by the correct design of the pack, and then to seek to recover as much of the rest of the material as possible. These requirements may soon be legally enforced, via EU environmental standards. Future design of packaging will require a means of disposal to be thought of during the design process; the package must be as simple as possible, for example, made from the minimum number of materials to assist recovery. It will also be necessary to consider the impact of packaging within the more complex system of distribution and recovery, perhaps through life-cycle analysis.

9.7 Clean technology and the food industry

This chapter has reviewed the environmental problems of the food industry. Of the five areas discussed, only the cleaning problem is unique to the food industry. The other four are essentially similar to the problems faced by other process industries. The contamination of ingredients is a similar problem to that faced by farming, that of product contamination to the pharmaceutical industry, that of waste to any process industry and that of packaging is similar to the problems of any retailer. Research in any of these industries will be of relevance to food problems.

In cleaning and hygienic design, the industry has developed efficient solutions. Some of these, by neglecting products from waste streams or designing processes in which all streams are used, fit the ideas of clean technology well. The industry has concentrated on the production of food with minimal contamination from the environment; this involves separating the food from the environment, and extensive cleaning.

It may be possible both to reduce the amount of effluent produced by cleaning, by correct selection of cleaning chemical and the length of the cleaning cycle, and to reduce the effluent produced by recyling and by extracting valuable products from waste streams before they are treated. The design of plants which minimise the waste they produce and which incorporate energy, water and product recovery is not as yet widespread; developments in both waste management and process integration will be important. More research is needed to optimise the cleaning process: such work will be cross-disciplinary, involving microbiologists and chemists as well as engineers. Computer-aided design and computational fluid dynamics can be of assistance in designing pipework and flow systems; however, hygiene requires an understanding of surface events, such as the rates of adhesion and removal of contaminants, which need more information than the computers will provide.

The problems discussed in this chapter will become more severe over the next decade, as the environmental constraints on the industry increase.[46] The ideas of clean technology have not spread into the food industry; outside the major food companies, there is little awareness of waste minimisation issues. There will be significant scope for technology transfer into the food industry of solutions developed for other industries.

Finally, it should be noted that there are areas of waste which have not been covered by the above discussion. Within the food processing factory, it is possible to minimise waste. Another, perhaps greater, environmental problem is the waste of food. Even in developed countries such as the UK, a large amount of the food grown is wasted through losses in distribution and storage, and through losses in the market, in shops and in the home. Studies from the Ministry of Agriculture, Fisheries and Food (MAFF) in 1980 showed that about 25% of food was lost between producer and final consumption. It is uncertain what the figure is in less developed countries; however, where it is possible to reduce this loss, the environmental benefits would be huge. It should thus be remembered that, in food production, as in farming, high technology solutions may not be only ones which produce environmental benefits.

Acknowledgements

The work on food fouling and cleaning described here was sponsored by Unilever and by the AFRC and was carried out by Drs M.R. Bird and M.T. Belmar-Beiny. The author is also grateful for discussions with a number of industrial companies.

References

1. Mayes, T. (1992 & 1993) Simple users guide to the HACCP concept for control of microbiological safety, *Food Control*, **3**, 14–20, and: The application of management systems to food safety and quality, *Trends in Food Science and Technology*, **4**, 216–219.
2. Sloan, A.E. (1993) Consumers, the environment and the food industry, *Food Technology*, 72–75, 91, Aug.
3. Nriagu, J.O. and Simmons, M.S. (eds) (1990) *Food Contamination from Environmental Sources*, Wiley, New York.
4. Packard, E.G. (1906) *Entomol. News*, **17**, 256.
5. Flegal, A.R., Smith, D.R. and Elias, R.W. (1990) Lead contamination in food. In: *Food Contamination from Environmental Sources*, eds Nriagu, J.O. and Simmons, M.S., Wiley, New York, Ch. 4, pp. 85–120.
6. Cappon, C.J. (1990) Speciation of selected trace elements in edible seafood. In *Food Contamination from Environmental Sources*, eds Nriagu, J.O. and Simmons, M.S., Wiley, New York, Ch. 6, pp. 145–195.
7. Kulkarni, A.P and Ashoke, M. (1990) Pesticide contamination of food in the US. In: *Food Contamination from Environmental Sources*, eds Nriagu, J.O. and Simmons, M.S., Wiley, New York, Ch. 9, pp. 257–293.
8. Phillips, J.D. and Griffiths, M.W. (1990) Pasteurised dairy products: the constraints imposed by environmental contamination. Nriagu, J.O. and Simmons, M.S. (eds), *Food Contamination from Environmental Sources*, Wiley, New York, Ch. 13, pp. 387–456.
9. Fellows, P. (1988) *Food Processing Technology*, Ellis Horwood.
10. Hasting, A.P.M. (1992) Practical considerations in the design, operation and control of food pasteurisation processes, *Food Control*, **3**, 27–33.
11. Holdsworth, S.D. (1993) *Aseptic Processing and Packaging of Food Products*, Elsevier.
12. Burton, H. (1988) *Ultra-high Temperature Processing of Milk and Milk Products*, Elsevier.
13. Vinson, H.G. (1990) Hygienic design of a dairy, *J. Soc. Dairy Techn.*, **43**, 39–41.

14. de Goederen, G, Pritchard, N.J. and Hasting, A.P.M. Improved cleaning processes for the food industry. In: *Fouling and Cleaning in Food Processing*, ed H.G. Kessler and D.B. Lund, Univ., of Munich, 198, pp. 115–130.
15. Romney, A.J.D. (1990) *CIP: Cleaning in Place*, Society of Dairy Technology, London.
16. Lalande, M., Tissier, J.P. and Corrieu, G. Fouling of a plate exchanger used in ultra-high temperature sterilisation of milk, *J. Dairy Res*, **51**, 557–568.
17. Belmar-Beiny, M.T, Gotham, S.M., Fryer, P.J., Paterson, W.R. and Pritchard, A.M. (1993) The effect of Reynolds number and fluid temperature in whey protein fouling, *J. Fd Eng.*, **19**, 119–139.
18. Belmar-Beiny, M.-T., and Fryer, P.J. (1993) Preliminary stages of fouling from whey protein solutions, *J. Dairy Res.*, **60**, 467–483.
19. Perlat, M.N. (1986) PhD Thesis, UST 1 Lille, France.
20. Graßhoff, A. (1989) Environmental aspects of the use of alkaline cleaning solutions. In: *Fouling and Cleaning in Food Processing*, (eds Kessler, H.G. and Lund, D.B.), Munich Univ., pp. 107–114.
21. Bird, M.R. and Fryer, P.J. (1991) An experimental study of the cleaning of surfaces fouled by whey proteins, *Trans. I ChemE C*, **69**, 13–21.
22. Bird, M.R. and Fryer, P.J. (1992) An analytical model for the cleaning and hygienic operation of food plant, *I ChemE Symp. Ser.*, **126**, 325–330.
23. Bird, M.R. (1993) PhD Thesis, Univ. of Cambridge.
24. Bird, M.R., Milford, B.G., and Tucker, B.J. (1994) The removal of carbohydrate based deposits from stainless steel surfaces using chemical cleaning agents, *IChemE Research Event*, **1**, 529–531.
25. Bird, M.R. and Espig, S.W.P. (1994) Cost optimisation of dairy cleaning in place (CIP) cycles, *IChemE Research Event*, **1**, 458–460.
26. Liu, Z.H. and Machietto, S. (1993) Cleaning in place policies for a food processing batch plant, *IChemE Research Event*, **1**, 25–27.
27. Graßhoff, A. (1992) Hygienic design – the basis for computer controlled automation, *Trans. IChemE C*, **70**, 69–77.
28. EHETG (1993) (European Hygienic Equipment Design Group) have produced a series of articles on the basic principles of hygienic design, in *Trends in Food Science and Technology*, **4**, 21–25, 52–55, 80–82, 190–192, 225–229, 306–310.
29. Ugerlu, A, and Forster, C.F. (1992) A comparison of the effect of oleate and palmitate on thermophilic and mesophilic anaerobic studies, *Trans. IChemE B*, **70**, 125–130.
30. Carawan, R.E. (1989) Environmental issues facing food processors in the 1990s. *Proc. Food Processing Waste Conf.*, Georgia Tech. Res. Inst., Atlanta, pp. 217–237.
31. Zaror, C.A. (1992) Controlling the environmental impact of the food industry: an integral approach, *Food Control*, **3**, 190–199.
32. Machietto, S. (1992) Automation research on a food processing pilot plant, *IChemE Symp. Ser.*, **126**, 179–190.
33. Knorr, D. (1983) Recovery of functional proteins from food processing waste, *Food Tech.*, **37**, 81–84.
34. Bottomley, R.C., Evans, M.T.A. and Parkinson, C.J. (1990) Whey proteins. In *Food Gels*, ed P. Harris, Elsevier, London, pp. 435–466.
35. Taylor, S.M, and Fryer, P.J. The effect of shear on the gelation of whey proteins, *Food Hydrocolloids*. (In press)
36. Asenjo, J.A. (1990) *Separation Processes in Biotechnology*, Marcel Dekker, New York.
37. Patel, M. (1976) *Processes for the Treatment of Food Trade Effluents*, The British Food Manufacturing Research Association, Scientific and Technical Surveys, Report No. 93, 1976.
38. Grant, R.A. (1978) in *New Processes of Waste Water Treatment and Recovery*, ed C. Mattock, Ellis Horwood, Chichester, pp 405–415.
39. Douglas, J.M. (1988) *Conceptual Design of Chemical Processes*, McGraw-Hill.
40. Douglas, J.M. Process systems for waste minimisation, *Ind. Eng. Chem. Res.*, **31**, 238–243.
41. SERC Clean Technology Unit (1993), *Stopping Waste Within the Production Process*, SERC, Swindon.
42. Statutory Instrument 3145 (1992) The plastic materials and articles in contact with food regulations, HMSO, London.

43. Cole, M. (1991) *Food Manuf*, 26 May.
44. Brown, D. (1993) Plastics packaging of food products: the environmental dimension, *Trends in Food Science and Technology*, **4**, 294–300.
45. ICI (1990) *Packaging – an introduction to ecological assessments*, Plastics Environmental Affairs Group, Welwyn Garden City, UK.
46. Hiranjan, K., Okas, M.R. and Rankowitz, M. (eds) (1994) *Environmentally Responsible Food Processing*, AIChE Symp. Ser. 300.

10 Clean synthesis

C.J. SUCKLING

10.1 Scope of chapter

10.1.1 Introduction

I remember being driven as a child through a major centre of the British chemical industry, the North of England town of Widnes, and noticing layers of brightly coloured sediments on plots at the roadside next to many of the chemical plants. Greens, presumably containing copper, yellows, perhaps chromates, and reds, probably from iron(III) salts, were all visible. The chemical industry also made itself known through the characteristic smells that were common on those banks of the Mersey. These features were characteristic of early and mid-twentieth century chemical operations which today would not be tolerated either by the local population or by the environmental protection agencies working on their behalf. It would be more common now for the smell of paint in D-I-Y operations to be stronger than the scent of the air even within a chemical works such has been the change in the competence of industry to contain its operations. But containment of gases is only part of the question of clean manufacture. Liquid effluents, both solvents and solutions, and solid waste also require attention. In many countries, rivers have had to accept the burden of aqueous effluent and landfills the solid waste; solvents have been incinerated. However careful, responsible, and professional companies are in dealing with waste, there remains public and political pressure, not all of which is rational and well considered, to prohibit the disposal of waste in these ways.

If there is no solution or only a limited accepted solution in the better handling of waste, then new methods of process operation and new chemical reactions that are, to a first approximation, waste free must be discovered and developed. If clean-up is not adequate, then intrinsically clean synthesis is the only alternative if a viable chemical industry is to remain to stimulate the economy and serve the public through its products. This chapter discusses some of the recent chemical discoveries that are contributing to the attainment of the goal of clean synthesis. Examples of discoveries already brought into production together with recent research with potential for development will be discussed. The challenge of clean synthesis has attracted the attention of public research funding agencies in many countries including the British Agricultural and Food Research Council's and Science and Engineering Research Council's Clean Technology Unit. A report to which I contributed identifies many of the key properties of clean synthesis, several of which I shall introduce now to provide a framework for the following discussion.[1] Many of the possibilities discussed are not accessible with currently available chemistry or would clearly be uneconomic in today's economic climate. The importance of clean synthesis to the planet is vital for future generations.

10.1.2 Clean synthesis

If the currently available chemical and biological science applicable to chemical manufacture is considered from a point on the horizon under a blue sky of anticipated new knowledge, a number of shortcomings become obvious. Let us take for granted the waste intrinsic in by-products and in the inefficient use of energy – which is beyond the scope of this chapter – and consider some common features of the industrial synthesis of organic compounds.

The scene for industrial synthesis is set by the feedstocks available. In this century, there has been a shift from coal, which afforded tars and acetylene, to oil, whose corresponding products are naphtha and ethylene. Through various hydrogenation, cracking, and isomerisation reactions, large quantities of intermediates have become available at very low prices. Modern high tonnage petrochemical operations are also clean in a great many respects. However, all of these operations consume fossil resources. One of the main pressures of the late twentieth century is to move to renewable resources because exhaustion of fossil hydrocarbons can be anticipated in the next century. The major consumer of such resources is not the chemical industry but the energy industry; even so, it is not easy to separate the two in political terms: both are targets for so-called carbon taxes along with the private motorist. Inevitably there will be strengthening forces to move the chemical industry to renewable carbon feedstocks.[2] Whilst for most purposes and in most countries, ethanol from fermentation of sugar is not economic, conceivable changes in fiscal regimes could alter the position rapidly. There is thus a strong coupling between the politics of the day and the pace at which clean synthesis develops.

Beyond the political arena and very much within the field of chemistry, there are several limitations of current practice that can be recognised, limitations that are inherent in chemistry itself. Given a range of feedstocks with reasonable chemical reactivity such as alkenes or alkynes, the principal task of chemistry is to construct larger chains and rings bearing the appropriate substituents of carbon, oxygen, nitrogen or other elements for the product to cause the desired effect when sold to and used by the customer. Traditional chemistry relies upon the introduction of oxygen or of halogens in particular to activate the hydrocarbon fragment to enable new bonds to be formed. Jentzsch has identified the problems from another direction,[3] pointing out that feedstocks are also major determinants of the technology to be used. Although there may be alternative primary carbon feedstocks (coal, oil, gas) the chemistry that converts them into synthetically useful compounds (steam cracking) tends to afford a similar mixture of compounds whatever the primary source. This poses an additional challenge for the development of clean chemistry.

Colleagues in industry have remarked that at one of their sites they import tens of tons of bromine each year, none of which ends up in the product sold. To put an extreme view, therefore, nucleophilic substitution and organometallic reagents in aliphatic systems are inherently unclean reactions. By this definition, intrinsically cleaner reactions for constructing carbon chains and rings would be catalysed aldol and Michael reactions. In particular α,β-unsaturated carbonyl compounds are important; for example, in the plastics industry methacrylic acid and its derivatives are important monomers. A plausible industrial synthesis of these compounds via acetone can be envisaged (Figure 10.1a); this route, however, uses the hazardous reagents hydrogen cyanide and sulphuric acid. Moreover, for

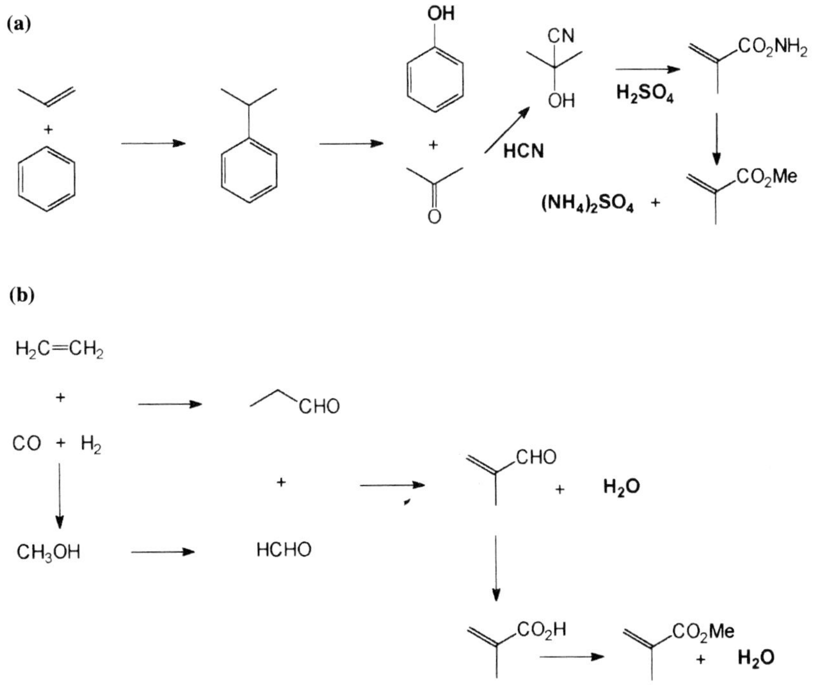

Figure 10.1 Industrial synthesis of α,β-unsaturated carbonyl compounds.

every mole of product, half a mole of ammonium sulphate is formed as waste. From any point of view, there is little to commend this route. In contrast, taking advantage of technology associated with the oxo-process for hydroformylation, BASF has developed a clean, catalytic route to methacrylates that exemplifies well the characteristics of clean synthesis (Figure 10.1b). It is also interesting to note that prior to the development of this process, the company was not in the methacrylates business.[3] Clean technology combined with entrepreneurial vision thus led to new business.

Similar chemical shortcomings can be found in aromatic chemistry where the generation of electrophiles dominates bond-forming processes of almost all types. For example, a Friedel Crafts acylation requires stoichiometric quantities of anhydrous aluminium chloride and a suitable inert solvent together with an acyl halide. Problems with chloride and aluminium waste are obvious. As will be discussed later, some solutions to these problems have been found. Nevertheless, the dependence of synthesis upon these types of chemistry limits severely the cleanness currently available. Ideally, but impossibly, only those atoms required in the product to create its effect should appear in the process. An asymptotic approach to such ideality would be to make every leaving group water, or perhaps a lower alcohol, and every activating group a carbonyl group that could be removed by reduction and elimination, or by decarboxylation.

This chemistry implies a greater emphasis upon carbocations than that used so far and this in itself identifies a need for catalysts. The importance of catalysis is not only to generate and control the chemistry of carbocations but also to initiate the synthetic sequence by providing suitably activated high tonnage building blocks for elaboration into larger compounds. A feature of the advancement of chemistry of the past 25 years has been the symbiotic development of biological catalysts and the rise of the process biotechnology industry together with the invention of increasingly sophisticated and selective non-biological catalysts. (The paper by Seebach[4] contains references to many topics relevant to this chapter; it is a landmark article for organic synthesis in the late twentieth century.) Many of the new nonbiological catalysts have been devised to introduce chirality in a controlled manner into substrates in response to the stereoselectivity of enzymes; many of these catalysts and enzymes will also find application in the fine chemicals manufacturing industry (see below). However, at the current level of the argument, a key question that arises is whether biological catalysis or even agriculture can furnish the defined building blocks for synthesis.

The major bulk chemical products of agriculture are sugars,[2] which are too highly functionalised for most synthesis, and lipids, which are too poorly functionalised.[5] There is currently no good agricultural process that leads to a bifunctional C4 to C8 building block that would satisfy the chemical criteria outlined above as methacrylates do. Thus a secondary operation at least must be carried out before a synthetic process can begin. Such an operation might be a fermentation leading to ethanol or a C3 or C4 ketone as has been known for 80 years (Weizmann acetone process using *Clostridium acetobutylicum* acting on starches[2]) but inevitably, any secondary process will lead to waste in the form of unwanted biomass such as dead cells or straw. This biomass contributes intimately to the economic viability and technical feasibility of a process. A conventional wisdom in some quarters holds that a process based upon a biological catalyst will always be cleaner than one using a non-biological catalyst. Such a view can be sustained unchallenged only by an emotional fixation. If we are to achieve the reality of clean synthesis, it is essential for scientists and engineers to make known all the relevant technical information and for politicians to be open and straightforward with the current and anticipated fiscal regimes under which companies will be expected to act. Politicians share in the economic success of the companies whose trading environments they manage; equally they must take their share of the blame for failures.

Despite these strictures, biological catalysis clearly has a place in clean synthesis as does non-biological catalysis. Even a good catalyst with respect to the selectivity of the reaction it catalyses is not necessarily clean. There are always problems associated with poisoning and regeneration or of leakage of the catalyst itself into the environment, especially where homogeneous catalysts are used. For these reasons, a major recent field of development has been in supported catalysts.[6,7] Such advances are built upon the chemistry of organic and inorganic polymers (especially zeolites) and have already provided ways to avoid the halogen problem outlined above. Electrophilic aromatic substitution catalysed by zeolites and by polymeric strong acids will be discussed later.

Supported catalysts provide encouragement for the field of clean synthesis. However, these catalysts provide substitutes for existing unclean catalysts or reagents; they arise from the same traditional chemistry. What scope is there for discovering new chemistry that will be intrinsically clean? Of course, no reactions can sidestep the inherent chemical reactivity of compounds. What is needed is to identify a new series of catalysts and arrays of functionality upon which they can act consistent with the clean concept. More easily said than done, but many examples of isolated reactions moving science towards the clean horizon are published each year (see Section 10.5). I have derived conceptually through many stages from some of our work[8] a scheme that approaches the requirements presented in this section (Figure 10.2). There is no direct experimental evidence for the entire sequence although readers will recognise precedents for individual steps. The strategy requires the production of a push-pull system to trap an intermediate carbanion as represented by **2.1** in Figure 10.2. In this compound, the amino nucleophile adds to the carbon source (acetylene) and the resulting carbanion (**2.2**) is trapped by the aldehyde

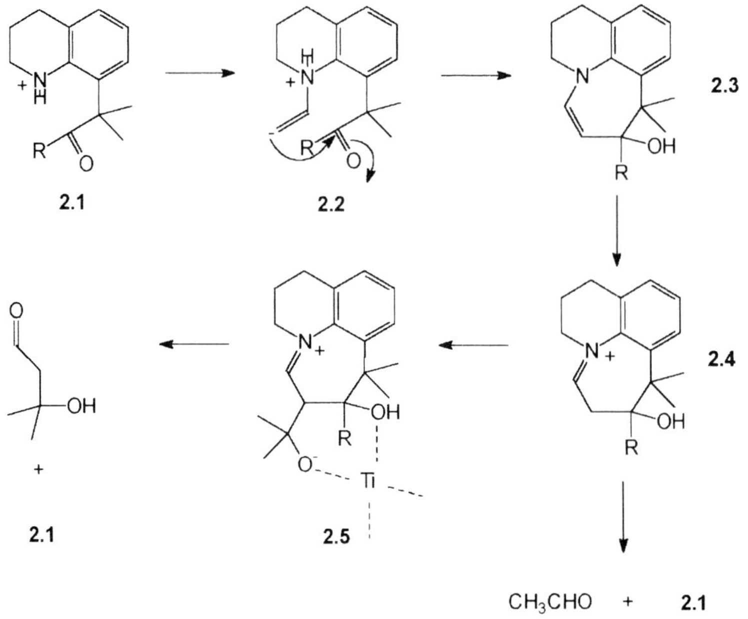

Figure 10.2 Experimental scheme for catalysts providing substitutes for unclean catalysts or reagents.

which is forced to remain in the plane of the aromatic ring by the geminal methyl groups which also prevent dehydration to give the intermediate **2.3**. This versatile intermediate is an enamine which could be hydrolysed to acetaldehyde via the corresponding iminium salt **2.4** thus regenerating **2.1**. This sequence would add water without resorting to heavy metal catalysts. To generate nucleophilic activity, **2.4** could be reacted in the presence of titanium alkoxides with carbonyl acceptors to form new carbon-carbon bonds as in **2.5**; hydrolysis could be envisaged to lead to bifunctional molecules in the C4–C8 range.

Even if this paper chemistry works, there must be an economic niche for its exploitation. However, it is the responsibility of academic research to create such possibilities. The ability of chemists, in particular process development specialists, to improve the performance of a chemical reaction from a prototype showing an important feature to a small extent, such as a new enantioselective reaction, to a commercially viable one must not be underestimated. Indeed, in the pharmaceutical industry, such feats are not uncommon; the asymmetric cyclopropanation reaction first introduced by Noyori in 1966[9] has now been improved to the extent that it forms a part of the synthesis of the enzyme inhibitor and drug component, cilastatin. The examples chosen in this chapter will therefore include both commercially exploited chemistry and reactions that, in my opinion, have features of interest with respect to the development of the clean synthesis of bulk chemicals.

10.2 Stoichiometric reactions

10.2.1 Does clean synthesis depend upon catalysts?

Catalysts are very often expensive. They have limited lifetimes. The disposal of spent catalysts must be considered. Traces of catalyst can leak into the environment. These statements imply that an ideal for clean synthesis would be a reaction in which all the reactants were converted quantitatively into one single product, an attractive but improbable situation. Approaches to clean synthesis in the absence of catalysts can nevertheless be considered in a more limited fashion if there is an improvement in the overall environmental impact of the reaction. The replacement of hazardous or toxic reagents by safer alternatives is a relevant context especially if single products can be obtained in high yield. Similarly the avoidance of odours would be an advantage. Two general chemical aspects of such a replacement can be recognised in addition: on the one hand, a conventional reagent that is too reactive may be modified so that its reactivity is reduced to provide selectivity and reduced hazard and, on the other hand, a conventional reagent that is too unreactive may interact in such a way that its activity is increased (see copper(II) halides below). A third possibility exists in which a reaction environment is provided to control the course of reaction without entering into the sequence of chemical transformations as would a catalyst. The main class of reactions in which these features have been demonstrated is electrophilic aromatic substitution.

10.2.2 Selectivity in aromatic substitution

The mixtures that typically arise in any electrophilic aromatic substitution charge a penalty on any process that uses them. Further, the involvement of highly oxidising and corrosive reagents such as halogens, fuming nitric acid or oleum, and heavy metal salts such as mercury(II), identifies the reactions as candidates for phasing out in the era of clean synthesis. For many reactions involving carbon–carbon bond formation, catalysis is required conventionally by Lewis acids such as aluminium chloride; examples of these will be discussed in the following section.

Over-reactivity of a reagent is particularly relevant in substitution reactions of aromatic compounds such as phenols and naphthols, themselves more reactive than benzene. Derivatives of these compounds are important intermediates in the synthesis of dyes. Chlorine itself is sufficiently

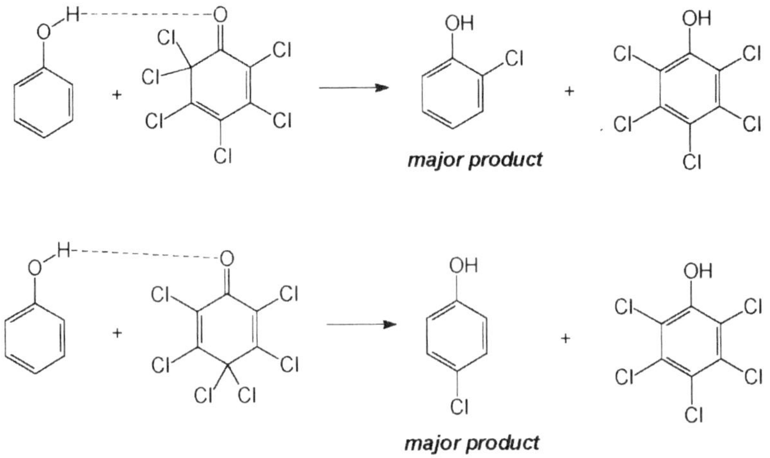

Figure 10.3 Selectivity in aromatic chlorination.

reactive to polychlorinate under some reaction conditions and is a prime candidate for reactivity reduction. Lemaire and colleagues showed that hexachlorocyclohexadienones, obtainable from pentachlorophenol by the addition of chlorine, are able to act as selective chlorinating agents for phenols and their ethers.[10]

Two isomers can be formed and each is isolable under suitable conditions (Figure 10.3). The 2,2-dichloro isomer favours 2-chlorination of phenol and the 4,4-dichloro isomer the complementary 4-chlorination. Similar selectivity can be obtained with 2-naphthol as substrate.[11] The mechanism leading to selectivity was suggested to involve the formation of a hydrogen bond between the hydroxyl group of the phenol and the carbonyl group of the reagent. In agreement with this hypothesis is the observation that the ethers of phenols are chlorinated in the position normally most reactive. Although the selectivity is modest between isomers, the key point is that monochlorination was obtained exclusively; the normally competing polychlorination was avoided by the choice of the special halogen transfer reagent. A similar reagent, derived from 2,3,4,6-tetrabromo-4-ethylphenol and nitric acid in acetic acid, was shown to be effective for mononitration of phenols.[12]

In related work, we found that selective 2-nitration of phenol could be achieved by a similar mechanism.[8] In the course of work on the effect of micelles on the course of aromatic subsitution, we had found that pyridinium salts form significantly strong charge transfer complexes with phenol. This and other results led to experiments in which the pyridinium salt contained a nucleophilic group that was capable of accepting a nitro group (Figure 10.4). The most convincing demonstration of the properties of these systems came with the discovery that nitrogen dioxide could serve

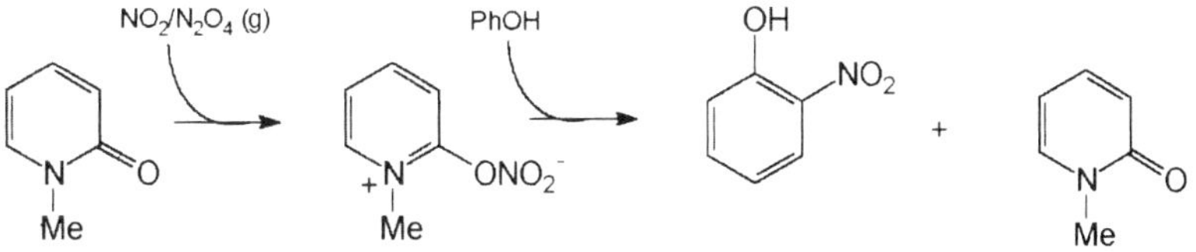

Figure 10.4 Nitration of phenol using *N*-methyl-2-pyridone.

as the donor of the nitro group, avoiding expensive reagents such as nitronium tetrafluoroborate, and that very simple pyridine derivatives such as *N*-methyl-2-pyridone could serve as nitro group acceptors. In order to bring the chemistry closer to a potential application, a styrene copolymer bearing 1% pyridone groups was synthesised and activated by passing nitrogen dioxide gas into a suspension of the polymer in an inert solvent. Phenol reacted with the activated polymer to give crystalline 2-nitrophenol, after evaporation of solvent, in greater than 95% yield. Both this reaction and the chlorination reactions mentioned above share mechanistic similarities and a disadvantage with respect to application. In the activation of the pyridone or cyclohexadienone, one equivalent of nitrite or chloride respectively are formed, a feature that reduces their value as clean synthetic options. On the other hand the lack of side reactions and the ability to prepare immobilised supported systems are positive features.

So far, such systems have been limited to the most reactive electrophiles; for carbon–carbon bond formation in phenols, transfer substitution in this way has not been demonstrated. An alternative approach using the ability of cyclodextrins to form complexes with many monocyclic aromatic compounds has been extensively investigated in the laboratory. Whilst the direct scale-up of these reactions to large scale is unlikely for the simple substrates such as phenols, it could be imagined that in the production of a pharmaceutical intermediate the ability to introduce a carbon substituent exclusively *para* to a phenolic hydroxyl group would be valuable. The reactions also demonstrate the potential of host–guest complexation in solution in controlling the course of well known reactions. Similar features are relevant to the properties of inorganic solid supports such as zeolites (see Section 10.3.5).

In cyclodextrins, the hydrophobic interior cavity is suitably sized to incorporate one benzene ring in such an orientation that substituents are excluded, leaving the guest molecule placed along the axis of the cyclodextrin ring. In this position, substitution can be encouraged to occur exclusively in the *para* position. Further, it is possible to modify the structure of the cyclodextrin by forming ethers and esters, especially with the primary hydroxyl groups, which can contribute to the selectivity of reactions. For example, it has been possible to mediate selective 4-hydroxymethylation of

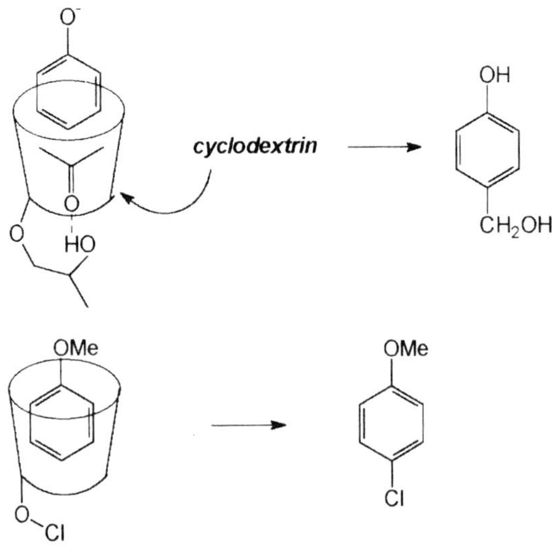

Figure 10.5 4-Hydroxymethylation of phenol in the presence of cyclodextrin.

phenol in the presence of a cyclodextrin;[13] the rate of 4-substitution was determined to be 16 times that of 2-substitution in the presence of the cyclodextrin (Figure 10.5). If copper salts additionally are included 4-carboxylation can be accomplished.[14] There are also examples in which phenyl ethers undergo selective 4-chlorination;[15] in this case, there is not only a protective effect of the cyclodextrin on the 2-position but also a transfer substitution reaction in which chlorine is exchanged from hypochlorous acid to one of the hydroxyl groups of the cyclodextrin sugar. To my knowledge, none of these reactions has been developed to any stage approaching a commercial process but the fact that they all occur at ambient temperature, with high yields, and in high selectivity commands attention from the point of view of concepts for clean synthesis.

10.2.3 Heterogeneous reactions

The best performance in the nitration of phenol discussed above was obtained when a reaction discovered in homogeneous solution occurring by a well understood mechanism was converted by simple polymer chemistry into a reaction upon a solid support. In general, heterogeneous reactions using supported reagents[16] have found much favour in clean synthesis because of the avoidance of extraction and solvent separation operations; ideally, a pure product can be isolated simply by evaporation of the reaction medium and the insoluble reagent or catalyst (see below) regenerated or reused. If the reagent itself has some environmental or other safety restrictions, the potential exposure of workers and the environment can be minimised by support on a solid for which containment is easy. In many cases, the reactivity of reagents supported on polymers is reduced compared with that in normal solution, that is they become more selective, but there are isolated examples of the reactivity of reagents being enhanced by absorption on to a solid support. Many supported reagents are commercially available. Most common oxidising and reducing agents have been explored and many nucleophiles (azide, thiocyanate, and even fluoride) can be used as supported reagents.[16]

As examples of improved selectivity and reduced hazard, the oxidation of alcohols by chromium(VI) compounds may be cited.[17a,b] Chromium(VI) compounds are well known as potentially carcinogenic and they are also such powerful oxidants that it is difficult to prevent over-oxidation of the substrate. For example, the direct oxidation of primary alcohols to aldehydes often occurs in low yield because of further oxidation to the corresponding carboxylic acid. Chromyl chloride chemisorbed on to silica–alumina has been found to be a clean oxidising agent for the conversion of octan-1-ol and benzyl alcohol to the corresponding aldehydes in yields well over 90%. An alternative is to support dichromate anions on quaternary ammonium ion exchange resins. The same type of reaction can be accomplished by a large number of metal nitrates absorbed on to silica with the reverse effect on the reagent.[18] In this case, the unsupported nitrates (copper and zinc were most practical) are essentially inactive but high yields without over-oxidation were obtained using the supported nitrates in non-polar solvents such as hexane or carbon tetrachloride at reflux. Aromatic halogenation in polymethylbenzenes can also be promoted by the absorption of the reagent, a copper(II) halide, on to neutral alumina.[19] Not only are the reagents inactive unless supported, but also halogenation only occurs in the ring leaving the side chain unchanged. Both bromination and chlorination reactions were demonstrated.

A completetly different reaction type, carbanion-mediated condensations, has also been accomplished using supported reagents.[20] Whereas fluoride under certain conditions, especially aprotic conditions, behaves as a nucleophile, with suitable substrates such as enolisable carbonyl-containing compounds, it behaves as a base. Potassium fluoride absorbed on alumina has been used in Michael, Knoevenagel, and Wittig reactions. It has the advantage of being insensitive to water, unlike conventional bases such as sodium methoxide. Further, such a material packed into a chromatography column can be used to catalyse reactions simply by percolating a solution of the substrates through the column.

The alert reader will have recognised that there are features of many of the reactions cited that must be modified if they are to be effective, clean, synthetic operations. For example, the use of chlorinated solvents proscribed by the Montreal Protocol must be avoided. Further, there are by-products in many cases deriving from the reagents used; even chloride or nitrite is a by-product with an environmental and process cost attached to it. Obviously one way to minimise reagent-derived by-products is to use catalytic reactions as will now be described.

10.3 Non-biological catalysts

10.3.1 The niche for non-biological catalysts

Fashions change in science and technology, more slowly than in *haute couture*, but perceptibly. Enthusiasm for biological catalysts on grounds of catalytic efficiency, selectivity, or apparent lack of environmental burden must be tempered by the lack of suitable biocatalysts for many reactions of interest to synthetic chemistry. Nature provides only a limited range of enzymes, with which much can be done but the introduction of halogens, nitro groups, sulphates, and the formation of carbon–carbon bonds in aromatic compounds are either unknown or very limited in scope. There is a clear role for non-biological catalysts in aromatic substitution, in oxidation and in hydrogenation reactions. Equally, there are several situations in which the synthesis of a commercially important compound has been shown to be possible using routes based upon biological or non-biological catalysis, for example β-blockers (see below). Commercial and practical considerations beyond the scope of this chapter determine what will be the method of choice.

10.3.2 Aromatic substitution

The SERC/AFRC report identified Friedel Crafts alkylation and arylation as a major reaction in which cleaner methods should be found to avoid the problems associated with aluminium chloride or other Lewis acids, which are required stoichiometrically in acylation reactions. Anyone who has carried out a Friedel Crafts acylation will be aware of the continuous evolution of hydrogen chloride gas from the reaction and the inevitable messy work-up in which the aluminium chloride complexes are carefully decomposed with ice-cold dilute hydrochloric acid. The need for dry, pure aluminium chloride is also obvious because of its ready hydrolysis in water. Further, the evolution of hydrogen chloride associated with aluminium chloride hydrolysis and use causes problems of corrosion. Work begun at the University of York and subsequently vigorously commercialised has shown that montmorillonite clays, suitably restructured, can act as efficient catalysts for Friedel Crafts alkylation such as benzylation.[21] The activity of the catalyst depends greatly upon the method of preparation.[22] Such materials, known felicitously as ‘Envirocats’, are free flowing, non-toxic powders. In a Friedel Crafts reaction using an alkyl or acyl chloride, the only hydrogen chloride evolved is that produced in the substitution itself. As with all supported reagents, the problem of reagent residues remaining in the product is minimised. In some cases, the material can be reused although eventually it must be disposed of by landfill. There are thus many advantages characteristic of Envirocats from the point of view of clean synthesis; in parallel with the reactions discussed in the preceding section, there are also chemical advantages in the selectivity of substitution.

Using suitable Envirocats, the benzoylation of benzene and mono-substituted benzene derivatives is catalysed at temperatures between 140° and 165°C; the *para* selectivity in the benzoylation of fluorobenzene is particularly remarkable (Figure 10.6). Other acylations such as with acetyl chloride are so far much less selective giving mixtures of products in moderate yields with substrates such as anisole. In many cases, the catalyst can be recovered and reused. Side reactions such as ether cleavage or halogen exchange in fluoroalkyl groups (common in agrochemicals) are not usually observed. Sulphonation with arylsulphonyl chlorides also gives mixtures but good yields are obtained. The reaction occurring under the mildest conditions is alkylation with alkyl halides which takes place at 80°C or below in good yields. In all of these reactions, a halogenated electrophilic reactant is required, a disadvantage from the purest view of clean synthesis. However, non-halogen containing electrophiles such as alkenes and formaldehyde can also be used to give good conversions to commercially significant products such as di-*t*-butylmethoxyphenol (anti-oxidant) and alkylphenols (surfactant components). This range of enhancements to Friedel Crafts chemistry is a major achievement.

Whilst Envirocats have been extensively promoted commercially, other supported reagents have been described for similar chemical reactions. 'Clayzics' are also montmorillonite clays with zinc chloride, a Lewis acid, absorbed upon them. These materials have been found to be effective catalysts for some Friedel Crafts reactions such as the alkylation of alkyl benzenes with benzyl chloride. Apart from smooth conversion to benzylated products at temperatures ranging from 40°–160°C, a number of surprising phenomena were observed which, if exploited, could form the basis of a further range of controls on the selectivity of this type of reaction. For example, if two substrates are used, synergistic rate accelerations can be obtained. Thus mesitylene in the presence of benzene reacts faster than mesitylene alone[23] and is preferentially alkylated with respect to benzene. The latter result runs contrary to the normal expectation of reactivity of a large electrophile interacting with a hindered versus a non-hindered substrate. Other observations in which mesitylene and anisole were competed suggest that different mechanisms may operate for each substrate on the clay support and that this might account for the again unexpected observation that mesitylene is more reactive than anisole with respect to benzoyl chloride under certain conditions.[24] Absorption phenomena and selective interactions of substrates with the support probably play a role; the exploitation of the selectivity caused by the supports and the avoidance of by-products of aluminium chloride are, as with Envirocats, opportunities for clean synthesis.

10.3.3 Oxidation

The importance of functionality in molecules to be used as synthetic building blocks was mentioned in Section 10.1. In much traditional organic chemistry, the primary way to introduce functionality into aromatic and saturated hydrocarbons has been halogenation, which, as has been discussed, is an intrinsically unclean process. With alkenes, epoxides are common activated derivatives which can be obtained in many cases by direct oxidation of the alkene with air. One way round these difficulties is to use oxidation, especially hydroxylation, to introduce the functionality; in this way, water becomes the eventually leaving group. The search for efficient oxidations has been a major problem for organic chemistry. Metal complexes have featured strongly in this chemistry and many industrial processes for easier reactions such as the formation of epoxides and the oxidation of *p*-xylene to terephthalic acid have been in operation for many years. More difficult reactions such as with saturated hydrocarbons and generally useful catalysts for epoxide formation have been targets for more recent research.

Envirocats have been found to be suitable for the oxidation of activated methylene groups in such substrates as ethyl benzenes. Although conversions are typically modest (30–75%), some interesting selectivities have been observed. For example, the oxidation of 4-aminoethylbenzene proceeds in 39% conversion apparently without competing oxidation of the amino group, a reaction that would be otherwise difficult to accomplish[21] (Figure 10.6). 'Clayniacs' contain nickel(II) acetylacetonate impregnated upon K10 montmorillonite have been shown to be effective in mediating the epoxidation of a wide range of alkenes using compressed air and *i*-butyraldehyde in a sacrificial reduction.[25] One of the reasons for investigating this type of material is to obviate the need to use reagents of limited stability such as 3-chloroperbenzoic acid. Clayniacs have been reported to achieve impressive results including 90–100% conversions and turnovers of better than 10^5. Although the best results were obtained using the environmentally undesirable solvent, dichloromethane, satisfactory performance was obtained in *t*-butanol. Whilst these results are encouraging, there are disadvantages: unreactive alkenes are not converted, yields are not quantitative, and a stoichiometric reductant is required to regenerate the active site of the nickel complex (cf. enzyme catalysed reactions in Section 10.4). Further research is being undertaken to overcome these problems.

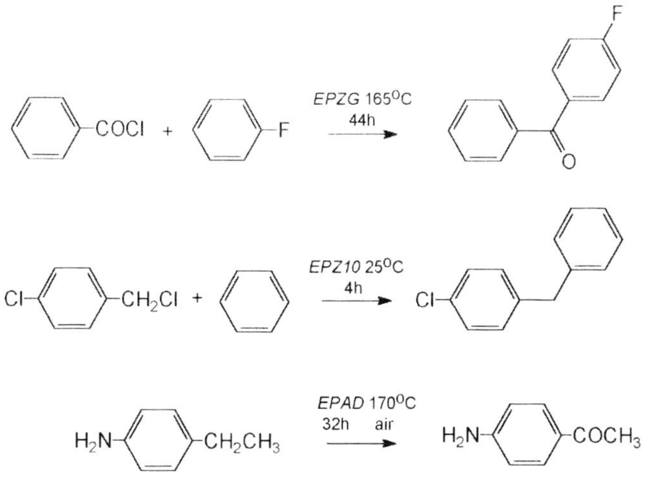

Figure 10.6 Reactions promoted by Envirocats.

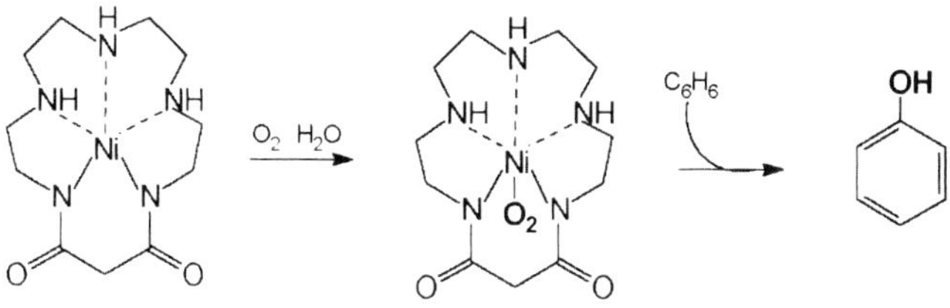

Figure 10.7 Hydroxylation of benzene by a nickel complex.

Hydroxylation reactions are particularly difficult to accomplish because low selectivity is usual. Much effort has gone into the modelling of enzymes that catalyse hydroxylations at non-activated carbon atoms in biosynthesis or in the metabolism of xenobiotic compounds. Many such enzymes are haemoproteins of the cytochrome P-450 class. If such model reactions could be made really efficient, there would be good opportunities for clean synthesis; efficiency in this case must be measured in terms of three factors: the selectivity of the reaction, the turnover of the catalyst, and the requirements for reducing cofactors.[26] A difficult reaction that has been achieved with a different class of metal complex is shown in Figure 10.7.[26b] One of the most stable and effective porphyrin derivatives has recently been obtained by nitrating the octachlorotetraphenyl porphyrin shown in Figure 10.8.[26d] Nitration with excess fuming nitric acid afforded a material that was capable of oxidising cyclohexane to cyclohexanol and cyclohexanone with oxygen, achieving turnovers of 370 and 270 respectively. Epoxidation of cyclo-octene was essentially complete within 30 min, using

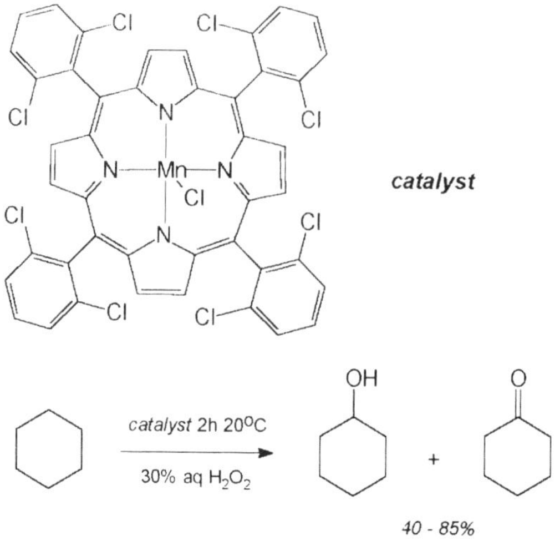

catalyst

catalyst 2h 20°C

30% aq H_2O_2

OH

O

+

40 - 85%

Figure 10.8 An effective cytochrome P-450 model.

dilute hydrogen peroxide as oxidant. Despite these advances, in no case have the problems of catalyst turnover and the introduction of reducing equivalents been completely solved.

An alternative readily available oxidant is hydrogen peroxide. For epoxide formation, the activation of hydrogen peroxide by a strongly electrophilic species would be a potential reagent. On solid supported materials, this possibility has been realised in polystyrene bound arsenic(V) oxyacids and selenium derivatives;[27a,b] but an especially stable polymeric support, polybenzimidazole, on to which molybdenum(VI) derivatives such as MoO_2(acac) can be bound has been described recently by Sherrington.[27c] The stability of this polymer is high at temperatures up to 100°C because it lacks the benzylic hydrogens found in polystyrene that are readily removed during oxidation reactions. Further, leakage of the supported catalyst was shown to be less than 0.2% in the epoxidation of propene using *t*-butylhydroperoxide as oxidant. Using the arsenic and selenium reagents, oxidation reactions including Baeyer Villiger conversion of ketones into esters, epoxidation of alkenes, and hydroxylation and further oxidation of diols can all be carried out with these reagents in direct analogy to the solution phase reactions. In solution chemistry, such reactions have been discovered additionally with perfluoroketones (Figure 10.9). Perfluoroketones exist as hydrates which undergo exchange with hydrogen peroxide under suitable conditions.[28] The perhydrate contains an activated hydroxyl group made electrophilic by the electron withdrawing effect of the fluorines; oxygen transfer to the alkene leads to epoxide formation and regeneration of the hydrate. It would be interesting to see whether a solid phase version of this reaction would be successful; perfluoroalkylsulphonic

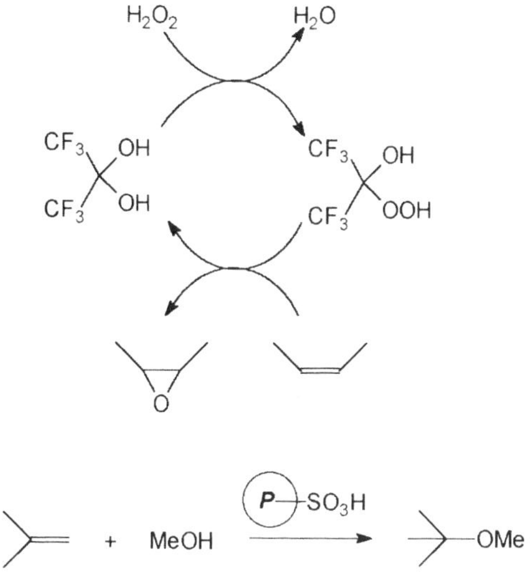

Figure 10.9 Clean reactions mediated by perfluorocarbons.

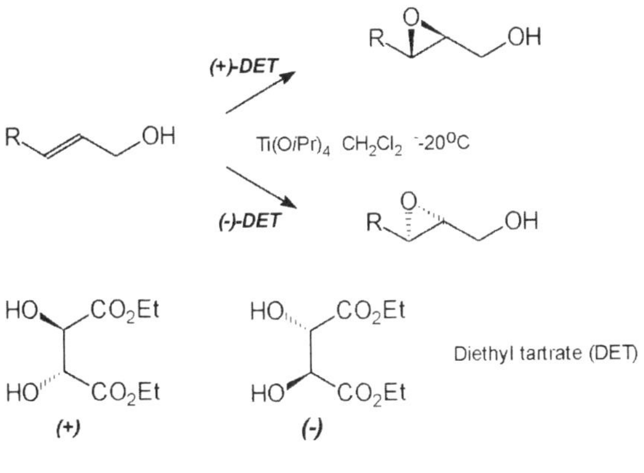

Figure 10.10 Sharpless epoxidation.

acids are well established solid phase strong acids useful for promoting reactions of carbocations and dehydration.[29] On a laboratory scale, one of the major advances of the 1980s was enantioselective epoxidation of allylic alcohols developed by Sharpless.[30] Although the original version of the reaction required stoichiometric quantities of the chiral auxiliary (diethyl tartrate), a catalytic version has been developed and chiral allylic alcohols are now commercially available (Figure 10.10).

10.3.4 Reduction

Broadening the concept of clean synthesis to include the applications, the importance of chirality in organic compounds becomes clear. Any compound that is to be introduced into the environment as a product for agriculture or medicine especially, will encounter a chiral environment; the proteins, carbohydrates, and nucleic acids that make up the functional components of biology are all chiral. Hence the interaction of a chiral molecule with the environment will be different for each enantiomer. This is because the complexes formed by associating each enantiomer with a chiral biological receptor made up of protein, carbohydrate, or nucleic acid, will have a *diastereoisomeric* relationship to each other; in such a situation, the properties of the two complexes will not be the same just as diastereoisomeric compounds have different physical and chemical properties with respect to each other. The significance of this is readily understood with reference to a few examples of biologically active compounds (Figure 10.11). For chemical synthesis today, it is therefore essential to have the

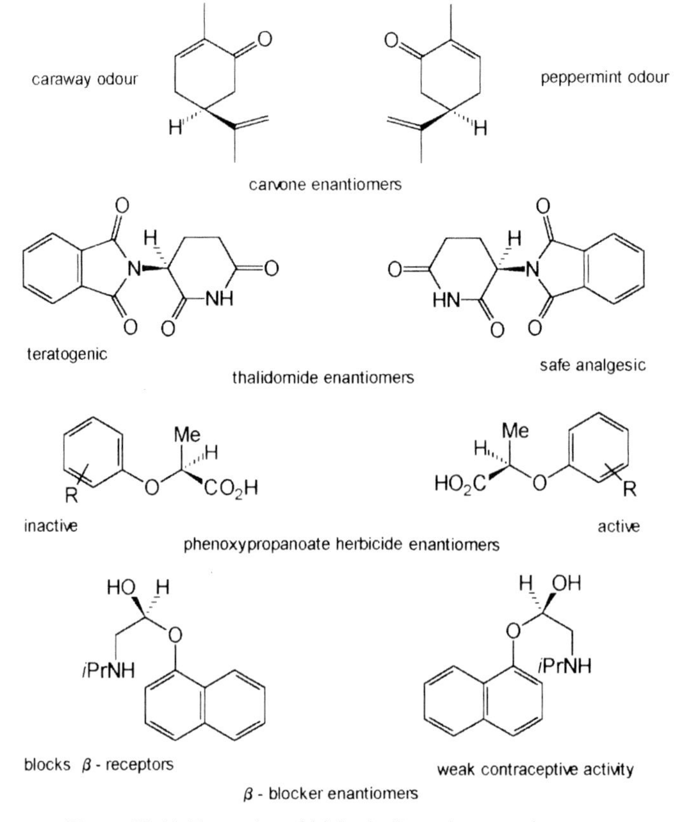

Figure 10.11 Examples of biologically active enantiomers.

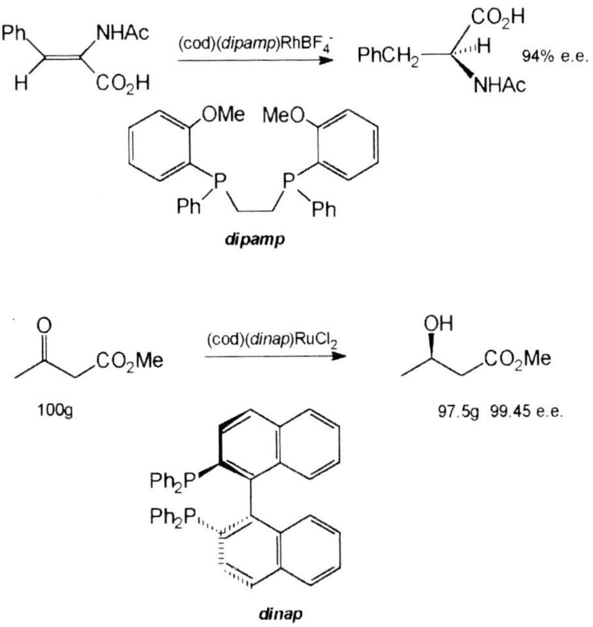

Figure 10.12 Hydrogenation by chiral metal complexes.

ability to synthesise either enantiomer of a chiral compound at will. This is one of the main drives for the use of biologically based catalysts (see below) and also is a stimulus for much new organic chemistry. Hydrogenation by chiral metal complexes was one of the first reactions to be studied and many variations are known (Figure 10.12).[31] The reactions, although sometimes tricky to establish effective reproducible conditions, are general. It remains to be seen whether economic conditions permit commercial applications.

10.3.5 Catalysis by zeolites

Zeolites hold a special position in selective chemical transformations because they were the first materials discovered with the ability to discriminate between organic compounds on the basis of their shape. Zeolites are aluminosilicates with regular well-defined structures the frameworks of which define pores of fixed size and shape.[7] Pore sizes in the range 4–15 Å can be obtained by suitable choice of composition and method of preparation. The shape of the pores is also significant: some zeolites present essentially a one-dimensional pore such that a molecule of *n*-hexane can travel down only in one direction (VPI-5) whereas others permit all directions of motion for the same substrate (NaX). The ability of zeolites to discriminate between the size and shape of molecules has been used in the petrochemical industry since the 1960s in conversions and separations involving the isomers of xylenes, for example. Recently, there have been major developments in the synthesis of zeolites with a range of pore sizes and with reactive sites included such that the potential of zeolites in clean chemical transformations has been greatly expanded.[7] They share with Envirocats the advantages of solid materials essentially built from innocuous components such as are found in natural rocks and clays; disposal of spent material into landfills is therefore a reasonable option.

The increased potential of zeolites arises from the ability to prepare materials designed to a precision of 10^{-10} m in length which is equivalent to the scale of the molecules that are substrates. Docking of substrates into designed pores of zeolites is as real a consideration as the docking of substrates into the active sites of enzymes. In addition to the shape of the pores of zeolites functional groups can be introduced. Zeolites are intrinsically cation exchangers and hence act as acid catalysts. The first report of shape selective acid catalysis by zeolites was in the dehydration of *n*-butanol in the presence of *i*-butanol, a reaction that under normal acid catalysed conditions would favour the opposite course. The kinetic diameters of the two alcohols differ by less than 1 Å yet a zeolite CaSA was able to differentiate between the two.[32]

Recent work has been concerned not only with expanding the range of pore sizes available but also with the inclusion of reactive sites. As was just explained, shape selective acid catalysis has been known for some time but shape selective base catalysis poses a greater challenge because of the intrinsic cation exchange properties of the anionic zeolites. The production of zeolite bases can be approached in principle in two ways. Either a positively charged framework element can be included or a neutral zeolite can be made host to an intrinsically basic molecule. There are experimental and theoretical reasons why the former is not likely to be achieved[7] but success has been obtained with the latter. Metallic sodium particles have been included in zeolites by the decomposition of sodium azide but although the materials are basic catalysts, they are air sensitive and difficult to prepare. A more practical solution is to include caesium oxide within the pores of the zeolite; these materials are capable of carrying out reactions typical of caesium oxide[33] but in addition can catalyse reactions that free caesium oxide does not. For example, but-1-ene has been isomerised to but-2-ene at 0°C.

The commercial potential of this type of chemistry has been demonstrated by scientists at Merck who have shown how base catalysis by zeolites can offer an environmentally sound preparation of 4-methylthiazole avoiding several unattractive reactions. This is an excellent example of clean synthesis. Not only are hazardous reagents such as chlorine and carbon disulphide avoided, wastes deriving from chlorine, ammonia and sulphur are all

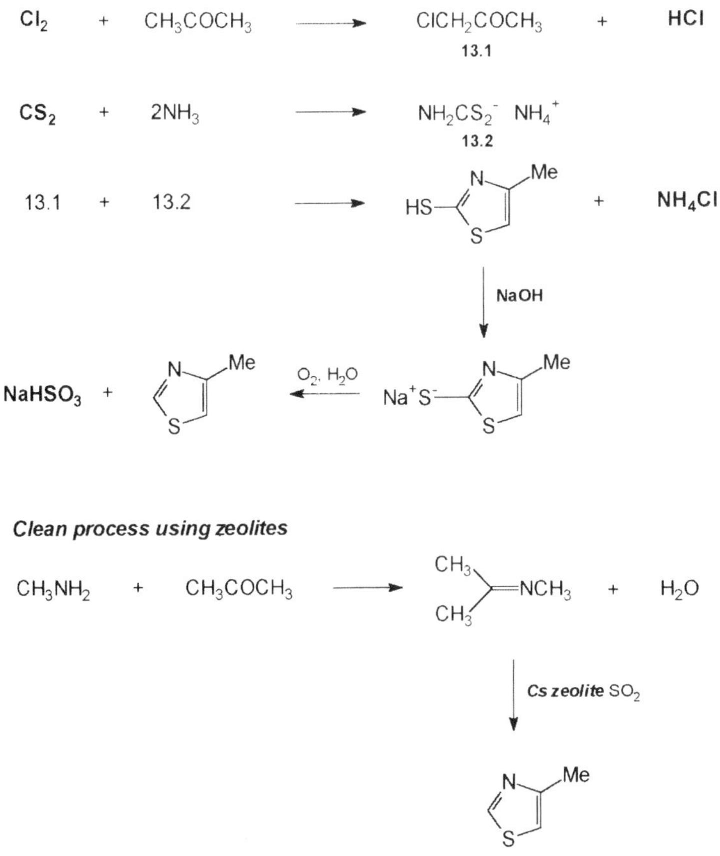

$Cl_2 + CH_3COCH_3 \longrightarrow ClCH_2COCH_3 + HCl$
13.1

$CS_2 + 2NH_3 \longrightarrow NH_2CS_2^- \ NH_4^+$
13.2

13.1 + 13.2 ⟶ HS–(4-methylthiazol-2-yl) + NH_4Cl

NaOH

$NaHSO_3$ + 4-methylthiazole ⟵ (O_2, H_2O) Na^+S^-–(4-methylthiazol-2-yl)

Clean process using zeolites

$CH_3NH_2 + CH_3COCH_3 \longrightarrow (CH_3)_2C{=}NCH_3 + H_2O$

Cs zeolite SO_2

4-methylthiazole (N, S, Me)

Figure 10.13 Preparation of 4-methylthiazole.

eliminated (Figure 10.13). In the potential process, the only leaving group is water and all other atoms of the reactants are found in the product.[7,34]

The potential of hydrogen peroxide as a clean oxidant was mentioned earlier. Titanium-containing zeolites such as TS-1 have become useful in the commercial manufacture of catechol and hydroquinone from phenol and aqueous hydrogen peroxide.[35] It is notable that titanium plays a key role in many modern reactions of synthetic organic chemistry including the Sharpless systems for epoxidation. Similarly, TS-1 is capable of oxidising alkenes to epoxides and alkanes to alcohols and ketones. The detailed mechanisms of reactions of this type are not yet well understood and reactions with later transition elements are even more complex. It is probable that exchange between hydroxide and hydroperoxide at titanium atoms occurs, as is believed to be the case with the fluoroketones discussed above. Nevertheless, the ability to use an environmentally acceptable oxidant such as hydrogen peroxide assures that these systems will be much more widely studied and developed in the future.

10.3.6 Electrochemistry

A simple definition of oxidation and reduction is, of course, the removal and addition of an electron respectively. Such processes occur at electrodes in electrochemical cells. An oxidised or reduced product is not the only result of an electrochemical reaction; electron transfer affords a radical which can undergo coupling and other reactions typical of organic compounds with unpaired electrons. Further oxidation by a one-electron step can lead to cations and reduction to anions. Chemistry of more general relevance to synthesis can therefore be considered. Electrochemistry is well known in the manufacture of bulk chemicals and materials including chlorine, hydrogen peroxide, and, of course, metals but its use in more general chemical manufacture has been limited. Nevertheless, there are many selective electrochemical processes described in the literature and these might reasonably be developed to a manufacturing scale although there are problems associated with heating effects of electric currents in solvents of relatively low dielectric constant and despite the use of organic electrolytes such as quaternary ammonium tosylates. A number of reactions interesting from the point of view of bulk chemicals have been described, for example, the electrochemical oxidation of benzene to phenol by hydrogen peroxide in the presence of iron(II),[36a] and a Michael addition as a key step in a lysine synthesis.[36b]

The absence of additional chemical reagents and the ability to avoid local toxicity gives electrochemistry advantages from the point of view of clean synthesis but if one broadens the whole operation to consider how the electricity is produced in the first place, electrochemistry's contribution to clean synthesis becomes more dubious. In economies with ready access to hydroelectricity, such as Norway, the opportunities for electrochemistry in chemical synthesis are greatest.

10.4 Biological catalysis

10.4.1 Introduction

The avoidance of hazardous or toxic reagents and of noxious effluents has been a major force in the development of new approaches to important synthetic reactions such as aromatic substitution. Similar arguments have been advanced in favour of biological catalysts, a range which includes various forms of live organisms, dead cells, and isolated enzymes, because biological catalysts operate under mild conditions, usually in water or other non-toxic solvents, and often have intrinsically high catalytic efficiencies. These reasons in themselves are sufficient to justify attention to biological catalysts in the context of clean technology. However, their attractiveness hides a number of problems that emerge if the whole operating system containing the biological catalyst is considered.

In economic and environmental terms, the preparation of the catalyst, its subsequent disposal and the isolation and purification of the product must all be evaluated. Because of the ease of production of large quantities, most biological catalysts derive from bacterial sources. The attendant risks in the growth and manipulation of what may be sensitive or even pathogenic cultures of organisms must be considered in the production and disposal of catalyst. Most biological catalysts operate in aqueous solution. Since many substrates of interest are only sparingly soluble in water, there are difficulties with large volumes of reaction and problems in the subsequent solvent extraction of the product. For these reasons, much emphasis in clean technology using biological catalysts is placed upon the effective engineering of the reactors and upon the downstream processing of the product, topics beyond the scope of this chapter. Equally, the disposal of a spent biological catalyst by incineration or landfill poses as many problems as that of a more conventional waste. Clearly, biological catalysis needs to be considered case by case on its merits and cannot unreservedly be accepted as the way forward for all synthetic processes.[37] Furthermore, biological catalysts are not available for all reactions that might be required in synthesis, aromatic substitution being a good example. It is therefore likely that clean synthesis of commodity and speciality chemicals will follow the path recently well established in organic chemistry research, namely mixed reaction sequences of biologically and non-biologically catalysed steps.

The impression should not be given from the above that biological catalysis is a new contributor to the process industries. It is reasonable to recognise an upsurge in systematic interest in biological catalysis in the last quarter of the twentieth century but biological products have been with us for many decades. The Weizmann process for acetone production using *Clostridium acetobutyliticum* has already been mentioned. Several amino acids (glutamic acid, lysine, phenylalanine and arginine) are produced on a kiloton per annum scale using microbiological methods and the dicarboxylic acids, citric, gluconic and itaconic acids on a similar or larger scale.[38] Most of the demand for these compounds has been in the food industry but the citric acid market is growing as its potential as a clean substitute for phosphate in detergents is being exploited. Ethanol, where national fiscal regimes permit, is being used as a fuel. Vitamin C is produced on the tens of kilotons per annum scale.

In addition to these products, there is clear potential for further contributions of microbial products should the political, economic and regulatory authorities make the appropriate decisions. Such compounds include tryptophan and aspartic acid, propanoic and butanoic acids, the manufacture of which is known to be technically feasible, and a resurrection of processes for acetone and butanol, each of which would be in the megaton per annum range. The drive for increased biological production would be in fuels, because no net increase in the global carbon dioxide burden would result from their combustion, and in materials, because biodegradable materials might alleviate waste disposal problems. Other views of the future recognise the potential for the production of new raw materials for fine chemicals manufacture from bulk carbohydrates such as xylose, rhamnose and arabinose, and for applications in biologically derived surfactants to replace those derived from fossil fuels.[39] It is notable, however, that little attention appears to have been given to the development of biologically catalysed preparations of bifunctional molecules with the general synthetic potential outlined in the introduction to this chapter. As will be described below, one window has been opened, if by remote control, through the production of biodegradable polymers.

10.4.2 Redox reactions

The problems of introducing functionality through oxidation reactions in organic chemistry have already been mentioned together with non-biological approaches to such reactions. Many enzymes that catalyse hydroxylation reactions are known and their application to clean synthesis has been considered. Of these many belong to the cytochrome P-450 class of enzymes which have an iron porphyrin active site; the P-450 hydroxylase requires the delivery of two electrons from a partner reductase protein as part of the oxidation cycle. Because of the requirement for recycling, these enzymes are usually exploited in intact cells, although the power of modern molecular biology is now being brought to bear on them so that purified preparations can be used. So far, enzymes of this class have not fulfilled the potential in synthesis in bulk chemicals that clearly exists; in the pharmaceutical industry, however, steroid hydroxylation has been used commercially for many years (Figure 10.14).[40] Such specific hydroxylation of non-activated methylene groups is not possible using non-biological catalysts despite many recent advances in organic reagents.

Whereas cytochromes P-450 introduce single hydroxyl groups into a substrate, a second class of oxidase, dioxygenases, introduces the two atoms of molecular oxygen into the substrate. A much publicised class of enzyme is benzene dioxygenase. This enzyme produced in the organism *Pseudomonas putida* was first shown to be useful in preparative organic chemistry through the production of a strain that lacks the ability to cleave the

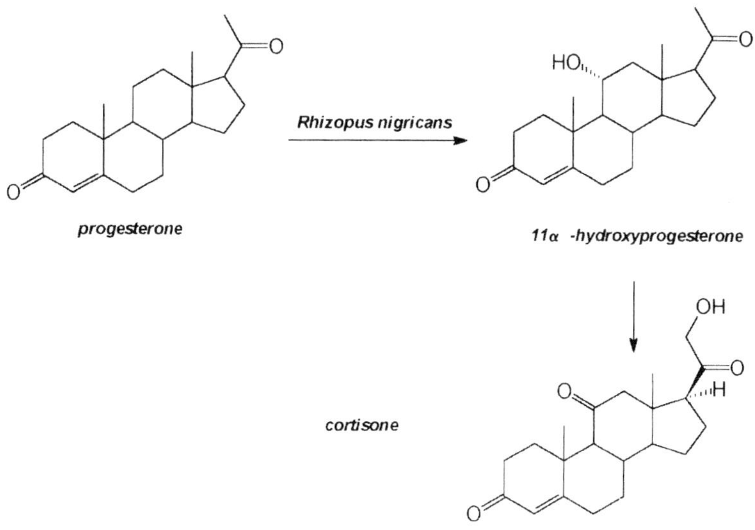

Rhizopus nigricans

progesterone

11α -hydroxyprogesterone

cortisone

Figure 10.14 Commercial steroid hydroxylation.

product (*cis*-dihydrobenzene-1,2-diol: Figure 10.15[41]) so that the dihydrodiol accumulates. Ironically, the significant strain of *Ps. putida* was itself discovered in a landfill. It was later shown that the diol could be converted via its acetate into polyphenylene, a polymer of interest in electrical applications. Scientifically, this work illustrated many attractive features of biotechnology in clean synthesis, the large scale potential, the discovery of effective reactions difficult by normal non-biological catalysis, but the commercial exploitation has been elusive.

One of the key features that characterises many biological catalysts is their ability to convert achiral compounds into chiral derivatives. This has also been demonstrated in many reactions catalysed by benzene dioxygenase. First, monosubstituted benzene derivatives can be oxidised effectively and, second, heterocyclic bicyclic molecules have also been shown to be substrates although with lower conversion.[41b,d] One can ask little more from a biological catalyst than this range of difficult, selective, reactions. Research scientists have readily taken up the products as the basis of stereochemically controlled syntheses but commercial applications remain to emerge.

Chirality has especial importance in drugs and agrochemicals and it is not surprising to find that many published applications of biological catalysts are directed towards these products, which have higher added value than commodity chemicals. For example, a mono-oxidase from *Pseudomonas* species that catalyses the stereospecific epoxidation of alkenes has been exploited in patented syntheses of *β*-blockers (Figure

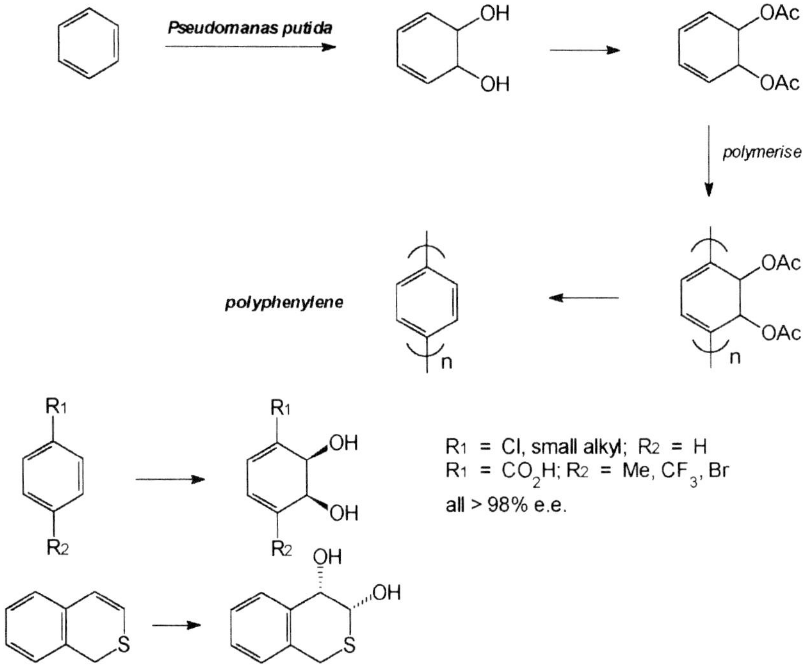

Figure 10.15 Production of *cis*-dihydrobenzene by *Pseudomonas putida*.

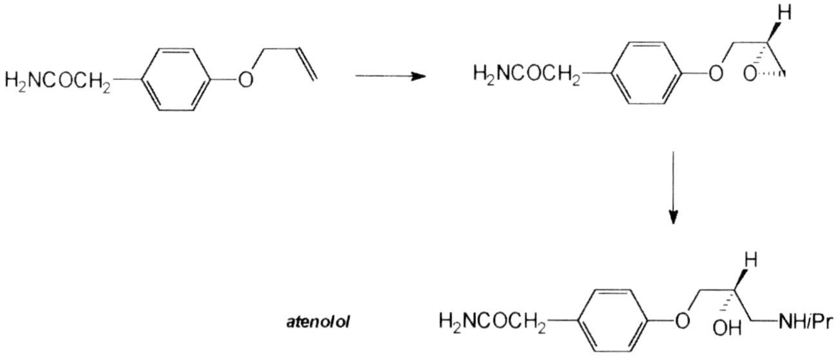

Figure 10.16 Synthesis of a β-blocker.

10.16[42]); the reactions have been scaled up to kilogram quantities and offer a competitor to non-biological processes such as those involving Sharpless epoxidation.[30]

One of the reactions that led the field in the development of the application of enzymes to organic synthesis and stimulated synthetic biotechnology was the stereospecific reduction of carbonyl groups catalysed

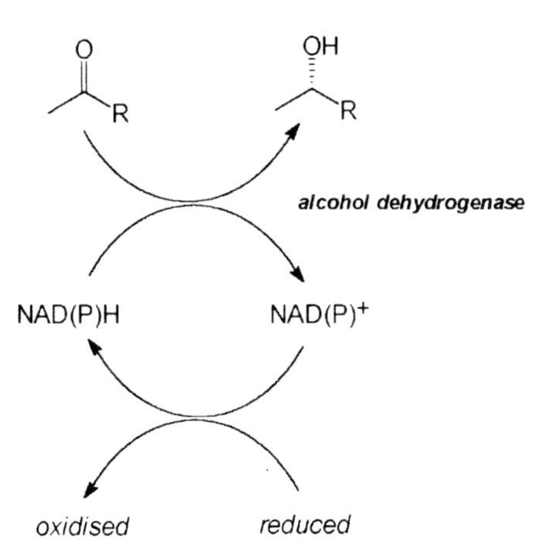

Figure 10.17 Stereospecific reduction with a cofactor, NAD(P) or NAD(P)H.

by alcohol dehydrogenases.[37] At the time of the original developments, there were no really efficient non-biological competitors although research over the last 25 years has provided alternative reagents.[4] The chief problem was that of recycling the expensive cofactor, NAD(P) or NAD(P)H, one equivalent of which is required for the transformation of one equivalent of substrate (Figure 10.17). Several alcohol dehydrogenases have been investigated for their general synthetic potential[43] and a range of strategies and technologies for recycling has been developed. Commercial development of such methods has included using simple recycling methods such as a cheap cosubstrate[44] as well as sophisticated solutions using membranes and modified cofactors[45] and electrochemistry.[46] In the case of the membrane systems, the cost of the cofactor has been reduced to less than 50 US cents per kg product for manufacture of amino acids with ammonia: ketoacid dehydrogenase. Much scientific and technological skill has been invested in studies in this field and the lack of widespread commercial application, like benzene dioxygenase products, is due to the lack of demand for the chiral secondary alcohols and not to a lack of technical sophistication or chemical specificity in the catalysts. As noted above, amino acids may prove to be a more viable commercial proposition.

Many such reduction reactions can easily be carried out using actively fermenting yeast for which the favoured substrates are ketoesters;[47] the reaction is sufficiently reliable to be described in *Organic Syntheses* and can be applied to a wide range of substrates. The chiral bifunctional compounds, typically (*S*)-3-hydroxybutyrates, are useful starting materials for synthesis. This reaction is a very suitable one through which the training organic chemist can appreciate the problems inherent in using unsophisticated biocatalysts such as fermenting yeast together with the benefits of mild conditions and chiral product: the work-up by solvent extraction is messy compared with the same procedure in normal laboratory reactions because of the large volume of dilute aqueous solution; the reaction takes two days typically; there is a large quantity of biological debris for disposal. Magnify these problems in scale-up to a commercial process and the importance of well-developed biotechnology becomes obvious. Today, many reactions using biocatalysts have been developed into useful and reliable laboratory procedures; they are recorded in a valuable new series, *Preparative Organic Biotransformations*.[43] Obviously, these reactions could be considered as prototypes for scale-up should a given product find a commercial niche.

10.4.3 Biopolymers

Both yeast-mediated reduction and benzene oxidation reactions are carried out using whole cells. A product that has attracted genuine commercial interest is the polymer of poly-3-hydroxyalkanoic acids which is produced by a number of bacteria including *Hydrogemonas eutropha* and *Alcaligenes eutrophus*. The bacterial polymerisation was discovered more than 60 years ago but its commercial potential was developed only in the 1980s.[48] Whereas animal cells and many other organisms store excess carbon as fatty acids, these bacteria synthesise the polymers as storage using a modified and attenuated biosynthetic sequence related to the biosynthesis of fatty acids. Under appropriate growth conditions, the organism accumulates up to 80% of the cellular mass as particles of polymer that can be harvested. Further, the nature of the polymer in terms of its composition and hence physical properties can be widely varied by suitable choice of the species and growth conditions. Materials with properties as varied as elastomers, films and fibres can be produced.

One of the most fully developed is a random copolymer of 3-hydroxybutanoic and 3-hydroxypentanoic acid in which the latter makes up 0–30% of the monomers and is known as BIOPOL. This material has properties similar to polypropylene. Interestingly, the configuration of the chiral centre in these monomers is *R*; thus by hydrolysis of the polymer, (*R*)-3-hydroxybutanoic acid becomes available through biotransformations to complement the *S* enantiomer available from yeast-mediated reduction of 3-oxobutanoate (see above).

Although BIOPOL has apparent environmental advantages in its biodegradability, it degrades in aerobic or anaerobic environments over periods from as little as a few weeks to several months, and its production required the development of new process technology. Consequently, the cost of the material is high, limited by production costs, and commercial success in today's business environment remains uncertain. The production and development of biodegradable polymers nevertheless remains an area of opportunity for clean synthesis.

10.4.4 Esterases

As can be seen in *Preparative Organic Biotransformations*, a great many selective reactions leading to useful chiral intermediates can be carried out using esterases such as that from pig liver. However, on a commercial scale the esterases that hydrolyse and transesterify fatty acids, the lipases, have come to the fore. These enzymes are particularly interesting in that they operate at an interface between an organic and aqueous phase; thus the use of organic solvents and water-immiscible substrates ceases to be a major problem. One major area of application is in the food industry where large quantities of triglycerides are manipulated in the production of edible oils and fats. The balance of nutritional properties, such as the proportion of polyunsaturates, physical properties such as spreading ability, and flavour are commercially crucial; they can be adjusted by varying the composition of the fatty acyl groups of the triglyceride through transesterification catalysed by lipases.[49]

Like all enzymes of major value in organic synthesis, lipases have been shown to catalyse reactions between compounds that are relatives of their natural substrates. There has been much interest in clean technology in defining efficient processes for the preparation of fully biocompatible detergents, typically non-ionic surfactants. An obvious candidate for such applications would be the fatty acyl ester of a carbohydrate. However, cheap carbohydrates such as glucose contain many esterifiable hydroxyl groups and good surfactant properties depend upon reasonable homogeneity in the chemical structure of the material. Simple chemical acylation would not only give a mixture of products, it would also suffer from the disadvantage of the need for activation of the acid (as an acyl halide or anhydride, for example) in order to produce the ester. It is therefore interesting to note that the lipase from *Candida antarctica* has been found to be effective in the selective acylation of glycosides at C-6 (Figure 10.18[50]).

Significant extensions of the applicability of enzymic catalysis have often arisen through arguments typical of organic chemistry rather than biochemistry. For example, the ability of lipases to catalyse transesterification has been extended to arrange for them to convert a carboxylic acid into the corresponding peracid essentially by exchanging water for hydrogen peroxide (Figure 10.18[51]). Both the lipase and the acid are used in catalytic quantities and the substrate, an alkene, is present in an organic phase. Conversion into the epoxide takes place in yields as high as 99%. This is a safe procedure that avoids the need for either isolated peracids or

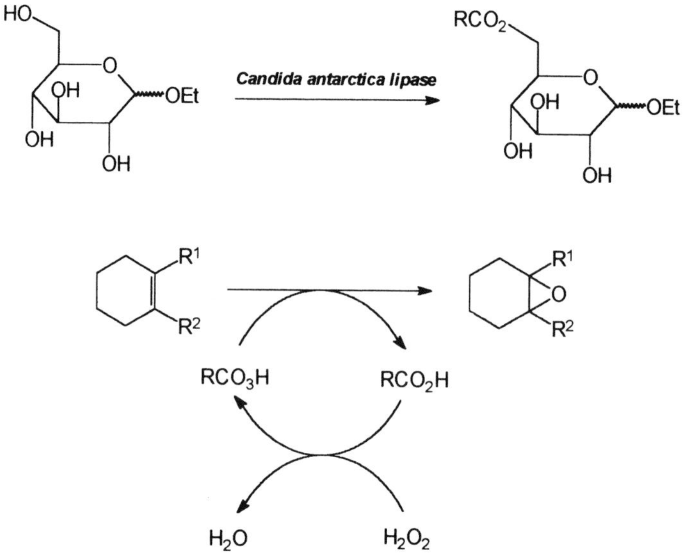

Figure 10.18 Clean reactions mediated by a lipase.

concentrated hydrogen peroxide, both of which are hazardous on a large scale.

10.4.5 Carbon–carbon bond formation

The major task of organic synthesis is the formation of new carbon–carbon bonds. It is ironic, therefore, that so few useful biological catalysts for this type of reaction have been developed. Two cases, both of which used the conversion of the product into β-blockers as an example of the utility of the reactions, have been described. Enzymes that catalyse the stereospecific addition of hydrogen cyanide to aromatic aldehydes affording cyanohydrins have been known for many years[52] and they have been demonstrated to be useful in the laboratory. However, the key to a commercial application was to obtain a sufficiently pure enzyme and, most importantly, to identify reaction conditions in which chirality was not lost through the reverse reaction followed by uncatalysed addition.[53] Having found suitable reaction conditions, the product had to be modified to obtain a stable, commodity chemical (Figure 10.19a); this was achieved through the standard chemical expedient of blocking the unwanted reaction by derivatisation of the product with *t*-butyldimethylsilyl chloride. The reaction could also be carried out in largely organic media thus improving the solubility of aromatic aldehydes. Organic media have proved to be the key to the large-scale usage of this enzyme.[54] An alternative sequence has been discovered using yeast as the biocatalyst (Figure 10.19b[55]). In this case, the aldehyde is

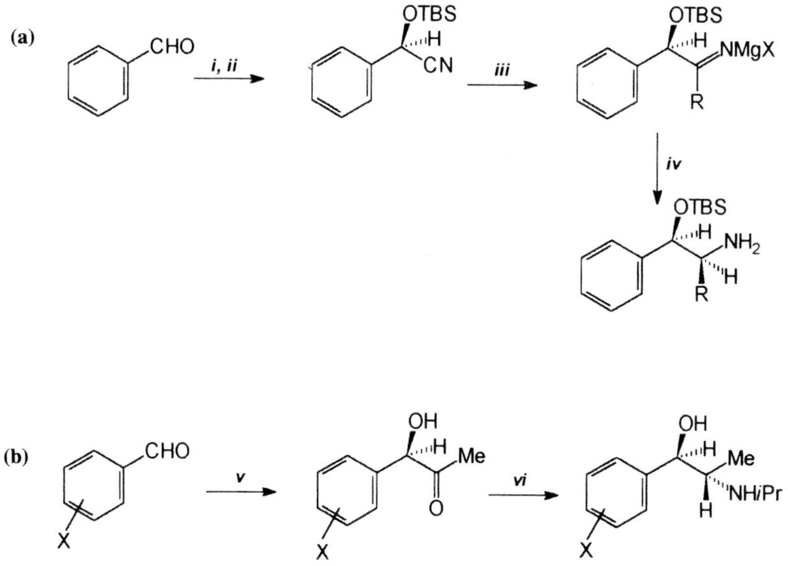

Figure 10.19 Production of β-blockers retaining chirality of starting material. (a) Stabilisation of product by derivatisation. (b) Production using yeast as biocatalyst.

converted by brewers' yeast into a chiral α-hydroxyketone, presumably by thiamine-dependent enzymes. This intermediate was then readily converted into the chiral drug.

10.4.6 Designing new biological catalysts

To conclude the discussion of biological catalysts, it is appropriate to outline some of the modern approaches to obtaining new catalysts. If the only catalysts that are available are those that are synthesised by nature, naturally occurring enzymes, the future of biocatalysts would be limited to those enzymes that can be identified and isolated from natural sources. Natural sources can be extended to include enzymes from micro-organisms grown in stressed environments so that special enzyme activities are induced or evolve over a number of generations and there is significant commercial scope there as has been shown by the development of a process for the stereoselective synthesis of phenoxypropanoate herbicides (Figure 10.20[56]). However, chemists will always consider the question: can a specific catalyst be found for their important problem reactions? For biological catalysis, the supplementary question of whether the key reaction type is catalysed by any known enzyme is also important. Two main approaches to this problem can be identified for biological catalysts. First, modifications to existing enzymes using the techniques of molecular biology can be used, and, second, and more speculatively, catalytic antibodies can be developed for reactions including many that are not catalysed in nature. The biological techniques by which these new proteins are obtained are beyond the scope of this chapter and the reader is referred to the original papers.

If a crystal structure is available for the enzyme that is to be modified, it is possible to use molecular modelling to design structural changes to improve the properties of the enzyme for synthesis. Although many plausible designs can be considered, our understanding of the detailed factors that control both protein structure and catalysis by enzymes is insufficient to be sure that an improvement can be made. In general, the modification of loops on the surface of an enzyme is unlikely to cause major structural changes but alteration of ionic groups in the interior, including the active site, may have a substantial effect. Knowledge is accumulating in these areas. As an example of a successful modification, the extension of the substrate specificity of lactate dehydrogenase to include larger alkyl chains than the normal methyl of the substrate is eloquent. Holbrook identified a loop region close to the active site and introduced a number of modifications to it.[57] One series of modifications increased the hydrophobicity and flexibility of the loop so that there would be stronger binding of alkyl chains. A second was designed to introduce greater plasticity into the helix on to which the loop folds (Figure 10.21). With these modifications, an enzyme that accepted branched C-4 alkyl substituents was obtained; on a millimolar scale, yields of 91% in 99%

Phenoxypropanoate herbicides

fluazifop

dehalogenase

R isomer active

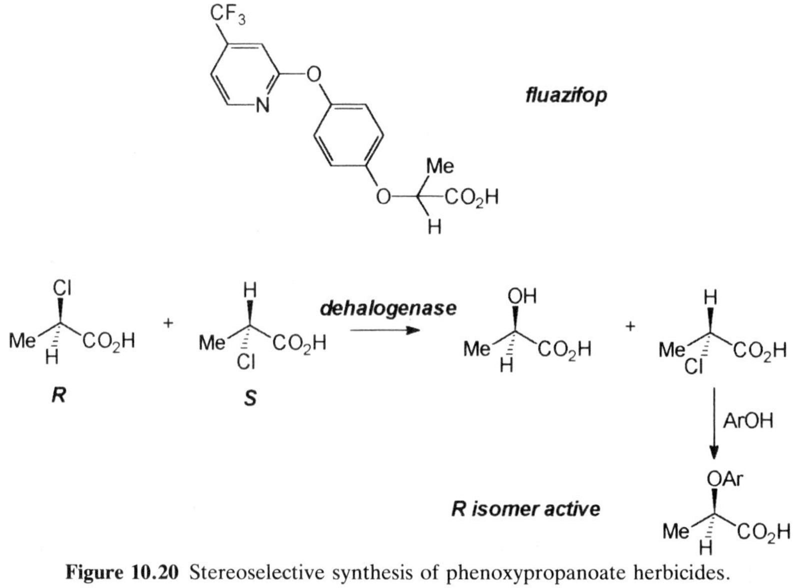

Figure 10.20 Stereoselective synthesis of phenoxypropanoate herbicides.

enantiomeric excess were obtained. There seems no reason why, if the demand existed, this enzyme could not be developed into a commercial catalyst.

If there is no known enzyme that catalyses the reaction in question, it has recently been found possible to prepare antibodies with the ability to catalyse reactions. The essential argument in the discovery of a catalytic antibody is as follows.[58] Enzymes in general achieve their catalytic effect by binding to the transition state for the reaction thus lowering its energy and increasing the reaction rate. Since antibodies can be raised to almost any molecule of molecular mass greater than 200, an antibody to a transition state would be expected to be a catalyst like an enzyme. However, transition states are not isolable molecular entities and the best that can be done is to synthesize a stable analogue of the transition state. Thus for a reaction involving an acyl transfer proceeding via a tetrahedral intermediate, a phosphonate, which is also tetrahedral and negatively charged, would be a suitable analogue. Indeed many antibodies that catalyse acyl transfer reactions have been obtained in this way.[58]

For preparative chemistry, however, adequate supplies of hydrolytic enzymes exist. With this in mind, we and others[59] investigated the possibility of preparing antibodies to the Diels Alder reaction, which is a mainstay of modern organic synthesis. The transition state was an analogue of the product in this case; the Diels Alder reaction is well known to have a late transition state. Antibodies were obtained with catalytic properties for several Diels Alder reactions and products were isolated and characterised

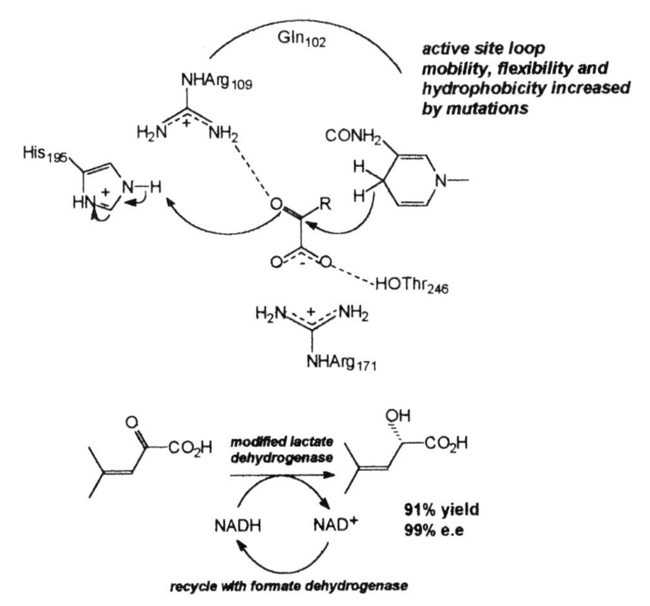

Figure 10.21 The Diels Alder reaction.

in each case. Since antibodies are large proteins (150 kD), it was important to establish that the whole molecule was not essential for catalysis; this was done by cleaving the whole antibody and isolating the antigen binding region which maintained the full catalytic activity of the protein. Animal cell fermentation has been used to produce hundred milligram quantities of the protein and the gene has been cloned and sequenced. There seems no reason in principle, therefore, why a catalytic antibody of suitable efficiency could not be developed using molecular biological techniques from the laboratory demonstrations that we have achieved to a practical catalyst for large scale production. As has been seen with several examples in this section, if the demand is there, it can be done.

10.5 Embryonic ideas in conclusion

There is no doubt that the research chemical community is well aware of the importance of clean technology and clean synthesis in general. To conclude, a few examples will be given of reactions that might become the basis of useful technology in the future. Aromatic substitution continues to be studied. The displacement of aryl halides by nucleophiles in non-activated cases usually requires temperatures in excess of 200°C. However, in the presence of copper(II) hydroxide and carbonate, yields approaching 90% can be obtained in an atmosphere of carbon dioxide.[60] The carbon dioxide atmosphere is the novelty raising the yield to a synthetically useful figure. Epoxidation has been mediated electrochemically using oxygen generated at an anode.[61] For example, cyclohexene epoxide is formed at the anode whilst hydrogen gas is produced at the cathode. There may well be potential for such double product generating systems.

New organometallic compounds with useful activity continue to be discovered. The conversion of cheap alkenes that can be obtained readily from oil or natural sources into valuable but inaccessible alkenes is a reaction that has been known for many years. However, the catalysts have typically been tungsten, molybdenum and rhenium alkyls which are themselves pyrophoric and have to be prepared from aluminium alkyls, which are also pyrophoric. Further, the metal alkyls also react with carbonyl groups by addition in a manner similar to Grignard reagents, and so the existing catalysts are limited compounds containing non-electrophilic functional groups. In the last few years, it has been discovered that methyl trioxorhenium is a stable and effective catalyst for the metathesis of alkenes.[62] It is stable to air and is soluble in water and organic solvents. Unlike conventional metal alkyl catalysts, methyl trioxo-rhenium catalyses transformations of alkenes such as 5-bromopent-1-ene and methyl oleate (Figure 10.22).

Finally, it is well to recall that the end fate of many organic compounds is

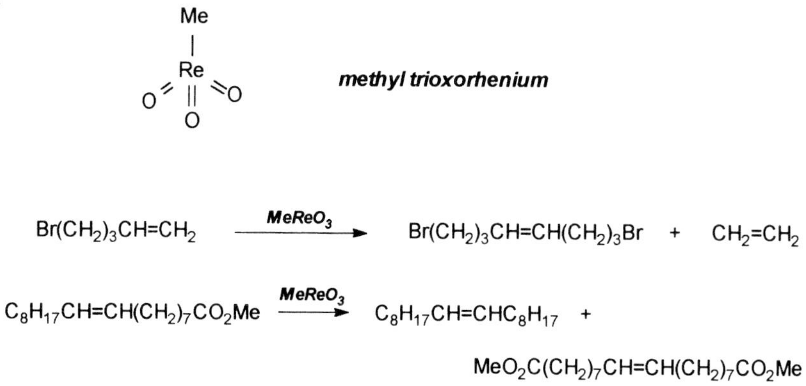

Me–Re(=O)₃ *methyl trioxorhenium*

$$Br(CH_2)_3CH{=}CH_2 \xrightarrow{MeReO_3} Br(CH_2)_3CH{=}CH(CH_2)_3Br + CH_2{=}CH_2$$

$$C_8H_{17}CH{=}CH(CH_2)_7CO_2Me \xrightarrow{MeReO_3} C_8H_{17}CH{=}CHC_8H_{17} + MeO_2C(CH_2)_7CH{=}CH(CH_2)_7CO_2Me$$

Figure 10.22 Catalytic transformation of alkenes using methyl trioxorhenium.

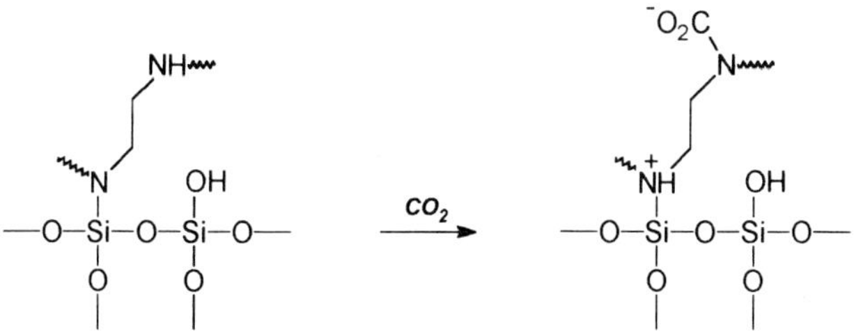

Figure 10.23 Reversible trapping of CO_2 with polyamine coated silica gels.

to be burnt, giving carbon dioxide emissions to the atmosphere. The emissions contribute to the greenhouse effect and also dispose of matter that might be well employed built into a further generation of useful molecules. An approach to the reversible trapping of carbon dioxide has been demonstrated using polyamine coated silica gels.[63] These materials will absorb substantial quantities of carbon dioxide to form carbamic acid salts (Figure 10.23) at temperatures between 20° and 90°C. The next stage would be to transfer the trapped carbon dioxide into a chemically useful reaction and even this problem is being tackled.[64] It has been shown that water-soluble rhodium–phosphane complexes are very efficient catalysts for the hydrogenation of carbon dioxide to formic acid in aqueous solution in the presence of amines (trimethylamine or dimethylamine). Thousand-fold turnovers have been obtained. Of course there are limitations: the maximum concentration of formic acid only reached the concentration of the amine present through salt formation but a clean reaction without further reduction or combination with the amine was observed. I have little doubt that technologically relevant solutions to carbon dioxide conversion will be discovered in the future.

References

1. Suckling C.J., Halling P.J., Kirkwood R.C., and Bell G., (1991) *Clean Synthesis of Effect Chemicals*, AFRC/SERC Clean Technology Unit, Swindon.
2. CEC (1990) *Towards a Carbohydrate-based Chemistry*, Commission of European Communities, Luxembourg.
3. Jentzsch W., (1990) *Angew. Chem. Int., Edn. Engl.*, **29**, 1229–1234.
4. Seebach D., (1990) *Angew. Chem. Int. Edn. Engl.*, **29**, 1320–1367.
5. Gunstone F.D. and Hersdoef B.G., (1992) *A Lipid Glossary*, The Oily Press, Ayr, Scotland.
6. Blossey E.C. and Ford W.T., (1989) *Comprehensive Polymer Science*, vol 6., eds. Allen G., Bevington J.C., and Eastmond G.C., Pergamon Press, Oxford, pp. 81–114.
7. Davis M.E., (1993) *Accts. Chem. Res.*, **26**, 111–115.
8. Pervez H., Onyiriuka S.O., Rees L., Rooney J.R., and Suckling C.J., (1988) *Tetrahedron*, **44**, 4555–4568.
9. Nozaki H., Moriuti S., Takaya H., and Noyori R., (1966) *Tetrahedron Lett.*, 5234–5235.
10. Guy A., Lemaire M., and Guette J.-P., (1982) *Tetrahedron*, **38**, 2339–2346.
11. Guy A., Lemaire M., and Guette J.-P., (1982) *Tetrahedron*, **38**, 2347–2354.
12. Roussel J., Lemaire M., Guy A., and Guette J.-P., (1986) *Tetrahedron Lett.*, **27**, 27–28.
13. Komiyama M., (1989) *J. Chem. Soc., Perkin Trans.* **1**, 2031–2053.
14. Komiyama M. and Hirai H. (1984) *J. Am. Chem. Soc.*, **106**, 174–178.
15. Breslow R. and Campbell P., (1969) *J. Am. Chem. Soc.*, **91**, 3085.
16. Clark J.H., Kybett A.P., and Macquarrie D.J., (1992) *Supported Reagents, Preparation, Analysis, and Applications*, Verlag Chemie, New York.
17. a SanFillipo J.J., and Chern C.-I., (1977) *J. Org. Chem.*, **42**, 2182–2183; b. Clark J.H., Kybett, A.P., Landen P., Macquarrie D.J., and Martin K., (1989) *J. Chem. Soc., Chem. Commun.*, 1355–1357; c; Frechet J.M.J., Darling P., and Farrall M.J., (1981) *J. Org. Chem.*, **46**, 1728–1730.
18. Nishiguchi T. and Asano F., (1989) *J. Am. Chem. Soc.*, **111**, 1531–1535.
19. Kodomari M., Sato H., and Yoshitori S., (1988) *Bull. Chem. Soc. Jpn.*, **61**, 4149–4150.
20. Texier-Boullet F., Villemin D., Ricard M., Moisson H., and Foucaud A., (1985) *Tetrahedron*, **41**, 1259–1266; Villemin D., (1985) *Chem and Ind.*, 166.
21. *Envirocats* information from Contract Chemicals, Warrington, England
22. Clark, J.H., Kybett A.P., Macquarrie D.J., Barlow S.J., and Landon P., (1989) *J. Chem. Chem. Soc., Chem. Commun.*, 1353–1354.
23. Cornelis A., Dony C., Laszlo P., and Nsunda K.N., (1991) *Tetrahedron Lett.*, **32**, 1423–1424.
24. Laszlo P. and Montaufier M.-T., (1991) *Tetrahedron Lett.*, **32**, 1561–1564.
25. Bouhlel E., Laszlo P., Levart M., Montaufier M.-T., and Singh G.P., (1992) *Tetrahedron Lett.*, **34**, 1123–1126.
26. a. Renaud J.-P., Battioni P., Bartoli J.-F., and Mansuy D., (1992) *J. Chem. Soc., Chem. Commun.*, 888–889; b. Kimura E., (1992) *Tetrahedron*, **48**, 6175–6271; c. Zhang W., Loebach J.L., Wilson S.R., and Jacobsen E.N., (1990) *J. Am. Chem. Soc.*, **112**, 2801–2803; d. Bartoli J.F., Battioni, P., DeFoor, W.R. and Mansuy D. (1994) *J. Chem. Soc., Chem. Commun.*, 23–24
27. a. Jacobson S.E., Mares F., and Zambi P., (1979) *J. Am. Chem. Soc.*, **101**, 6938–6946; b. *idem., ibid.*, 6946–6950; c. Miller, M.M. and Sherrington, D.C. (1994) *J. Chem. Soc., Chem. Commun.*, 55–56.
28. Heggs R.P., and Ganem B., (1979) *J. Am. Chem. Soc.*, **101**, 2484–2846.
29. Waller F.J., (1986) *ACS Symp Ser.*, **308**, 42.
30. a. Gao Y., Hanson R.M., Klunder J.M., Ko S.Y., Masamune H., and Sharpless K.B., (1987) *J. Am. Chem. Soc.*, **109**, 5765–5780; b. Hanson R.M. and Sharpless K.B., (1986) *J. Org. Chem.*, **51**, 1922–1925.
31. a. Kagan H.B. (1982) Asymmetric synthesis using organometallic catalysts in *Comprehensive Organometallic Chemistry* ed. Wilkinson G., Stone F.G.A., and Abel E.W., Pergamon Press, Oxford, vol. 8., pp. 463–498; b. Noyori R. and Kitamura M., (1989) *Modern Synthetic Methods*, **5**, 115–198.

32. Weisz, P.B. Frilette, V.J. Maatmann, R.B. and Mower, E.B. (1962) *J. Catal.*, **1**, 307–312.
33. Tsuji, H. Yagi, F. and Hattori, H. (1991) *Chem Lett.*, 1881–1884.
34. Reuben B.G. and Wittkoff, H.A. (1989) *Pharmaceutical Chemicals Perspective*, Wiley, New York.
35. Bellusi, G. Clerici, M. Buonomo, F. Romano, U. Eposito, A. and Notari, B. EP 86 200 603.
36. a. Wellman J. and Steckhan, E. (1977) *Chem. Ber.*, **110**, 3561–3571; b. Baizer, M.M. (1964) *J. Org. Chem.*, **29**, 1670–1676.
37. Suckling C.J. (1990) in *Enzyme Chemistry, impact and applications*, 2nd edn., ed. Suckling C.J., Chapman and Hall, London, pp. 95–170.
38. Monsan P.F. in ref. 2, pp. 119–140.
39. Barker S.A. in ref. 2, pp. 77–98.
40. a. Charney W. and Herzog, H.L. (1967) *Microbial Transformation of Steroids*, Academic Press, New York; b. Fieser, L.F. (1959) *Steroids*, Reinhold, New York, pp. 672–678; c. Peterson D.H. Murray, H.C. Eppstein, S.H. Reinecke, L.M. Weintraub, A., Meister, P.D. and Leigh, H.M. (1952) *J. Am. Chem. Soc.*, **74**, 5933–5936; d. Hogg, J.A. Beal, P.F. Nathan, A.H., Lincoln, F.H. Schneider, W.P. Magerlein, B.J. Hanze, A.R. and Jackson, R.A. (1955) *J. Am. Chem. Soc.* **77**, 4436–4438.
41. a. Ballard, D.H.G. Courtis, A. Shirley, A.M. and Taylor, S.J.C. (1983) *J. Chem. Soc., Chem. Commun.*, 954–955; Taylor, S.J.C. Ribbons, D.W. Slawin, A.M.Z., Widdowson, D.A. and Williams, D.J. (1987) *Tetrahedron Lett.*, **28**, 6391–6394; c. Boyd, D.R. Sharma, N.D. Boyle, R. McMurray, B.T., Malone, T.F. Chima, J. and Sheldrake, G.N. (1993) *J. Chem. Soc., Chem. Commun.*, 49–51; d. Boyd, D.R. Sharma, N.D. Hand, M.V. Groocock, M.R. Kerley, N.A. Dalton, H. Chima, J. and Sheldrake, G.N. (1993) *J. Chem. Soc., Chem. Commun.*, 974–976.
42. EP 86 193 227, 86 193 228, 88 256 586.
43. *Sigma* and *Fluka* catalogues list several useful dehydrogenases. Other sources are detailed in *Preparative Organic Biotransformations*, (1992) ed. Roberts, S.M. Wiley, 1992, with annual updates, under relevant preparations.
44. Keinan, E. Sett, K.K. and Lamed, R. (1987) *Ann. N.Y. Acad. Sci.*, **501**, 130–149.
45. Wandrey C. and Wichman, R, (1985) *Biotechnol. Ser.*, **5**, 177.
46. Simon, H. Bader, J. Gunther, H. Neumann, S. and Thanos, J. (1985) *Angew. Chem. Int. Edn. Engl.*, **24**, 539–553.
47. Mori K. and Mori H., (1989) *Org. Syn.*, **68**, 56–63.
48. Holmes, P.A. Wright, L.F. and Collins, S.H. EP 82 52 459.
49. P. Critchley (1987) In *Chemical Aspects of Food Enzymes* ed. Andrews, A.T. Royal Society of Chemistry, London, pp. 150–155.
50. a. Bjorkling, F. Gotfredsen, S.E. and Kirk, O. (1989) *J. Chem. Soc., Chem. Commun.*, 934–935; b. Adelhorst, K. Bjorkling, F. Gotfredsen, S.E. and Kirk, O. (1990) *Synthesis*, 112.
51. Bjorkling, F. Gotfredsen, S.E. and Kirk, O. (1990) *J. Chem. Soc., Chem. Commun.*, 1301–1303.
52. Becker, W. Freund, H. and Pfeil, E. (1965) *Angew. Chem. Int. Edn. Engl.*, **4**, 1079.
53. Kruse, C.G. Geluk, H.W. and van Scharrenburg, G.J.M. (1992) *Chimia Oggi*, 59.
54. a. Effenberger, F. Ziegler, T. and Forster, S. (1987) *Angew. Chem.*, **99**, 491; b. Brussee, J., Roos, E.C. and Van der Gen, A. (1988) *Tetrahedron Lett.*, **29**, 4485–4488.
55. Kren, V. Crout, D.H.G. Dalton, H. Hutchinson, D.W. Konig, W. Turner, M.M., Dean, G., and Thomson, N. (1993) *J. Chem. Soc., Chem. Commun.*, 341–342.
56. Taylor S.J.C. (1990) In *Opportunities in Biotransformation* eds. Copping, L.G. Martin, R.E. Pickett, J.A. Bucke, C. and Bunch, A.F. Elsevier Applied Science, Barking, pp 170–176.
57. Casy, G. Lee, T.V. Lovell, H. Nichols, B.J. Sessions, R.B. and Holbrook, J.J. (1992) *J. Chem. Soc., Chem. Commun*, 924–926.
58. Schultz, P.G. (1989) *Accts. Chem. Res.*, **22**, 287–294; (1993) **26**, 391–395.
59. a. Suckling, C.J. Tedford, C.M. Bence, L.M. Irvine, J.I. and Stimson, W.H. (1993) *J. Chem. Soc., Perkin Trans.* **1**, 1925–1929; b. Gouverneur, V.E., Houk, K.N., Depascualteresa, B., Beno, B., Jerda, K.D. and Lerner, R.A. (1993) *Science* **262**, 204–208.

60. Nobel, D. (1993) *J. Chem. Soc., Chem. Commun*, 419–420.
61. Otsuka, K. Yoshinaka, M. and Yamanake, I. (1993) *J. Chem. Soc., Chem. Commun*, 611–612.
62. Herrmann, W.A. Wagner, W. Flesser, U.N. Volkhardt, U. and Komber, H. (1991) *Angew. Chem. Int. Edn. Engl.*, **30**, 1636–1638.
63. Tsuda T. and Fujiwara, T. (1992) *J. Chem. Soc., Chem. Commun*, 1659–1660.
64. Gassner F. and Leitner, W. (1993) *J. Chem. Soc., Chem. Commun.*, 1465–1466.

11 Clean energy supply and use

J.W. TWIDELL

11.1 Introduction

This chapter seeks to establish principles of clean energy, to give facts and figures and to propose strategies of energy use and supply for a sustainable, and therefore 'clean', future. The supply and use of energy, especially of fossil fuels, is a dominant activity of industrial and post-industrial society, strongly linked to pollution at the local and global scale.[1] There are intrinsic connections with world problems of sustainable development, climate change and biodiversity. The message of this chapter is that harm from energy supply and use is not necessary; we have engineering and institutional choices to make between options with different environmental impacts. Energy is necessary for development, but pollution is not.

At the most basic level, energy is needed to maintain all forms of life; to communicate, to heat, to move and to manufacture goods. However, the characteristics and amounts of energy for these functions vary greatly, showing significant differences in environmental impact. In use, some forms of energy and methods of energy supply cause more pollution and more environmental harm than others. However, there is always a choice between different energy processes to perform the same effect, each having different environmental impacts. The market place costs of these methods will vary also and will depend on whether or not the external costs of environmental impact are included, e.g. whether or not the polluter pays.[2]

In planning future sustainability, a critical factor is to know the 'carrying capacity' of each locality, nation and, indeed, the whole earth. Carrying capacity is the amount and number of biological organisms that can live within the ecological environment in a continuing and sustainable manner. Our own interest is in the 'biological organism' of mankind which is no different from any other organism in having to live harmoniously in an ecological framework. The ecological environment is defined as a system of many parameters such as nutrients, food, temperature, material resources and social behaviour. Each parameter will have maximum and minimum limits to allow sustainability. A major parameter will always be energy supply, whether this be the natural supply of sunlight for the photosynthesis of plants and food for animals or of heat and electricity for industry. Therefore energy is a necessary, but not sufficient, requirement for all life, be it biological, domestic, industrial or commercial. However, many other parameters are not necessary in the ecological context, and their presence may cause harm; in which case they are 'pollutants' with a usually small, maximum level of acceptability and a minimum level that may well be zero.

The analysis of energy supply and use should therefore be conducted within the debate on sustainable development, defined here as 'the continuation and enriching of human society in a manner that: (i) strengthens its ecological framework, (ii) improves the long term quality of life and the wealth of all humanity, and (iii) does not deprive future generations of necessary resources. This definition recognises our human environment as an ecological environment and our moral duty to consider the future as well as the present. These two aspects of sustainable development are unfortunately not pronounced in the commonly used definition of 'development that meets the needs of the present, without compromising the ability of future generations to meet their own needs'.[3]

By the standards of sustainable development, the energy supplies used today for human development are too intensive and are related to a multitude of pollutants and other adverse effects at unacceptable levels. Therefore the carrying capacities related to human activity seem to be at or beyond their limit within present modes of technology; *in other words our present technology and lifestyle are not sustainable*. Our choice is either to accept limited carrying capacity by, say, accepting a lower quality of life, or to seek other forms of technology that allow sustainability. This chapter seeks to discover what these technical options are, whilst not accepting any scenarios producing a lower quality of life.

The environmental impacts of energy relate to the capture of the energy, its storage and its utilisation. They are the impacts of the energy process *on* the environment, and not vice versa. The impacts are:[4]

- physical (e.g. hydro dams, noise);
- chemical (e.g. air pollution);
- biological (e.g. genetic damage from ionising radiation);
- ecological (e.g. climate change);
- aesthetic (e.g. electricity grid pylons, wind turbines).

The most serious impacts are those that affect the human body and the earth's ecology, usually due to chemical and ionising pollution. The presence of pollution and adverse environmental impact is relatively easily known within the local environment, but in recent years the increasing world population and increase in per capita energy use have produced environmental impacts definitely measurable on international and global scales. Obvious global examples related to energy supply are depletion of forests, atmospheric carbon dioxide increase, acid rain and mining.

The technology of energy supply and use has evolved throughout civilisation, with most rapid advance in the last 200 years through the Industrial Revolution and now into the post-industrial age. The major thrust of the Industrial Revolution was to increase energy supply from fossil fuels with little economic regard for the pollution of air, water and land from emissions or from the physical intrusion of engineering structures. Such adverse impacts of energy generation on the environment have been realised for over 300 years[29] but only recently have economists attempted to calculate the costs paid by society as a result of such pollution.[5] Once calculated, these 'external costs' can be charged to the polluter and so become internalised. In this way the overall costs of different energy supplies and processes can be compared in like manner without distortions.

Consequently the full costs of environmental pollution are beginning to be charged to energy suppliers with a resulting increase in the price of energy. Examples are the threefold increase in the price of nuclear electricity when full costs were applied in the UK in 1990, and the additional price of cars when catalytic exhaust converters are fitted. The overall effect is for economic emphasis to move from supply to the efficient *use* of energy. For instance, electricity utilities discover they can increase profit by encouraging low-energy lighting rather than installing more generating plant. However, despite the ethical fairness and technical advances resulting from the 'polluter pays' principle, despite bodies such as the Organisation for Economic Co-operation and Development supporting the principle as long ago as 1972, despite the good sense of adopting 'precautionary' and 'no regrets' principles, and despite Sweden and Denmark leading the way in having taxes for carbon, sulphur and nitrogen oxide emissions, very little action is being taken by countries as a whole. The UK Government is typical, in being strong on rhetoric[7] yet weak on action.

In this chapter, pollution has been defined in its *ecological* context, but it is reasonable to presume that pollution is also *economically* harmful at all scales. Realising this, governments and industry are encouraging new concepts of clean energy (usually interpreted as 'cleaner than before') and of sustainable technology (technology that allows economic development without limiting human and other ecological carrying capacities).[7] If such concepts become requirements, they present formidable challenges and opportunities for the next millennium.

11.2 Principles of clean energy

Energy processes are needed for all our activities and are listed below in approximate order of priority for human sustainability. With each process are associated various environmental impacts and choices of technology. It is therefore possible to propose principles for clean energy technology, which will be stated as they arise, and then consider examples of how these principles may be applied.

11.2.1 Solar energy

The radiation from the sun is at the highest temperature obtainable by humanity from any source. This is important, because high-temperature energy sources can be the most efficient and therefore potentially less polluting. At a high temperature (about 6300°C), the radiated energy by Planck's Law is 9% in the ultraviolet wavelengths (0.25–0.4 μ), 45% in the visible (0.4–0.7 μ) and 46% in the near infrared (>7.0 μ). The ultraviolet radiation would be harmful to life on earth if it were not absorbed by ozone as a natural component of the stratosphere at a height of 10–50 km.

The radiation that reaches the earth's surface therefore has a 'photon temperature' of about 6300°C that is far above temperatures otherwise available for engineering plant. Unlike nuclear power on earth, the *solar* nuclear fusion reactions produce no terrestrial pollution. Consequently solar radiation processes, such as photochemistry and photovoltaic electricity, have the potential for operating at high thermodynamic efficiency without harmful environmental impact. In addition, the received solar radiation is at intensities compatible with biological life. Such solar-radiation-driven engineering is the ultimate state for sustainability.

The intensity of bright sunshine at sea-level and under a clear and clean atmosphere is 1000 W m^{-2} (1 kW m^{-2}) onto a surface perpendicular to the beam. During a daylight period of about 12 h the absorbed radiation onto a dark horizontal surface is about 30 MJ/(m^2 day), equal to 8.3 kWh/(m^2 day). For the whole earth, the flux of solar radiation is about 1.2×10^{17} Watt, equivalent to about 20 000 kW per person, i.e. about 10 000 times more than the 2 kW per person needed within a nation having high quality life styles with high technology and the efficient use of energy.

Principle 1. There will never be a shortage of clean energy supply on earth; the challenge for science and technology is how to harness the abundant energy that we have freely available.
Application. Long term economic investment and markets in solar technologies.

11.2.2 Photosynthesis

The photon energy of solar radiation transmitted through the atmosphere is sufficiently energetic to initiate photosynthesis in plants, but not energetic enough to cause ionisation damage to molecular structure. The intensity of solar radiation peaks at a photon energy of about 1 eV (1.6×10^{-19} J) which 'by chance' is the optimum for these photochemical reactions.

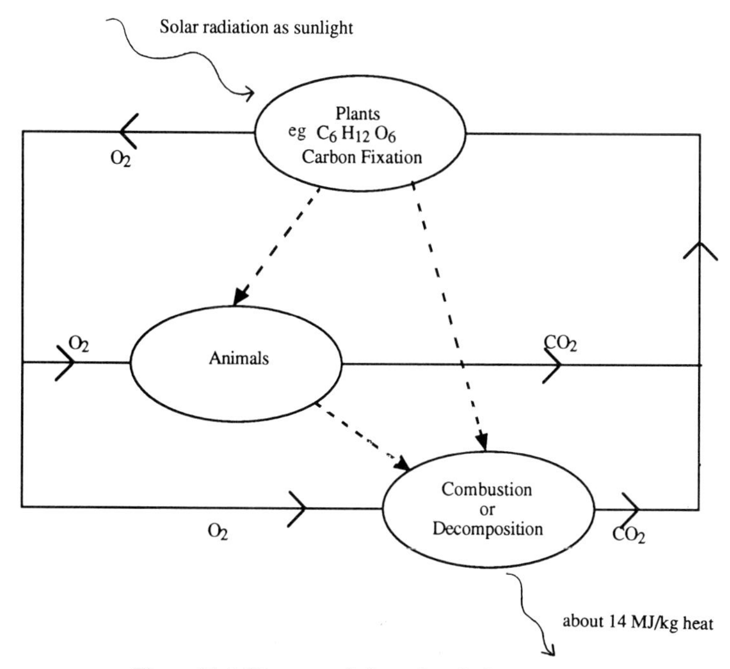

Figure 11.1 Diagram of photochemical processes.

When absorbed in photosynthetic molecules, this energy is able to initiate the biochemical reactions leading to the absorption by plants of carbon dioxide gas, CO_2, the formation of the organic structural material of the plant and the emission of oxygen (Figure 11.1). The oxygen is a reactive gas containing the chemical energy obtained from the solar radiation through photosynthesis; this chemical energy is released when oxygen is utilised in metabolic reactions or in combustion (the energy is not 'in' the fuels, it is in the oxygen – a pedantic, yet perceptive, fact). The absorbed carbon becomes the main chemical component of the plant material, biomass. When this biomass eventually decomposes in the presence of oxygen by microbial action or by burning, energy from the originally absorbed photosynthetic light is released as heat and the carbon returns to the environment as CO_2. Thus the carbon absorbed into plants is recycled but the energy continues its progress from solar radiation to become heat lost to space as long-wave infrared radiation.

The importance of the earth's atmosphere is understood by realising (i) that all the energy released from the combustion of fuels and from metabolism is obtained from oxygen and (ii) all plant and animal organic material is obtained from carbon in CO_2. Therefore the energy and materials of life are obtained from gases in the atmosphere. *The atmosphere is the delivery system of economic prosperity.*

Photosynthesis and metabolism are processes within a natural system of recycling materials interacting with a flow of energy from high temperature (the sun at 6300°C) to low temperature (space at −270°C). The materials are recycled and so conserved; the energy passes through the system and is lost. Natural ecology is driven by the *income* of solar radiation and not by the *capital* deposits of fossil or nuclear fuels.

Principle 2. Sustainability and development are dependent upon irretrievable flows of energy and the cycling of materials.
Application. Every economic activity should be assessed by its role in full life-cycle processes.

11.2.3 Production of biomass and fossil fuels

The material of plants is transformed into the organic body material of the animals that digest them and eventually this biomass decays under microbial action to mostly CO_2 and water, releasing heat equal to the heat of combustion of the same biomass. Thus whether biomass decays naturally or is burnt, the same energy is released, being about 35–55 MJ/(kg of carbohydrate) or 10–15 MJ/(kg of biomass). It is clearly reasonable to utilise this energy as a fuel rather than have it dissipated to the environment without benefit – hence the ecological justification for utilising biofuels in preference to fossil fuels.

If the biomass is in a condition where complete microbial decay does not occur, then fossil fuels are formed from the biomass together with inorganic matter. Peat is an early form of fossil fuel remaining at ground level with heat of combustion of about 15 MJ/kg of dry material. When the biomass deposits are covered and trapped anaerobically within the earth, they become coal, oil or gas with heats of combustion of about 30–55 MJ/kg. When extracted, these fuels may be burnt in air to produce oxides of carbon, other pollutant emissions and solid residue (Table 11.1). However, the carbon content of the atmosphere is in a natural balance without these anthropogenic activities and so every instance of the combustion of fossil fuels perturbs the equilibrium. Today, of the excess carbon added to the atmosphere, mostly from fossil fuels and partly from rapid forest burning,

Table 11.1 Gaseous emissions (kg) from the extraction, consumption and waste disposal of fuels (i.e. the life cycle of the fuels) as used to produce 1.0 GJ (280 kWh) of heat (based on Mortimer[9]).

Fuel system	CO_2	CH_4	N_2O	CO	volatiles	NO_x	SO_2
Coal	82	1.2	0.01	0.02	0.01	0.4	1.9
Oil	75	0.01	0.04	0.03	0.05	0.2	2.3
Natural gas	35	0.6	0.001	0.05	0.01	0.07	0.07
Nuclear (PWR)	3	0.04	0.0001	0.001	0.0004	0.02	0.08
Wind, hydro	0	0.00	0.0000	0.000	0.0000	0.00	0.00

only half is reabsorbed naturally by photosynthesis. The remainder stays in the atmosphere. For instance, the UK emits 160 Mt/y of carbon in CO_2 from fossil fuels, but an urgent policy for reafforestation is only absorbing an extra 1.5 Mt/y of carbon (HMSO[8]). Moreover there is no plan to use the resulting biomass to substitute for fossil fuels, so the reafforestation is only of short term effects. Therefore despite a government strategy for sustainable development (HMSO[7]), 158.5 Mt/y of excess carbon from fossil fuels are being deposited into the global atmosphere with no attempt being made to recycle the carbon.

The formation of fossil fuels via photosynthesis is the biological way to remove CO_2 from the earth's atmosphere. Since there is a danger of there being too much CO_2 in the atmosphere, the best ecological location for fossil fuels is to remain underground and not to be utilised by people. Nature took a long time forming a low-CO_2 atmosphere, and we should not perturb our carrying capacity by reversing the process. We have only to consider the inhospitable atmospheric and thermal conditions of CO_2-rich Venus and Mars to appreciate how important our atmosphere is to us. We are so woefully ignorant of how life for humanity became sustainable in the first place, that our presumption of understanding how to live now is amazing. Lovelock makes such points strongly from scientific argument.[10]

Principle 3. Ultimate sustainability depends on carbon not entering the atmosphere and being trapped in carbonates and fossil fuels. For long-term sustainability, the correct ecological location for fossil fuels is underground and we should not bring them up.
Application. Biomass and wastes should be utilised in preference to fossil fuels.

11.2.4 Processes in the atmosphere that maintain life

The unpolluted atmosphere transmits only the wavelengths of solar radiation that initiate photosynthesis and vision, and do not cause excessive ionisation. However, removal of natural ozone in the stratosphere, e.g. by reactions arising from chlorofluorocarbon compounds (CFCs), allows the harmful ultraviolet radiation to reach the earth's surface.[11] The excess ultraviolet causes skin cancers and harms primary photosynthesis and hence growth in marine phytoplankton and land plants, as recognised by the Montreal protocols on ozone protection.[12,13] Likewise a clean and unperturbed atmosphere allows heat to leave the earth as infrared radiation, so maintaining a temperature compatible with life processes. Several gases emitted from human industry absorb this infrared heat on a scale to perturb the temperature-balancing mechanisms of the atmosphere, leading to the risk of global climate change; examples are carbon dioxide, nitrogen oxides, methane and CFCs. The more atoms there are in a gaseous molecule, the more strongly that molecule absorbs infrared radiation, causing warming. Therefore methane is about 30 times more effective per molecule at absorbing infrared radiation than CO_2. At the present time there is overwhelming scientific reason for being concerned about the possibility of global warming and resulting global change from anthropogenic gaseous emissions, mostly from CO_2 from thermal combustion processes.[14,15,16,17] Consequently nations have conducted audits of their CO_2 emissions and developed policies to ameliorate the amounts produced from fossil fuels.[18] For UK anthropogenic activities, the 'forcing' of global warming is 87% from fossil fuel CO_2, 8% from methane and 4.4% from NO_x (mostly nitrous oxide emissions from organic chemical industry). The UK methane is 32% from agriculture (ruminant cattle enteric emissions account for over half), 39% from landfill gas (mostly anaerobic rotting food and usable as a fossil fuel substitute), 16% from coal mining, 10% from leaks of natural gas at distribution and extraction.[8] The UK NO_x emissions as a whole are 56% from transport petroleum, 28% from fossil fuel power stations, and about 10% from other energy sources. Thus over 93% of UK anthropogenic global warming potential is from fossil fuel-related processes.

An unpolluted atmosphere at ground level contains only oxygen as a reactive gas which maintains life processes. All other reactive gases are pollutants, especially those arising from fossil fuel combustion, producing adverse effects on biological life. Examples are ozone, and nitrogen and sulphur oxides.

The atmosphere is the source of the water that also maintains life. Condensed water as rain or mist becomes polluted from gaseous emissions already in the atmosphere and from polluted ground through which it flows. Examples of international importance are acid rain, radioactive aerosols as from the Chernobyl accident and nuclear waste.

Principle 4. Sustainability is endangered when air and water are polluted because these are the essential media controlling conditions for life.
Application. All chemical emissions need to be rigorously monitored and limited according to life-cycle analysis.

11.2.5 Ecological control systems – the danger zones

Nowhere is our general ignorance of the environment more pronounced than in our lack of appreciation that the earth's ecology is a thermodynamic non-equilibrium system, maintained far from its ground state by a sophisticated control system which maintains the earth within the narrow band of temperatures for water to be liquid and biochemical reactions to proceed. Whilst the earth itself is too large to be significantly influenced by humanity, the control systems are more delicate and can certainly be perturbed. We appreciate that a nuclear power station or a jumbo jet demands great sophistication of control, and that their control systems must be protected at all costs, but we do not apply the same care to the environment. The danger of excess CO_2 in the atmosphere is not that the balance of gas concentrations of the atmosphere will be changed, but that the control systems will be damaged. The easiest way to damage a nuclear power plant is to subtly perturb the control switches; the same is true for the environment.

Principle 5. The earth is controlled in a thermodynamic state far from thermal equilibrium by a natural and sophisticated control system strongly involving the atmosphere; it is extreme stupidity for us to perturb this control system.
Application. Teach and require an understanding of natural and human ecology for all students at secondary and tertiary level.

11.2.6 Energy subsidies to enhance food production and industry

Animals obtain energy from metabolising the organic carbon compounds of plants. For instance, our bodies metabolise at an average power of about 100–150 W from the food we eat (9–15 MJ/day). The total energy from food is therefore 3 to 5 GJ per person per year. Vegetable food is obtained from plants that grow with about 1–5% efficiency in sunlight that has a maximum intensity of 1000 W/m^2 and an average intensity day and night through the year of about 100 W/m^2. Consequently if we eat only plant food and allowing for non-digestible plant material, in terms of energy, each person requires a minimum of about 30–100 m^2 of closely planted fertile arable land for sustainable food production.

For the UK, with about 10^5 km^2 of arable crop land, this approximate calculation predicts a human carrying capacity of about 1000 million people – far in excess of the present 57 million. What is wrong with the calculation? The error has been to neglect many other energy factors relating to sustainability, principally energy supplies for food production and for domestic, commercial and industrial life, and, of course, many non-energy factors.

In industrialised countries the per capita energy consumption is about 50–100 times greater than metabolic requirements, e.g. USA 10 kW, Germany 6 kW, UK 5.5 kW. In the non-industrialised tropical rural economies, energy is needed for cooking, construction and transport with average per capita levels at 0.5–1 kW per person. Where does this extra energy above metabolicc requirements come from? Figure 11.2 shows the primary energy supply mix for several countries. The main impression is the dominant role of fossil fuels with the consequent impacts on the environment.

Two major factors concerning the energy analysis of food production are (a) the 'sequestered energy' or 'energy subsidy' needed for labour,

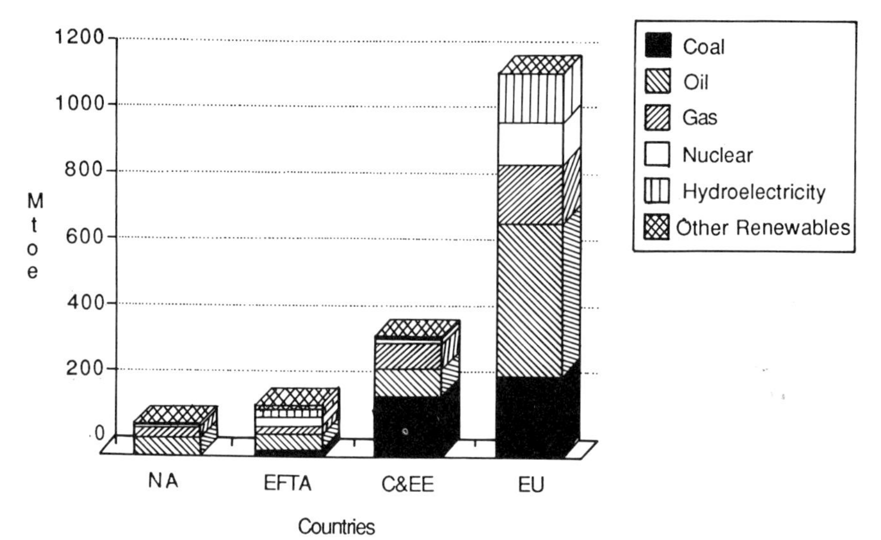

Figure 11.2 Primary energy mix of various countries.

machines, fertilizers and pesticides, and (b) the reduction in energy passed from plants to humans by eating meat rather than plants.

(a) On average in modern agriculture the energy stored from solar radiation in the photosynthesis of an arable crop is about 5 times the energy consumed for the machinery, fertilisers and pesticides used for that crop. Without such industrial energy expenditure crop yields would be drastically reduced. However, the energy of human labour in modern farming is negligible. Consequently the great majority of energy sequestered as a subsidy for agriculture arises because of the manufacture of fertilisers and pesticides from fossil fuels, mostly oil. Therefore mechanisms for increasing crop and cattle yields are related to environmental pollution from the fossil fuel subsidies.

(b) The energy obtained by humans from eating meat and fish is only about 10% of the energy metabolised by those animals from their plant feed. If the animals are raised on arable crops or on arable land, about 10 times the land area per unit of human food energy is required for meat than for plant food. Consequentially for such intensive husbandry, the sequestered energy subsidy for meat may be more than the metabolic energy obtained from the meat. If the animals are raised on non-arable land, the sequestered energy subsidy is much less. Because of the present-day large numbers of cattle worldwide (1.5 billion, doubled since 1950) and because methane molecules have 30 times the global warming impact of CO_2, the global amount of methane produced from intestinal fermentation in cattle is significant as a forcing agent for global climate change.

Principle 6. Human food yields may be increased significantly by technologies dependent on subsidies from fossil fuels, but the price paid is in the external costs of pollution, soil degradation and the danger of climate change.
Application. Policies are needed for less intensive agriculture and for less consumption of meat.

11.2.7 Sources of domestic, industrial and transport energy

There are five primary sources of energy available to humanity, of which the first three are renewable sources and the last two are finite sources:

Renewable sources of energy. These are obtained from the repetitive and continuing energy fluxes of the natural environment,[19] i.e.

1. From solar radiation that produces photosynthesis (biomass, biofuels), photovoltaic power (solar cells), wind, waves, and hydropower (falling water).
2. From tidal power (from the loss of kinetic energy of the earth as it slows from the friction of the tides).
3. Geothermal heat, from the heat of radioactive decay of natural elements within the earth.

Principle 7. By definition, renewable energy supplies are sustainable and, being associated with natural processes, they are unlikely to cause chemical and biological environmental harm or pollution.
Application. Financial credit should be allocated to renewable energy supplies in recognition of the low external costs of pollution.

Finite sources of energy. These are natural energy stores related to:

4. The combustion of the fossil fuels of gas, oil and coal in air for heat and power, especially electrical and motive power.
5. Energy released from nuclear power stations, either as fission or fusion of elements.

Principle 8. By definition as energy stores, fossil fuels and nuclear power are not sustainable. Nevertheless the lifetimes of the resources may be longer than usual planning horizons.
Application. Policies are needed for the steady phasing out of finite energy sources and for the increased limitation of chemical and ionising emissions.

The instinctive reaction after the 1973 and subsequent OPEC oil price increases was for countries to fear that other energy supplies would soon become insufficient. It is now obvious that no such dramatic endings will occur and that the world total of accessible fossil fuel deposits is more than sufficient for several hundred years at present extraction and discovery rates. The limiting constraints are now clearly environmental, as foreseen by a number of earlier studies.[20,21] Not only must we allow the natural control systems and processes of global and local ecology to proceed, but we must prevent ecological irreversibilities occurring.

> *Principle 9*. The sustainability of fossil and nuclear fuels is likely to be limited by harm to ecological systems rather than from the absolute quantities of resource.
> *Application*. Considerable scientific investigation and monitoring is required to trace and understand the effects of pollutant emissions.

Additional sources of energy. These can be classified as either renewable or finite depending on the net supply and consumption of the local resource, namely:

(a) Peat: a renewable resource only if allowed to restore itself over hundreds of years.
(b) Firewood: only renewable if the mass and variety of natural woodland and forests are not depleted.
(c) Wastes: effectively renewable, and often so classified (used as energy sources from direct combustion or from methane gas arising naturally from landfill tips of urban food waste).
(d) Hot rock geothermal: finite over many tens of years as subterranean granite rock is cooled by heat removal, but not a sustainable resource at economic extraction rates.

11.3 Consequences of the laws of thermodynamics and matter

There are well established physical laws of mass and energy which may be interpreted in terms of environmental impact and sustainability.

11.3.1 Conservation of mass

Every atom absorbed into a chemical reaction remains within the products. If such a reaction is not part of an established natural ecological cycle or linked to one, then the final products are pollutants. Thus the combustion of carbon and sulphur in coal produces carbon, sulphur and nitrogen oxides as pollutants. The CO_2 from fossil fuels is a pollutant despite it being chemically indistinguishable from the CO_2 of the natural cycles of photosynthesis and metabolism. This definition of a pollutant includes the products of all nuclear reactions.

Principle 10. Atoms not involved in natural systems of recycling are, by definition, pollutants.
Example. Life cycle assessment of all emissions into the environment should be the basis of pollution control.

11.3.2 The first law of thermodynamics

This law is valid in all energy transfer processes. It states that energy in a transformation is conserved and cannot be destroyed.

The relevance for environmental impact is that energy used for processes at high temperature can be passed on for further use at lower temperatures without the need to generate additional energy from other sources. In this way the same energy can be used several times as it 'cascades' down a temperature gradient. There is no additional environmental impact from this same energy used again for the subsequent processes. Examples are (a) the beneficial heating of a building from lights and computers, (b) the use of waste heat from a thermal power station for a fish hatchery.

Principle 11. The same energy can be used several times with no additional environmental impact when passing sequentially though cascading processes.
Application. Modern offices and many industrial buildings in cold countries are well insulated, and should need no special heating plant since the building is heated from thc 'frcc' and 'casual' gains of lighting, computers and other electrical devices.

Boilers for heating, heat exchangers and calorifiers are all examples of the transfer of heat from one form or one temperature to another. Simple fossil fuel boilers are about 65% efficient in transforming the heat of combustion into heating water from cold to hot. The remaining energy passes with the emission gases, including water vapour, from the flue. If these gases pass across the pipes of the cold water intake, as a heat exchanger, the overall efficiency increases to about 75% as in gas heaters with balanced flues. If this heat exchange process condenses the water vapour to liquid (a condensing boiler) the overall efficiency to useful heat is about 90%.

Principle 12. Heat recovery from the exhaust of energy processes increases overall efficiency without added pollution.
Example. Air extraction fans are available so the outgoing stale air exchanges heat with the incoming fresh air; consequently the energy loss is reduced by over 50%, for both heating and cooling systems.

11.3.3 Nuclear reactions

Changes in the nuclei of atoms are always accompanied by the creation of heat energy, E, an annihilation of mass, m, and the emission of radiation.

The relationship between the energy gain and mass loss is Einstein's relationship, $E = m c^2$ where c is the speed of light. The relationship is such that extraordinary amounts of heat are released for very small changes in mass. However, the First Law of Thermodynamics is still valid. After the nuclear energy is released from the annihilation of mass, this energy is eventually absorbed as heat into other forms of matter. At absorption, there is a total increase of mass that equals the initial loss. Finally, both mass and energy are conserved.

The initial and new nuclei formed in the nuclear release of energy may themselves be radioactive, releasing nuclear radiation capable of harming biological organisms. The intensity and lifetime of such radiation varies greatly, but the integrated effect is always harmful. No radioactive process is part of any natural ecosystem. To date, when the full internal and external costs have been assessed, the price of nuclear energy has always been higher than alternative supplies.

> *Principle 13.* By its very essence of requiring radioactive processes, nuclear power is intrinsically polluting and distinct from ecological sustainability, unrelated to any form of material recycling.
> *Application.* Nuclear power is distinctive in providing a capital intensive and diverse energy supply and not for providing cheap power; consequently governments have to be involved in nuclear power policy and funding.

11.3.4 The second law of thermodynamics

This law applies to engines powered from heat, e.g. a car engine, a steam or gas turbine, a fossil fuel or nuclear power station. The consequences are that heat cannot be entirely transformed by an engine into shaft work or shaft power. There is always a release of heat at lower temperature. In a power station the engine work is used to generate electricity. The hotter the initial energy, the greater the proportion of initial heat that can be transformed into work and hence electricity. The ratio of the electricity output to the energy input is the conventional 'efficiency' (of electricity generation) of a power station.

Engines, refrigerators and heat pumps. Heat engines take in heat $(A+B)$ to produce shaft work (A) as 'power' and rejected heat (B). For steam turbine power stations with no use of waste heat, A is about half B, so the majority of the input energy goes to waste with the efficiency being $A/(A+B)$, e.g. about 30%. In a combined heat and power (CHP) plant a fraction fB of reject heat is utilised, so the efficiency becomes $(A+fB)/(A+B)$, e.g. about 80%.

Principle 14. For any heat engine, the proportion of shaft work obtained is limited by thermodynamic laws, but the overall energy efficiency is not so limited if beneficial heat output is included.
Application. Combined heat and power should be the norm, rather than having power-only or heat-only installations. In practice this means having smaller scale and local generating plant so the heat can be piped as steam or hot water to consumers.

A refrigerator (i) takes in energy (A') from shaft power (e.g. from electricity) and (ii) removes heat (B') from the cold temperature of a refrigerated container and (iii) lifts this to a higher temperature outside the refrigerator, where the total energy ($A'+B'$) becomes waste heat to the environment.

Heat pumps are effectively refrigerators in reverse, using electricity (A') to take heat (B') from the outside environment and lifting this energy in temperature to give useful heat ($A'+B'$) to an inside environment. The environmental benefit of heat pumps is that B' equals approximately $2A'$ so three times the heat is given out as compared with a direct electric heater. The environmental impact of the electricity generation is therefore reduced by a factor of about three by using heat pumps instead of direct electrical heating, but only if the heat pumps themselves do not add to the environmental burden, e.g. by involving CFC refrigerants.

Principle 15. The heat output of a heat pump within specified ranges of temperature is obtained for significantly less primary energy expenditure than utilising that primary energy directly.
Application. A strong policy to encourage the technology and markets for heat pumps with environmentally acceptable refrigerants would lead to greater energy efficiency and less pollution.

Both CHP and heat pump technology is well established, for instance with CHP providing 20% of heat in the Swedish economy.

Electricity power stations and plant. The maximum mechanical efficiency of heat engines is given by the Carnot equation, which explains why high temperatures of combustion are needed. The calculation gives a theoretical efficiency of about 70% for a thermal power station. However, the Carnot theory is unrealistic in requiring processes that are infinitely long. Practical engineering plant should be designed to maximise power output, i.e. energy, in a finite and not infinite time. For single stages of operation the maximum power efficiency of plant is about half the Carnot efficiency (i.e. about 35% for a thermal power plant).

Nuclear power stations produce steam at slightly lower temperature than other thermal plant and so they have lower efficiency for electricity generation at about 33%; in addition they consume about 3% of this electricity on site for the complex operations. Thus about 70% of the input energy is emitted wastefully, usually as hot water into a river or the sea.

Since nuclear power stations have to be long distances from towns and cities because of the inherent danger of radiation release, the large amounts of waste heat cannot be easily utilised.

Modern coal-fired power stations produce steam at higher temperatures than nuclear plant and so have electricity efficiencies about 35%. The combustion temperature in gas turbines is higher than the temperature of the steam of nuclear, coal or oil plant, and so gas turbines have electricity efficiencies of about 40%. Moreover the output gases from a gas turbine are hot enough to produce steam for a steam turbine following after the gas turbine; such 'combined cycle plant' produces electricity from the two stages together at about 55% electricity efficiency. In addition gas turbines are relatively small, the gaseous emissions are not chemically or biologically harmful (being nearly 100% CO_2 and water), and so the waste heat can be used beneficially in 'combined heat and power', for instance in heat for a building. The overall efficiency of combined cycle electricity generation followed by the use of the reject heat may be 90%, i.e. combined cycle generation plus combined heat and power. This is a *three times* improvement in first law efficiency over a traditional thermal power station.

If the electrical generation or direct mechanical output of such a combined system powers a heat pump, then the total heat output can be more than the energy input, i.e. the efficiency for heating can be much more than 100%. The total output heat is from the engine, the exhaust and the heat raised in temperature from the outside environment by the heat pump.

> *Principle 16.* Single stage energy plant is intrinsically less efficient and more polluting than integrated and sequential plant having multiple purposes.
> *Application.* The market for high technology, low environmental impact, energy plant has to be encouraged.

Exergy. One reason for the failure of the UK and other nations to advance efficient heat and electricity generation is the failure to analyse the parameter *energy* in terms of its temperature and work related aspects. In practice we use the First Law of Thermodynamics for economic assessment, but not the Second Law. Arising from the Second Law, a better parameter than energy is available, that of *exergy*. Exergy may be defined as the work potential of an energy source within a specific environment, and is therefore related to the absolute thermodynamic functions of 'free energy' and 'availability'. Exergy has the same units as energy. A quantity of energy has higher exergy (i) if it is at higher temperature, and (ii) if the environmental temperature decreases. In exergy terms, 100 GJ of energy at 1000°C has far more work potential, and hence more economic value, than 100 GJ of energy at 100°C, yet in energy terms the two quantities are equal. Exergy, and the related function of second law (of thermodynamics) efficiency, enable much improved analysis for greater efficiency of engineering plant. By further including holistic, life cycle, analysis of the whole process together with the assessment of external costs, design of significantly cleaner energy systems will be possible.

Principle 17. Energy has value (i) depending on form (e.g. electricity is more valuable than hot water) and (ii) proportional to temperature (e.g. the temperature of burning gas in a gas turbine is higher than the temperature in a gas boiler). This value is recognised in exergy analysis, but not in energy analysis.
Application. Standards for energy equipment should be defined by reference to both the first and second laws of thermodynamics, rather than just the first.

One has to ask why society has been content with electricity generation and other efficiencies as low as 30–35% when 90% and higher efficiencies are obtainable? There are many aspects of an answer: 'inefficient plant is the easiest to construct with the cheapest capital cost', 'obtaining very high efficiency requires multidisciplinary design', 'institutional factors discourage innovation', etc. Yet *increasing efficiency always decreases environmental impact*, so efficiency improvements are vital. However, the converse is not true; decreasing environmental impact does not necessarily increase efficiency (e.g. desulphurisation of coal combustion emissions in power plant lowers efficiency). Therefore enlightened environmental legislation should lead to improvements in technology and consequently increased economic activity. This is one of the main mechanisms for economic advancement. In a modern world, *we become richer by using less, not more.*

Principle 18. Environmental care is a scientific task capable of producing advanced technology of high economic value, and, to the contrary, lack of environmental care protects outmoded vested interests of decaying value.
Application. Strict environmental legislation to prevent pollution and to enforce clean technology is a means to increase economic activity, despite the protests of conservative industry.

11.4 Case studies

11.4.1 Transport

The total energy used for transport in every region throughout the world is increasing significantly. The dominant form of primary energy for transport is oil, followed by electricity generated from fossil fuels. Consequently the global environmental impact of transport, especially with regard to air pollution, is increasing at a rate far higher than for any other class of energy use. Transport fossil fuels account for nearly 50% of world anthropogenic NO_x pollution. Power stations account for an equal quantity of NO_x and about 90% of all the approximately 70 million t/y of SO_2.

The problem is well illustrated by considering the United Kingdom, where transport is the only energy use classification requiring increasing primary energy supplies. Moreover a number of surveys, backed by street-side measurements, show that air and noise pollution and traffic congestion are the prime environmental concerns of the public. Of total UK CO_2 emissions in 1993, 20% arose from road transport. However, the amount grew in the 10 years from 1983 by 35%. This trend was due to (i) a 40% increase in the number of road vehicles, (ii) a 22% increase in the use of each vehicle, and (iii) a 35% greater average journey length per vehicle. Thus although the fuel efficiency of new vehicle engines is expected to increase by about 30% in the next 10 years, with a consequent reduction in CO_2 emitted per unit distance, the overall CO_2 emission from transport will significantly increase.

The impact of the car is clearly seen from the latest available statistics in the UK (Tables 11.2 and 11.3). In 20 years car use increased 600% by

Table 11.2 Billion passenger miles travelled per year in UK (i.e. excludes freight traffic).

	1970	1980	1990
All modes	210	300	700
All road	170	260	660
Cars and vans	100	200	600
Buses and coaches	70	50	45
Rail	40	35	40
Air, motorcycles, cycles	10	15	15

Source: UK Dept of Transport.

Table 11.3 Average distance (miles) travelled by car per person per year in the UK.

	1975	1990
To work	520	800
Shopping	110	420
Social	420	680
Holiday, other	200	220

Source: UK Dept of Transport.
Note that the increases are in regular activities, predominantly in regular journeys to work and shops. These are the short journeys best suited to electric vehicles.

500 billion passenger miles, and equated to the entire increase in passenger distance travelled by all methods. The same trends continue.

The air pollution from road vehicles continues to create concern for health, especially of the urban population. Over the last 30 years, there has been a growing list of pollution effects considered to cause chronic ill health or physical or mental disability.[22] The pollutants include:

- *Lead*. Organic lead compounds such as lead tetraethyl can be used in petroleum spark ignition engines to prevent pre-ignition, but lead breathed from the emissions accumulates harmfully in the body, for instance retarding mental ability, especially in children.
- *Benzene*. Used in place of lead, benzene combustion produces 1,3-butadiene which is probably carcinogenic and a genotoxin.
- *Hydrocarbons*, HC. Fuel, especially petroleum spirit, evaporates from vehicle tanks, escapes at filling stations, and may be only partially burnt in the engine, especially at starting, to add HC to the atmosphere. These volatile and gaseous compounds cause respiratory illness and react in sunlight with NO_x and other gases to form petrochemical smog and low level ozone, difficulties well known in Los Angeles, Athens and Santiago for instance.
- *Carbon monoxide*, CO. There is incomplete combustion in spark ignition engines, so the direct emissions include CO, a most poisonous gas. CO emissions are proportionally greatest, as with most other pollutants, during variable engine speeds in urban stop-go traffic.
- *Carbon dioxide*, CO_2. Due to the greenhouse effect, CO_2 has to be considered a pollutant. It is a necessary emission if fossil fuels are used.
- *Nitrogen oxides*. NO_x (nitric oxide NO, nitrogen dioxide NO_2 and nitrous oxide N_2O). NO_x is formed at the high temperatures in the engine as the nitrogen and oxygen of air react, with the amount increasing at high speeds. The pollutant causes acid rain, respiratory and cardiac illness and other unwanted atmospheric effects.
- *Particulates* (soot). Partial combustion, especially in old and badly maintained engines, produces exhaust smoke causing ill health, dirt within and damage to buildings, and general unpleasantness. In practice with lax regulation, particulate emission is a particular disadvantage of standard diesel engines.

Emission control. There are various strategies to decrease vehicle pollution, including the obvious but unheeded options of group travelling, zero emission vehicles, cycling and walking. For spark ignition engines the options are improved fuel consumption efficiency, lean-burn engines or catalytic converters.

Catalytic converters are fixed within the exhaust system of spark ignition vehicles. Two-way converters decrease HC and CO emissions by about

50%. Three-way converters decrease HC, CO *and* NO_x by about 90%. Unfortunately catalytic converters (i) do not begin to function until the engine and exhaust are hot, so the particularly offensive pollution from cold and choked engines is not decreased, and urban pollution is not decreased entirely, (ii) fuel consumption increases due to a decrease in engine performance, and (iii) there is a minor annoyance of hydrogen sulphide, 'rotten eggs', smell if SO_2 is present in the emissions.

Lean-burn engines have the stoichiometric fuel to air ratios of 14.7:1, instead of the more usual 15–16:1. This gives improved fuel consumption and therefore less pollution, but cannot usually be combined with catalytic converters.

The diesel engine is an alternative to the spark ignition engine, giving better fuel efficiency per unit volume of fuel and so potentially less mass of emissions per unit distance travelled. In addition the diesel engine burns its fuel completely to CO_2, whereas the spark ignition engine has incomplete combustion to toxic carbon monoxide, CO. Both engines produce unburnt particulates, usually small amounts of carbon with absorbed chemicals. When breathed, the particulates carry the pollution deep into lungs – the smaller the particulate size, the deeper the penetration and potential damage. Recent reports indicate concern about particulates of diameter less than 10 μm (PM10 particulates) as produced by diesel engines (77% of the total) and petrol engines (33%). UK cities average 30 $\mu g/m^3$ of PM10 in main streets, with peaks to danger levels of 150 $\mu g/m^3$. Individuals may be in high concentrations for extended periods of time because of their work, e.g. at bus stations and taxi ranks. The consequent pollution has been estimated to cause 10 000 accelerated deaths per year in the UK.

11.4.2 Clean coal

Coal is a fossil fuel of carbon composition usually much less than 100% (Table 11.4) and therefore the impurities provide ample opportunity for creating pollution from combustion, especially smoke (particulates) and acid rain from the oxides of sulphur and nitrogen.[24]

A series of national and international regulations has made it obligatory to include at least some forms of pollution control whenever coal is burnt. During the closing decade of the twentieth century, there has been remarkable success in producing a wide range of techniques for reducing and abating pollution from emissions. When pollution reduction is added to conventional plant, as with flue gas desulphurisation (FGD), there is a decrease in the net energy efficiency because the substantial extra plant needs to be powered and cooled flue gases may have to be reheated to form a thermal plume. However, with new plant, the discipline of providing cleaner emissions usually produces innovation of design that leads to

Table 11.4 Composition of coals, which are named by proportion of carbon but otherwise have very variable composition. The sulphur and ash content especially varies greatly between sources, with English coal having high sulphur content at about 2.5–3%.

By mass	Peat	Lignites	Bituminous	Anthracite
% moisture as found	70–90	30–50	1–20	1.5–3.5
% dry mass element				
Carbon	45–60	60–75	75–92	92–95
Hydrogen	3.5–7	4.5–5.5	4.0–5.5	3–4.0
Oxygen	20–45	17–35	3–20	2–3
Nitrogen	0.8–3	0.8–2	0.8–2	0.5–2
Sulphur	–	1–3	1–3	0.5–1.5
Volatiles	45–75	45–60	10–50	3.5–10
Ash	20	25	10	9

improvements in energy efficiency which alleviate somewhat the added costs of the clean technology. Such new plant will incorporate modern construction materials and complex processes under microelectronic control that usually allow the plant to be both less polluting and more efficient than previously.

Drax 4000 MW coal power station flue gas desulphurisation (FGD). This is the largest coal fired power station in Western Europe. Situated in North Yorkshire, it burns typical English coal with relatively high sulphur content (1.4–2.8% by mass) and previously succeeded in 'exporting' nearly 300 000 t/y of SO_2 eastward across Europe from its high stack (about 0.1 g/kWh of electricity produced). For many years no attempt was made to remove SO_2 from the emissions despite the long experience of FGD, as for instance at Battersea Power Station in central London from 1930. However, the European Community Large Combustion Plant Directive of 1988 forced the UK Government to take action, and so since 1994 a £680 million add-on plant uses a limestone ($CaCO_3$) water slurry process to remove about 90% of the SO_2 as well as some NO_x and particulates from the stack gases. The 800 000 t/y of gypsum ($CaSO_4$) is removed and sold, mostly for building materials. However, the mining of the limestone and the use of river water to form the slurry sprayed down the stack emissions, creates its own form of environmental intrusion, albeit not exported air pollution. The final stack emission meets European Union regulations for the UK at less than 400 mg $(SO_2)/m^3$ of cooled gas and less than 140 g (NO_x)/GJ (combustion heat), equivalent to 650 mg (NO_x/m^3). Limestone is mined, delivered to the power station, crushed dry, then milled with water to form a slurry. This is passed through hydro cyclones to separate the fine particulate slurry which is sprayed down and from the sides of the vertical absorber up which the flue gases are passed.

Flue gas desulphurisation is a retrospective measure added at great expense to old combustion plant. The SO_2 removal at Drax adds about

25% (0.6 p/kWh) to the generation cost of the electricity, a most significant amount.

Fluidised bed combustion. Fluidised bed combustion (fbc) is a technique for burning coal with both improved efficiency and less pollution using new plant. A bed of limestone at about 500°C is fed with a regulated stream of coal that mixes in the bed. Preheated compressed air is directed into the bed, which is controlled to burn at a constant temperature around 400°C. At this relatively low combustion temperature, nitrogen from the air forms less oxides, NO_x, than in normal combustion. SO_2 is removed from reaction with the hot limestone, so the final stack gases are low in both SO_2 and NO_x.

Diesel engines fuelled by coal dust. The diesel cycle engine was originally designed to burn a range of fuels, including heavy oils and coal dust. It is a high compression engine with no spark ignition. Commercial coal-burning 6 MW diesel engines use the fuel as approximately a 50:50% water:coal-dust mix injected directly into the cylinders with some top-up from conventional fuel. The coal is precleaned to remove impurities and some sulphur, ground with water to particles of average diameter 12 μm and no particle larger than 88 μm, and then injected as the fuel. The fuel has about half the heat of combustion per unit volume of diesel oil, but nevertheless electricity generating efficiency is about 47%.

Exhaust gases emerge from an engine under high pressure, and therefore can be passed easily through a series of stages: a turbocharger to increase engine efficiency, a cyclone to remove smoke particulates, selective catalysts to reduce NO_x concentration, sodium bicarbonate to remove sulphur and finally heat exchangers for preheating air and to provide useful heat.

Combined cycle generation from coal gasification. Coal may be converted into a gaseous fuel after mining, although there are proposals to gasify coal *in situ* underground without mining. The coal gas is mainly producer gas (also called town gas) of hydrogen and carbon monoxide formed from passing water over heated coal. Once in gaseous form, the fuel can be used in combined cycle generation of electricity at much increased efficiency. A conventional coal, oil or nuclear thermal power station is about 30–35% efficient, whereas a combined cycle station reaches 50–55% efficiency from a gas turbine with the exhaust gas generating further electricity from a steam turbine.

11.4.3 Combined heat and power from integrated fuels

There is no better example of the rapid changes taking place in engineering for improved efficiency and environment than in the generation of

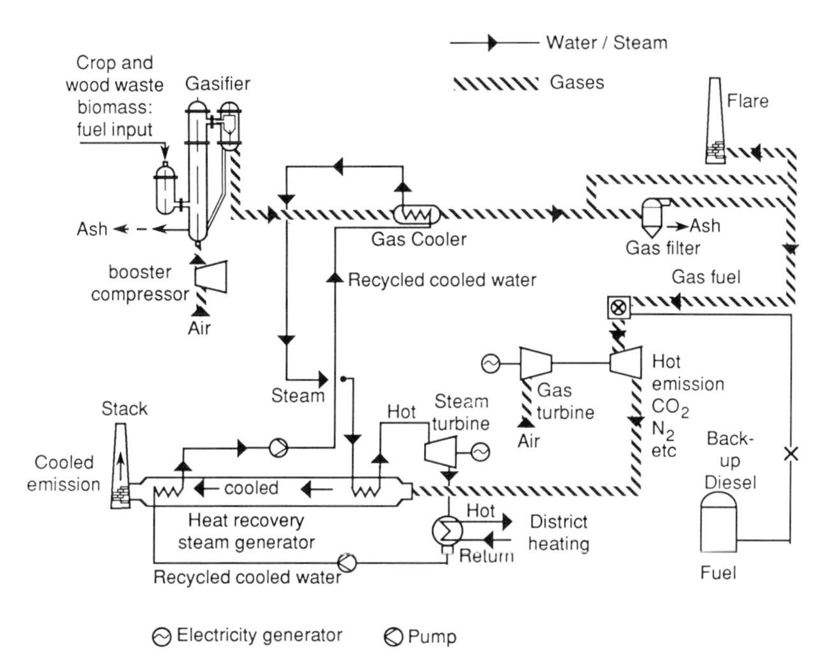

Figure 11.3 Schematic of Swedish combined heat and power plant utilising biomass.

electricity and heat. The crude and wasteful large scale methods inherited from the first industrial revolution are moribund as the commercial and economic benefits of clean and smaller scale technology are appreciated. Britain is an example of how a state electricity authority closed its corporate mind to generate electricity predominantly by steam turbines powered by coal without pollution control and without combined heat and power. This was done in contradiction of the 1947 Act of Parliament nationalising the industry. Consequently 65–70% of the energy was emitted as waste heat and the efficiency of generating electricity was limited to about 30–35% by the moderate temperatures of the steam. The process was inherently polluting and inefficient.

In contrast, as an example of a modern process, consider the type of technology of the demonstration 120 MW heat and 40 MW electricity power plant of Vårnamo in southern Sweden[27] (Figure 11.3). The modern features are:

- The feedstock fuel is biomass, usually wood grown as fuel or obtained from forestry and milling waste. The combustion does not add to the ecological burden since the carbon dioxide emitted is part of a natural cycle. There is negligible sulphur in the biomass, negligible CO emission, and NO_x is controlled within the plant. Combustion is complete, so organic emissions are negligible.

- The feedstock is converted to a combustible mixed gas (hydrogen, carbon monoxide and methane, together with CO_2 and water vapour) at high pressure in a gasifier, rather than being burnt directly. Consequently there is better pollution control and the secondary fuel is more manageable as a gas than solid fuel.
- The gas is cooled (the heat passes to a steam preheater), filtered and then burnt in gas turbines connected to electricity generators. Operating at higher temperature than a steam turbine (about 950°C compared with, say, 450°C) the efficiency of electricity generation is around 40% rather than 30%.
- The combustion gases from the gas turbine are hot enough to raise steam for a secondary steam turbine, also generating electricity. The 'combined cycle' leads to a total electricity generation of about 55% efficiency.
- The final waste heat passes to a district heating scheme. Adding the useful heat to the electricity production, the combined heat and power efficiency is 80–90%, i.e. three times that of an old fashioned steam turbine, electricity-only, power station.
- The electricity output of such combined cycle stations is about 4 MW per gas turbine at about 80% total efficiency. This should be compared with the approximately 33% efficiency of traditional 500 MW steam turbines as used in large fossil and nuclear power stations. Thus the fossil fuel technology of the 1990s is smaller, more adaptable, more efficient and much cleaner than the technology of the 1960s. In addition it is in a form to utilise biomass sources of renewable energy

11.4.4 Wind-generated electricity

The technical and commercial growth of renewable energy supplies has been outstanding in the last 20 years. In the 1970s the aim was to provide an insurance against the loss of oil supplies and in the 1990s there has been the further need to utilise the clean, non-polluting characteristics of renewable energy. National and international reviews of policy for renewables are regularly available, e.g. Twidell and Brice,[25] Johansson *et al.*,[26] DTI.[23] Hydro power having been always accepted for electricity generation, the major recent development has been in windpower. In the UK this has been assisted by the Non-Fossil-Fuel-Obligation, which compels electricity distribution utilities to purchase and then supply electricity from mainly nuclear, but also renewable, sources.

The wind farm at Blyth Harbour, near Newcastle-upon-Tyne in north east England, is a particularly interesting development selling electricity to the grid without any form of chemical emissions.[28] This is in contrast to the historic use of the harbour for coal exports to standard UK thermal and polluting power stations in the south of England. The general area of

Figure 11.4 The wind farm at Blyth Harbour, near Newcastle-upon-Tyne.

the harbour is industrial and residential, and therefore not green belt for planning purposes. A general consensus is that the line of turbines have added aesthetically to the view, are of popular interest to residents as well as visitors, do not give acoustic annoyance and are an outstanding example of modern, clean technology.

A total of nine wind turbines, each of 300 kW capacity, were installed along an established pier in 1992. The site is obviously at sea-level and therefore does not benefit from the increase in wind speed with height utilised by wind farms in hilly and upland regions; nevertheless winds off the sea are unperturbed by obstructions. The total installed capacity of 2.7 MW produces grid-connected electricity according to the wind conditions and has averaged 30% capacity factor over the full year, so providing clean and sustainable electricity for the equivalent of about 1000 homes.

The key factors from the development for future clean energy supplies are:

- No chemical emissions.
- By substituting for electricity from coal thermal power stations, the abatement each year of 1000 t CO_2, 100 t ash, 30 t SO_2, 15 t NO_x.
- Other potentially negative impacts, e.g. noise, electromagnetic interference, disruption of birds, have not caused difficulty.
- A renewable, sustainable and clean energy supply is produced in the locality of the consumers.
- The cost of the electricity is in line with expectations, cheaper than nuclear power and economic within the terms of the programme to

decrease CO_2 and other emissions. Subsequent developments after this first experience will be cheaper and competitive with fossil fuels.

- For planning purposes and public acceptance, the development is in harmony with the area and the other activities of the harbour.
- The machines are reliable (97% availability under remote and microprocessor control).
- Installation and maintenance are by local companies.
- Eventual decommissioning would leave the site as before, but with a stronger pier.

11.5 Conclusions

The aim of this chapter has been to show that despite the present developed world's lifestyle being unsustainable, there are strong scientific and technical arguments to explain how sustainability can be obtained. An essential aspect is to have clean energy supplies and to use energy efficiently, for which there are many opportunities. The appropriate methods frequently involve advanced and sophisticated technologies that promote economic growth. There should be no regrets in ardently pursuing such aspirations. The world can grow richer by using less.

References

1. Johansson, A. (1992) *Clean Technology*, Lewis Publishers, CRC Press, Florida.
2. Economist (1991), *The Environment – Industry, Energy and Economics,* The Economist Newspaper Ltd., London.
3. WCED (1987) *Our Common Future (The Bruntland Report)*, Report of the World Commission on Environment and Development, OUP, Oxford.
4. Twidell, J.W. (1994) The environmental impacts of wind and water power. In *Proc. Environmental Impacts of Energy Technology*, UK Solar Energy Society, Franklin Co., Birmingham.
5. Hohmeyer, O. (1986) *The Social Costs of Energy Consumption*, Springer-Verlag, Berlin.
6. POST (1992) *Costing the Environmental Impacts of Electricity Generation*, Parliamentary Office of Science and Technology, London.
7. HMSO (1994a) *Sustainable Development, the UK Strategy*, Cmd 2426, HMSO, London.
8. HMSO (1994b) *Climate Change, the UK Programme*, HMSO, London.
9. Mortimer, N. (1994) The role of energy efficiency in reducing environmental impacts. In *Proc. Environmental Impacts of Energy Technology*, UK Solar Energy Society, Franklin Co., Birmingham.
10. Lovelock, J. (1991) *Gaia*, Gaia Books Ltd, London.
11. Garcia, R. (1994) Causes of ozone depletion, *Physics World*, April, 49–55.
12. UNEP (1987) *Montreal Protocol on Substances that Deplete the Ozone Layer*, UNEP.
13. HMSO (1992) *Amendment to the Montreal Protocol on Substances that Deplete the Ozone Layer*, Cmd 2367, HMSO, London.
14. IPCC (1990) *Intergovernmental Panel on Climate Change First Assessment Report*, Cambridge University Press.
15. Mohen, V.A. Goldstein, W. and Wang, W. (1990) The scientific challenge of measuring climate change, *Energy Policy*, Sept., 641–651.

16. Houghton, J.T. Callender, B.A. and Varney, S.K. (eds) (1992) *Climate Change 1992: The Supplementary Report to the IPCC Scientific Assessment*, Cambridge University Press.
17. CSTI (1992) *The Greenhouse effect: Fact or Fiction*, Council of Science and Technology Institutes, London.
18. DoEn (1989) *An Evaluation of Energy Related Greenhouse Gas Emissions and Measures to Ameliorate Them*, Department of Energy, Energy Paper 58, HMSO, London.
19. Twidell, J.W. and Weir, A.D. (1986) *Renewable Energy Resources*, E. & F.N. Spon (Chapman & Hall), London.
20. SCEP (1970) *Man's Impact on the Global Environment*, Report on the Study of Critical Environmental Problems, MIT Press, Cambridge, Mass.
21. Meadows, D.H., Meadows, D.L., Randers, J. and Behrens, W.W. (1972) *The Limits to Growth*, Earth Island Ltd, London.
22. Eide, A. (1992), *An Enquiry into Energy Efficiency and Interrelated Environmental Issues of Automobiles*, MEnvSt report, Graduate School of Environmental Studies, University of Strathclyde, Glasgow.
23. DTI (1994) *UK Renewable Energy Progress*, ETSU Report, Department of Trade and Industry, HMSO, London.
24. Park, C.C. (1987) *Acid Rain*, Routledge, London.
25. Twidell, J.W. and Brice, R. (1992) Strategies for implementing renewable energy, *Energy Policy*, **20**, 464–479.
26. Johansson, T.B., Kelly, H., Reddy, A.K.N. and Williams, R.H. (1994) *Renewable Energy – sourcces for fuels and electricity*, Earthscan Publications, London.
27. Caddet (1994) Energy in Finland *Renewable Energy News*, Caddet, Energy Technology Support Unit, AEA, Harwell, UK.
28. Grainger, W. (1993) Blyth Harbour Wind Farm, in *Proc. British Wind Energy Association, Progress in Wind Power*, I Mech E, London.
29. Tickle, A. (1993) *The Acid Test for Plants (acid rain and British plants)*, Plantlife, Natural History Museum, London. (Tickle refers to the 17th century tract, *Fumifugium*, on London's air pollution by J. Evelyn, and to R.A. Smith, Britain's first inspector of pollution who coined the phrase 'acid rain'.)

Index